Essentials of
Advanced Composite Fabrication and Repair

THIRD EDITION

Louis C. Dorworth   Ginger L. Gardiner

AVIATION SUPPLIES & ACADEMICS, INC.
NEWCASTLE, WASHINGTON

Essentials of Advanced Composite Fabrication and Repair
Third Edition
by Louis C. Dorworth and Ginger L. Gardiner

Aviation Supplies & Academics, Inc.
7005 132nd Place SE
Newcastle, Washington 98059
asa@asa2fly.com | 425-235-1500 | asa2fly.com

Copyright © 2024 Abaris Training Resources, Inc.
Third edition published 2024 by Aviation Supplies & Academics, Inc. First edition published 2009 by
Aviation Supplies & Academics, Inc.

Abaris Training Resources, Inc.
5401 Longley Lane, Suite 49 • Reno, NV 89511
+1-775-827-6568 • www.abaris.com

   CompositesWorld.com

See the Reader Resources at **asa2fly.com/composite** for additional information and updates related to
this book.

All rights reserved. No part of this publication may be reproduced, stored in a retrieval system, or
transmitted in any form or by any means without the prior written permission of the copyright holder.
While every precaution has been taken in the preparation of this book, the publisher, Abaris Training
Resources, Louis C. Dorworth, and Ginger L. Gardiner assume no responsibility for damages resulting
from the use of the information contained herein.

ASA-COMPOSITE3
ISBN 978-1-64425-414-1

Additional formats available:
eBook EPUB   ISBN 978-1-64425-415-8
eBook PDF   ISBN 978-1-64425-416-5

Printed in the United States of America
2028 2027 2026 2025 2024   9 8 7 6 5 4 3 2 1

Cover photos: *Front (center left):* Phruet/Shutterstock.com. *(Bottom right):* Tarnero – stock.adobe.com.
(All other images): Abaris Training Resources, Inc. *Back cover:* Composite_Carbonman/Shutterstock.com.

Photography note: Credit and copyright information for photographs other than those owned by Abaris
are acknowledged throughout the text, stated adjacent to the printed photographs, and also in the
Bibliography and Acknowledgments section at the back of the book. All photos not so noted are copyright
Abaris 2024. Illustrations note: All drawn illustrations created by ASA, unless otherwise noted.

Library of Congress Cataloging-in-Publication Data
Names: Dorworth, Louis C., author. | Gardiner, Ginger L., author.
Title: Essentials of advanced composite fabrication and repair / Louis C. Dorworth, Ginger L. Gardiner.
Description: Third edition. | Newcastle, Washington : Aviation Supplies & Academics, Inc., [2024] | Includes
    bibliographical references and index.
Identifiers: LCCN 2024029310 (print) | LCCN 2024029311 (ebook) | ISBN 9781644254141 (hardback) | ISBN
    9781644254158 (epub) | ISBN 9781644254165 (pdf)
Subjects: LCSH: Airplanes—Materials—Handbooks, manuals, etc. | Composite materials—Handbooks, manuals, etc.
Classification: LCC TL699.C57 D67 2024 (print) | LCC TL699.C57 (ebook) | DDC 629.134—dc23/eng/20240723
LC record available at https://lccn.loc.gov/2024029310
LC ebook record available at https://lccn.loc.gov/2024029311

Contents

CHAPTER 1
Composite Technology Overview

CHAPTER 2
Matrix Technology

CHAPTER 3
Fiber Reinforcements

CHAPTER 4
Nanocomposites

CHAPTER 5
Sandwich Core Materials

CHAPTER 6
Basic Design Considerations

CHAPTER 7
Molding Methods and Practices

CHAPTER 8
Liquid Resin Molding Methods and Practices

CHAPTER 9
Introduction to Tooling

CHAPTER 10
Inspection and Test Methods

CHAPTER 11
Adhesive Bonding and Joining

CHAPTER 12
Machining, Drilling, and Fastening Composites

CHAPTER 13
Repair of Composite Structures

CHAPTER 14
Health and Safety Considerations

Preface

Updated from the second edition, this third edition of **Essentials of Advanced Composite Fabrication and Repair** covers a wider range of contemporary technical material and is designed to function as a textbook both for Abaris Training and for other technical schools teaching composites. It is an "essential" resource for everyone, from novice to professional, involved in the advanced composites industry.

Initially produced as a spiral-bound composition of excerpts from various technical documents gleaned from the Lear Fan, Ltd. Training Academy and other pioneer thought leaders of the early 1980s, the original text was used to support a single Abaris course, "Inspection and Repair of Composite Structures." Over the next twenty years, the book expanded and developed as more innovative materials and technologies emerged and as Abaris added relevant new courses. By 2005, the authors realized the need for a "real" textbook and began a collaborative effort with the publisher to formalize the content. In 2009, the first edition textbook was published and quickly became useful throughout the industry.

The composites industry moves fast, and by 2015 the authors once again realized the need for a thorough update of the content, thus beginning a new journey to identify what had changed in the past decade and how to include it in the second edition. It turned out that much had changed and much had remained the same. Over countless days and nights, the authors spent time taking it all in from industry sources, purging content no longer relevant, and weaving together new content and illustrations in a logical order that the reader can easily follow and understand. The second edition was published in 2019, and by 2023 it became clear that a third edition of the textbook was needed to further update the industry.

Like the second edition, this book starts with an introduction to composites and then takes a deep dive into the constituent materials such as fibers, matrix resins, nanocomposites, core materials, and curing or processing them. This is followed by chapters that cover basic design considerations, molding methods and practices, tooling, testing, bonding, machining, drilling, repair, and much more. In addition, online Reader Resources on the publisher's website at asa2fly.com/composite provide additional information and future updates relating to this book's content.

It is the sincere desire of the authors that readers gain a deeper knowledge and a better understanding of the subject and are empowered to put this information to use immediately in the workplace, on their projects, or to further advance their careers.

About the Authors

 **LOUIS C. DORWORTH** is the Direct Services Manager for Abaris Training Resources, Inc., where he currently manages all marketing and training activities worldwide. By trade, he is a composite materials and process (M&P) specialist with experience in research and development (R&D), manufacturing engineering, tool design/engineering, tool fabrication, and repair. Louis has been involved in the advanced composites industry since 1978, starting in aerospace as a toolmaker and part-time M&P technician at Lear Fan in Reno, Nevada, after which he has continued in the industry for more than four decades. Louis has been associated with Abaris since its inception in 1983 and began his teaching career at Abaris in 1989.

Louis has been a professional member of the Society for the Advancement of Material & Process Engineering (SAMPE) since 1982 and a senior member of the Society of Manufacturing Engineers (SME) since 1997. He currently serves as an advisor to the SAMPE Technical Excellence Subcommittee for Bonding and Joining, is a member of the Thermoplastics Technical Committee, and is an advisor to conference and workshop steering committees within these technical organizations.

 **GINGER GARDINER** has worked in the composites industry since 1990. She has a degree in Mechanical Engineering from Rice University and began her career as a technical marketing representative in DuPont's Composites Division for KEVLAR and NOMEX products in aerospace and marine applications. After leaving DuPont, Ginger formed Vantage Marketing Services, providing market and product development consulting for companies such as Hoechst-Celanese and Ciba-Geigy, and also developed and marketed technical conferences. She also wrote articles for several magazines, including *Professional BoatBuilder*, and began writing for *Composites Technology* and *High Performance Composites* magazines in 2006. She has worked as Senior Editor and Senior Technical Editor of the now combined *CompositesWorld* magazine since 2013.

Editor's Note

The design of this textbook takes advantage of visual elements to aid the reader's navigation through the narrative: a yellow "dot" helps identify the numbered figures referred to in the text, as well as, in most cases, a gray "bar" along the outside edge of the page to differentiate between illustration content and the narrative content. Tables are numbered separately to distinguish them from the drawn and photographic illustrations. Footnotes are contained at the bottom of pages where they fall in the text, and further bibliographic references are listed at the back of the book. In addition, a short main-topic contents list is added to the chapter-start pages.

1

Composite Technology Overview

CONTENTS

Composites vs. Advanced Composites

Composites are comprised of two or more materials working together, where each constituent material retains its unique identity within the composite and contributes its own structural properties, yet upon combination the resulting material has superior properties to those of its constituents. A good example of an everyday composite material is concrete. Concrete is made with select amounts of sand, aggregate, and perhaps even glass fiber mixed with cement to bind it together. If the concrete were broken open, the individual constituents would be visible. The type and quantities of these individual constituents can also be adjusted to give the resulting concrete different properties depending on the application.

This textbook is focused on composite laminates, which combine fibers and a matrix material that binds the fibers together. There are many different types of composite materials in use today. One example is fiber-reinforced polymer (FRP) composites made with short glass fibers in a polymer resin matrix. These materials are used in bath tubs, showers, pools, doors, car fenders, and a variety of construction materials including wall panels, corrugated sheet, profiles, and skylights. (Figure 1-1)

FIGURE 1-1. Fiber-reinforced composite.

 +  =

Fibers
- Tensile strength
- Flexural stiffness
- Somewhat brittle

Matrix
- Compressive strength
- Interlaminar shear
- Controls shape
- Governs temperature
- Low density

Composite
- Increased strength
- Increased stiffness
- Increased toughness
- Lightweight

Highly loaded composite structures typically use continuous or long-fiber reinforcement that transfers load along bundles or layers (**plies**) of fibers arranged to run the length and width of the structure, much like the layers in a sheet of plywood. This type of composite laminate is used in the manufacture of boats, bridges, snowboards, bicycle frames, race cars, aircraft and spacecraft structures, to mention a few.

❱ ADVANCED COMPOSITES

"Advanced composites" are generally considered to be those that use advanced reinforcements such as carbon, aramid and S2 glass fibers that exhibit high strength-to-weight ratios.[1] They are typically more expensive, with more precisely tailored properties to achieve a specific objective.

Fiberglass vs. Advanced Composites

Some composites are generally referred to as "fiberglass" due to their use of randomly-oriented, chopped glass fiber (E-glass) and polyester resin, whereas most aerospace structural parts are made using precisely-laid plies of carbon fiber/epoxy **prepreg**, an example of advanced composites. (Figure 1-2)

Examples of Typical Applications

Large components of commercial airliners—such as the vertical and horizontal tail plane (stabilizer) on the Airbus A320, A330/340, A380 and Boeing 777, the wing, center wing box and fuselage for the Boeing 787 Dreamliner and Airbus A350, and various structures on many smaller craft such as the wings for the Bombardier C Series airliners. (Figure 1-3)

Large primary structures on military aircraft—such as the wing and cargo doors for the Airbus A400M transport, fuselage/wing for the B-2 Spirit Stealth Bomber, rotor blades and aft fuselage for the V-22 Osprey tilt-rotor, as well as the most of the fuselage and wings for the F-22 Raptor and F-35 Joint Strike Fighter. (Figure 1-4)

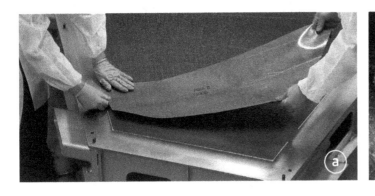 

FIGURE 1-2. Fiberglass vs. advanced composites.

[a] To fabricate an aerospace structural component, technicians carefully lay down each ply of carbon-fiber prepreg prior to vacuum bagging and autoclave cure. Green "templates of light" are accurately projected from a 3D laser projector to ensure precise positioning of the ply. *(Photo courtesy of Aligned Vision)*

[b] Chopped fiberglass and resin are sprayed onto a gel-coated mold to form the outer shell of a Class 8 truck hood. However, this is a more advanced and higher performance example of spray-up fiberglass because the shell is cured in an oven at 130°F and then reinforced with structural members made using RTM, which are secondarily bonded in-place using methacrylate adhesive. *(Photo courtesy of Marine Plastics Ltd.)*

---

1   ASM Handbook Volume 21, *Composites* (pg. 1113) defines "advanced composites" as: "Composite materials that are reinforced with continuous fibers having a modulus higher than that of fiberglass fibers. The term includes metal matrix and ceramic matrix composites, as well as carbon-carbon composites." Material Park, Ohio; ASM International, 2001.

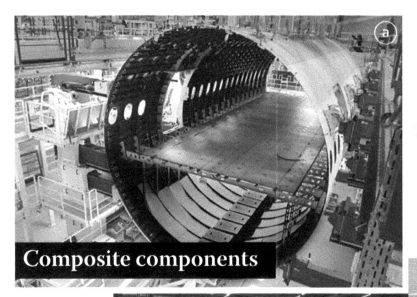

Composite components

FIGURE 1-3. [a] A350 forward fuselage; [b] prototype center wing box for future A320-type aircraft; [c] A350 lower wing cover. *(Photos courtesy of Airbus)*

Bottom diagram: Boeing 787 Dreamliner composite components.

Boeing 787

- ▮ Carbon laminate
- ▮ Carbon sandwich
- ▮ Fiberglass
- ▮ Aluminium
- ▮ Aluminum/steel/titanium pylons

FIGURE 1-4. The F-35 Lighting II Joint Strike Fighter features vertical tail feathers and horizontal stabilators made from carbon fiber-reinforced bismaleimide composite. *(Photo courtesy of Lockheed Martin)*

Many other components on modern airliners—such as radomes, control surfaces, spoilers, landing gear doors, wing-to-body fairings, passenger and cargo doors, trailing edges, wingtips and interiors. (Figure 1-5)

Large marine vessels and structures including hulls, decks and superstructure of military and commercial vessels, as well as composite masts (one of the largest carbon fiber structures in the world is the *M5* sailing yacht's 290-foot mast), wing masts and foils, rigging, propellers and propeller shafts. (Figure 1-6)

Primary components on helicopters and other vertical takeoff and landing (VTOL) vehicles including rotor blades and rotor hubs have been made from carbon fiber (CF) and glass epoxy composites since the 1980s. Composites can make up 50 to 90% of a rotorcraft's airframe by weight, including radomes, tail cones and large structural assemblies. (Figure 1-7) For example, Bell Helicopter Textron's 429 corporate/EMS/utility helicopter features composite structural sidebody panels, floor panels, bulkheads, nose skins, shroud, doors, fairings, cowlings and stabilizers, most made from CF/epoxy.

FIGURE 1-5. Airliner components.

[a] J-nose thermoplastic composite leading edge for the Airbus A380 made by Stork Fokker. Note the stamped thermoplastic stiffeners, which are attached using resistance welding. *(Photo courtesy of Stork Fokker)*

FACC is a leading Tier 1 supplier with composite structures on every commercial aircraft in production, including [b] carbon fiber/epoxy flaps for the Airbus A321, [d] translating sleeves for the Boeing 787, and [c] bypass ducts for Rolls-Royce aircraft engines. *(Photos upper right and bottom, courtesy of FACC)*

FIGURE 1-6. Marine structures.

Left to right, top to bottom:

- The *M5* (previously the *Mirabella V*) is the world's largest composite ship. *(Photo courtesy of Select Charter Services)*

- The Visby class of corvettes is built using carbon-fiber-reinforced composite sandwich construction. *(Photo courtesy of Kockums AB)*

- Placid Boatworks' 3.6m long, 8-kg *Spitfire Ultra* canoe uses carbon fiber biaxial (±45°) and quasi-isotropic (0°/+60°/-60°) braided fabrics in the hull and biaxial braided carbon and aramid sleeving wrapped around Divinycell foam in the gunwales. *(Photo courtesy of Placid Boatworks)*

- The lightweight composite construction for CMN Group's 43.6m *Ocean Eagle 43* Ocean Patrol Vessel (OPV) was built by H2X using glass fiber/epoxy/foam core sandwich and resin infusion. *(Photo courtesy of H2X)*

- The ECsix composite rigging on this sailing yacht is made from a bundle of 1-mm pultruded carbon fiber/epoxy rods encased in an abrasion-resistant braided synthetic fiber jacket. Able to cut rigging weight by 70%, ECsix has been used on over 500 yachts, sailing more than 1 million miles without a single failure. *(Photo courtesy of Composite Rigging/Southern Spars)*

- The 48m Palmer Johnson SuperSport yacht features all-carbon fiber composite construction made using vacuum infusion. *(Photo courtesy of Brødrene Aa)*

FIGURE 1-7. Airbus helicopter components.

Airbus Helicopter's H160 twin-engine medium helicopter features an all-composite airframe, new composite Biplane Stabilizer™, Blue-Edge® composite main rotor blades with double-swept tips, a CF/PEKK thermoplastic composite rotor hub and a double-canted tail incorporating the largest-ever composite Fenestron® shrouded tail rotor *(inset). (Photos courtesy of Airbus Helicopter)*

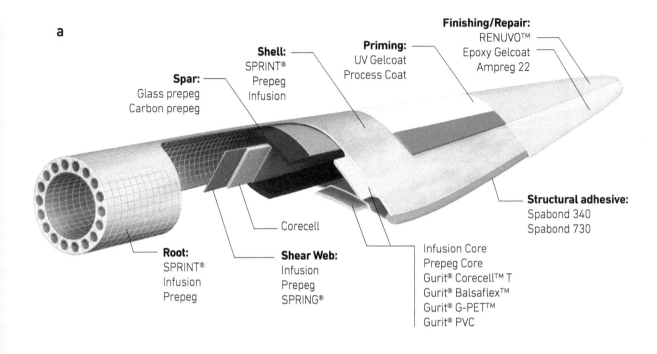

a

Spar:
Glass prepeg
Carbon prepeg

Shell:
SPRINT®
Prepeg
Infusion

Priming:
UV Gelcoat
Process Coat

Finishing/Repair:
RENUVO™
Epoxy Gelcoat
Ampreg 22

Root:
SPRINT®
Infusion
Prepeg

Corecell

Shear Web:
Infusion
Prepeg
SPRING®

Infusion Core
Prepeg Core
Gurit® Corecell™ T
Gurit® Balsaflex™
Gurit® G-PET™
Gurit® PVC

Structural adhesive:
Spabond 340
Spabond 730

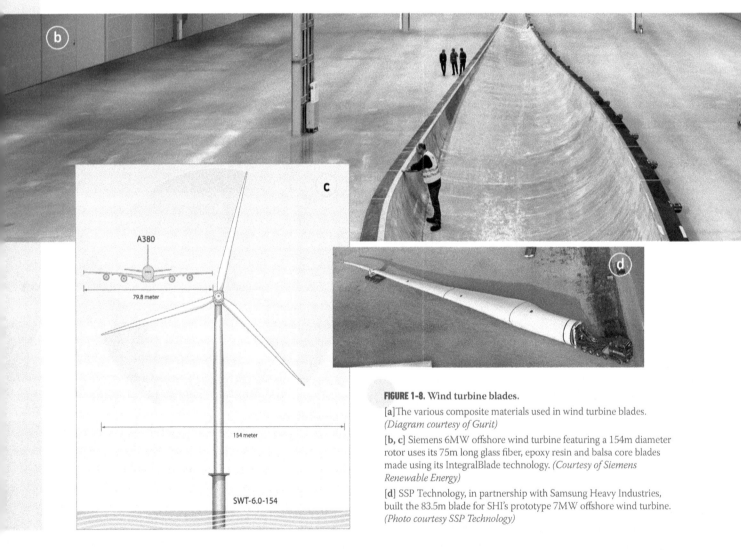

b

c

A380

79.8 meter

154 meter

SWT-6.0-154

d

FIGURE 1-8. Wind turbine blades.

[a] The various composite materials used in wind turbine blades. *(Diagram courtesy of Gurit)*

[b, c] Siemens 6MW offshore wind turbine featuring a 154m diameter rotor uses its 75m long glass fiber, epoxy resin and balsa core blades made using its IntegralBlade technology. *(Courtesy of Siemens Renewable Energy)*

[d] SSP Technology, in partnership with Samsung Heavy Industries, built the 83.5m blade for SHI's prototype 7MW offshore wind turbine. *(Photo courtesy SSP Technology)*

Wind turbine rotor blade manufacturers use glass fiber and resin—epoxy, polyester or polyurethane are often used—plus a core material such as balsa wood or foam to create a lightweight yet stiff sandwich structure. The blades are made using **resin infusion** or prepreg; carbon fiber is increasingly used in spar caps as blades get longer. For example, LM Wind Power makes the 88.4-meter-long blades used on the Adwen 8 MW turbine and is planning a 107-meter-long blade for GE's Haliade-X 12 MW turbine. Blades have also become more slender and may feature aeroelastic tailoring, with airfoils optimized for specific wind conditions. (Figure 1-8)

Missiles and space vehicles rely heavily on carbon fiber composite construction due to its high strength- and stiffness-to-weight ratios as well as its negative linear coefficient of thermal expansion (CTE), which gives dimensional stability in the extreme temperatures of space. The Delta family of launch vehicles have used carbon fiber and epoxy construction in more than 950 filament-wound motor cases and are in their sixteenth year of production. The Pegasus rocket is the first all-composite rocket to enter service. Its payload fairing, which is 4.2 feet/1.3 meters in diameter and 14.2 feet/4.3 meters in length, is comprised of carbon/epoxy skins and an aluminum honeycomb core.

The Chandra X-ray Observatory spacecraft features numerous composite structures, including the optical bench assembly, mirror support sleeves, instrument model structures and telescope thermal enclosure. The next generation James Webb Space Telescope (JWST) uses

FIGURE 1-9. Space structures.
The James Webb Space Telescope (artist's rendition) uses a carbon-fiber-reinforced cyanate ester matrix backplane support [inset]. *(Photos courtesy of NASA)*

FIGURE 1-10. A robot places composite fibers on an 8-ft diameter cryogenic fuel tank, tested as part of development of even larger tanks for future heavy-lift launch vehicles. *(Photo courtesy of NASA)*

a carbon fiber reinforced cyanate ester backplane structure, deployable sunshield booms, and other composites in the design of its primary structures. (Figures 1-9 and 1-10)

Automotive applications range from carbon fiber reinforced polymer (CFRP) monocoque chassis and driveshafts in race cars and high-end sports cars, to selected structural and aesthetic carbon composite applications in lower-volume luxury models, to applications in higher-volume models such as glass-fiber-reinforced polyurethane leaf springs and GF/epoxy coil springs. (Figure 1-11)

Selected applications in production automobiles include roofs for BMW's M3 and M6 models, the hood and fenders for Corvette's LeMans Commemorative Edition Z06, the inner deck lid and seats for the Ford GT, composite coil springs for the 2015 Audi A6 Avant 2.0 TDI ultra, and front and rear axles (actually termed "multi-function suspension blades" because they integrate suspension, steering, anti-vibration/noise and anti-roll) for the Peugeot 208 FE. There is also

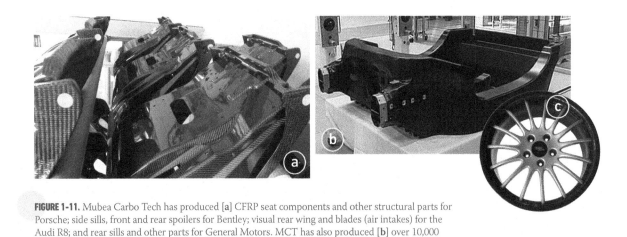

FIGURE 1-11. Mubea Carbo Tech has produced [a] CFRP seat components and other structural parts for Porsche; side sills, front and rear spoilers for Bentley; visual rear wing and blades (air intakes) for the Audi R8; and rear sills and other parts for General Motors. MCT has also produced [b] over 10,000 monocoques, including serial production for the McLaren P1/MP4, Volkswagen XL1, and Porsche 918 Spyder—as well as [c] all-CF and CF/aluminum hybrid wheels. *(Photos courtesy of Mubea Carbo Tech)*

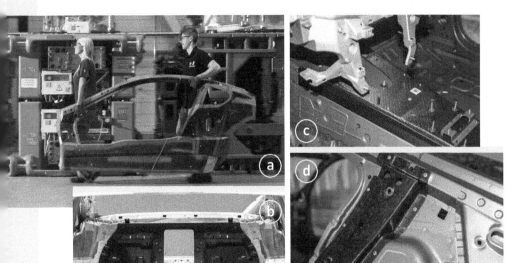

FIGURE 1-12. [a] The BMW *i3* and *i8* feature CF/epoxy passenger cells made with non-crimp fabrics, RTM and HP-RTM (BMW *i8* sideframe).

[b] Audi uses a CFRP rear wall in the A8 luxury sedan, made by Voith Composites.

[c, d] The BMW *7 Series* uses 16 different CFRP parts in its Carbon Core body in white (BIW), including tunnel and C pillar.
(Photos courtesy of BMW, Audi and Voith Composites)

a significant market in carbon fiber composite parts sold as aftermarket enhancements to production automobiles, including a number of companies mass-producing all-carbon-fiber wheels. (Figures 1-12 and 1-13)

Sporting goods, such as bikes, tennis rackets and golf clubs have taken advantage of carbon fiber for decades. The majority of tennis rackets, golf club shafts and hockey sticks are made using carbon fiber. Composites are also common in bats, arrows, snowboards and skis, and gaining widespread use in helmets and protective shoulder and shin pads. Latest trends in sporting goods include use of spread-tow reinforcements, nanomaterials and biomaterials, as well as added multi-functionality such as vibration damping for increased performance and control. (Figure 1-14)

FIGURE 1-13. [a] The glass fiber/polyurethane leaf spring for Volvo's Scalable Product Architecture is used in its S60, S90, V60, V90, XC60 and XC90 models, requiring up to 500,000 parts per year.
[b] This hybrid carbon fiber/aluminum suspension knuckle increases stiffness by 26% vs. all-aluminum.
(Photos courtesy of Volvo, SGL Carbon, Saint Jean Industries)

FIGURE 1-14. Sporting goods applications.

[a] Dylan Groenewegen of the Netherlands won two stages of the 2018 Tour de France racing this Oltre XR4 engineered with Bianchi's CV vibration-canceling system: Countervail®'s unique fiber architecture and viscoelastic resin, and CV tuning process that reduces cyclist muscle fatigue and increases control, for top performance on bumping roads and cobblestone sections. *(Source: Bianchi)*

[b] Innegra high modulus polypropylene fiber provides toughness, durabilty and vibration damping in carbon fiber composites such as this standup paddleboard and paddle, and BAUER hockey sticks and goalie masks. *(Photos courtesy of Innegra Technologies and BAUER*)*

[c] Lingrove's Ekoa biocomposite materials reduce mass and increase vibration damping in skis, skateboards, paddles and other sporting goods. They also improve sustainability, including drastic reductions in energy usage and greenhouse gas emissions during manufacture. *(Photo courtesy of Lingrove)* The TeXtreme Warrior 107 tennis racquet by Prince uses TeXtreme spread-tow carbon fiber fabric. *(Photo courtesy of TeXtreme)*

[d] Cevotec's carbon fiber patch preforms (the black arcs in the right-hand photo) improve board control via asymmetric torsion stiffness and board flex in a North Kiteboarding freestyle board. *(Photos courtesy of North Kiteboarding)*

Medical devices continue to spur the advancement of new composites, such as ENDOLIGN™ continuous carbon fiber/PEEK thermoplastic biomaterial from Invibio Biomaterial Solutions, which is being used as an alternative to metals in the development of implantable load-bearing applications in orthopedic, trauma, and spinal implants. Other carbon fiber composites are used in medical imaging tables and accessories because they offer high stiffness and light weight, while helping to minimize imaging issues such as signal attenuation. Composites are also gaining applications in prosthetics, orthotics and surgical tools via 3D printing. (Figure 1-15)

Civil engineering structures are progressing steadily in their use of composites. Carbon fiber wraps for repair and strengthening of columns, beams, concrete slabs and bridge structures have become increasingly common. They offer an alkaline-resistant repair, which is quicker and less costly to install due to its light weight. Carbon fiber in bridge cable stays is also progressing, offering high strength and stiffness at minimal weight as well as excellent resistance to temperature contraction and expansion due to its negative coefficient of thermal expansion. (Figure 1-16)

Composite structures using a wide range of additive manufacturing processes are offering a cost-effective means to produce structural parts with complex geometry using thermoplastic compounds reinforced with short fibers or continuous filaments. (Figure 1-17)

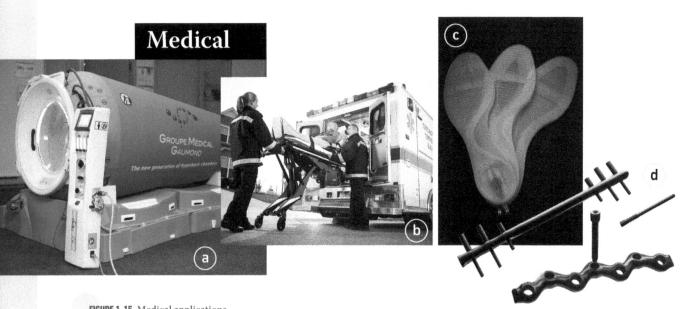

FIGURE 1-15. Medical applications.

[a] The HematoCare portable hyperbaric chamber weighs only 125 kg (275 lbs), thanks to a composite pressure vessel made using filament winding and Kevlar aramid fiber but with a special resin that provides flame resistance and high strength to meet pressure loads and flexibility. This enables the device to be folded like an accordion for easy transport. *(Photo courtesy of Groupe Médical Gaumond)*

[b] Ferno's iNX patient transport cot uses carbon fiber/SMC legs to minimize weight and reduce risk of injury to EMTs. Each leg is designed for a load of up to 590 kg, twice the load rating of the entire cot. *(Photo courtesy of Ferno)*

[c] These aramid fiber-reinforced nylon orthotics produced by the Mark One 3D composite printer enable customization to the individual without a cost or time penalty, as well as added functionality via embedded pressure sensor and/or RFID tag inserts. *(Photo courtesy of Markforged)*

[d] Invibio's ENDOLIGN composite is used in the development of orthopedic, trauma and spinal implants, such as these from icotec AG. Icotec develops and produces parts using high continuous fiber content composites and thermoplastic matrices. *(Photo courtesy of Invibio)*

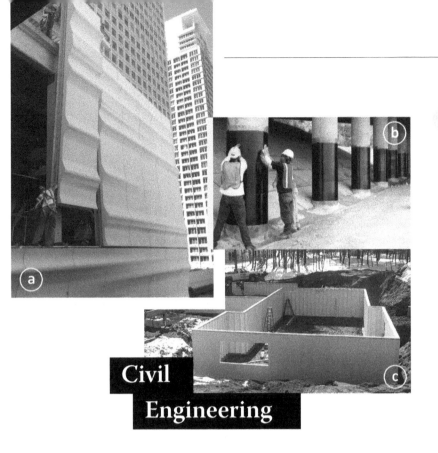

FIGURE 1-16. Civil engineering and construction.

[a] The San Francisco Museum of Modern Art (SFMOMA) expansion uses over 700 fiber-reinforced polymer (FRP) composite panels totaling 7,804 m² on a 10-story curved facade. The lightweight integrated water barrier/insulation/façade panels—which passed the rigorous NFPA 285 fire test, allowing their use on a high-rise exterior—saved three construction passes around the building and 1 million pounds of steel secondary structure. *(Photo courtesy of Felix Weber, Arup and SFMOMA)*

[b] The Utah Department of Transportation (UDOT) strengthened 76 columns using carbon fiber-reinforced polymer wraps. *(Photo courtesy of Sika Corporation)*

[c] Composite Panel Systems' EPITOME brand composite foundation wall system offers residential and other construction applications thinner walls (7 inches vs. 12 for traditional stud-framed and concrete) for more usable space, energy efficiency (R-16.5 insulation value) and fast, easy installation (a foundation can be constructed in less than 2 hours). *(Photo courtesy of Composite Panel Systems)*

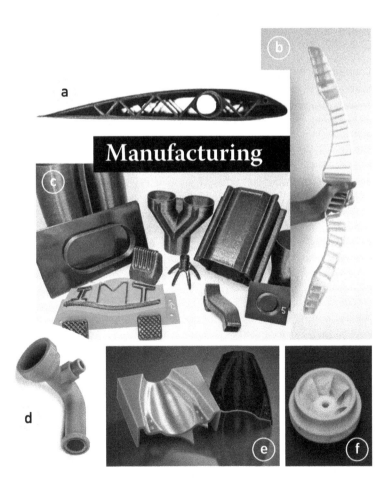

FIGURE 1-17. [a] A continuous carbon-fiber reinforced wing (length ≈20 cm) is 3D printed using 9T Labs' additive fusion technology which enables high fiber volume, low-void content, continuous fiber composites. *(Photo courtesy of 9T Labs)*

[b] Continuous glass-fiber-reinforced epoxy archery bow riser, printed using moi composites' CFM technology, weighs 700 g vs. 1.5 kg for conventional aluminum risers and 800–900 g for Olympic composites versions, yet is simpler to manufacture and customized to the athlete's needs. *(Photo courtesy of Moi Composites)*

[c] The continuous fiber-reinforced 3D printing technology developed by McNAIR Center for Aerospace Innovation and Research, funded by TIGHITCO and available through Ingersoll Machine Tools, enables highly complex structures difficult to achieve with other composites manufacturing methods. *(Photo courtesy of McNAIR Center)*

[d] Arevo Labs uses a 6-axis robot to control a carbon fiber-reinforced high-performance thermoplastic FDM nozzle with state-of-the-art thermal management technology and software control, to produce truly 3D reinforced composite parts such as this crane component. *(Photo courtesy of Arevo Labs)*

[e] Stratasys is 3D-printing tools using Ultem 1010 PEI for making composite prepreg parts that can deliver very fast turnaround and complex shapes not economically possible in traditional tooling materials. *(Photo courtesy of Stratasys)*

[f] Markforged's Mark One and Mark Two printers use continuous carbon, glass or aramid filament-reinforced nylon in a modified fused deposition modeling (FDM) process to produce a wide range of complex-shaped composite parts with the same or better strength vs. aluminum, such as this aramid-fiber-reinforced impeller. *(Photo courtesy of Markforged)*

Advantages of Composites

Strength and stiffness. Composites typically exhibit high strength and stiffness-to-weight ratios. Composite structures can attain ratios 4 to 10 times better than those made from metals. However, lightweight structures are not automatic. Careful engineering is mandatory, and many tradeoffs are required to achieve truly lightweight structures.

Optimized structures. Fibers are oriented and layers are placed in an engineered stacking sequence to carry specific loads and achieve precise structural performance. Matrix materials are chosen to meet the service environment for which the structures are subjected. (The matrix generally determines the temperature capability of the part.)

Multi-functionality. Composite materials are being developed that not only provide lightweight, load-bearing structures, but also integrate additional functions such as structural health monitoring, thermal and/or electrical conductivity, energy harvesting/storage, impact resistance, acoustic damping, self-healing, and morphing or shape-changing.

Fatigue resistance. Composite structures do not suffer from fatigue like their metallic counterparts. High fatigue life is one reason composites are common in helicopter rotor blade construction; however, composites do exhibit some fatigue behavior, especially around fastener and pin locations. Careful design and good process controls are required to ensure long service-life of both adhesively bonded and mechanically fastened composite joints.

Corrosion issues. Composites themselves do not corrode, hence their popularity within the marine industry. This is also pertinent to chemical plants, fuel storage and piping, and other applications that must withstand chemical attack.

Part geometry. Composite materials are easily molded to shape. Composites can be formed into almost any geometry, usually quite easily and without costly trade-offs in structural properties. Truly monocoque structures are possible with proper tooling.

Reduced parts count. Integrated composite structures often replace multi-part assemblies, dramatically reducing part and fastener count as well as procurement and manufacturing costs. Sometimes adhesively bonded or welded thermoplastic assemblies can almost completely eliminate fasteners, further reducing part count and production time.

Lower tooling costs. Many composites are manufactured using one-sided tools made from composites or Invar, versus the more expensive multi-piece metal closed cavity, machine-tooling or large two-sided die sets that are normally required for injection molding of plastics and metal forming processes.

Aerodynamically smooth surfaces. Adhesively bonded structures offer smoother surfaces than riveted structures. Composite skins offer increased aerodynamic efficiency, whereas large, thin-skinned metallic structures may exhibit buckling between frames under load (i.e. "oil-canning").

Low observable (LO) or "stealth" characteristics. Some composite materials can absorb radar and sonar signals and thereby reduce or eliminate "observation" by electronic means. Other materials are "transparent" to radar and work well as a radar "window" in radome applications.

Disadvantages of Composites

Material cost. High performance composite materials and processes typically cost more than wood, metal and concrete. The cost of oil and petroleum-based raw-material products often drives the price of these materials.

Storage and handling. Many materials such as film adhesives and prepregs have a limited working time (out-time), usually measured in days. Most prepregs also require frozen storage, and may have a limited shelf life of a few months to a year. Material management is critical with these materials. Some fibers and core materials also require special storage to prevent moisture absorption and/or must be oven-dried before use to prevent volatization of moisture during processing.

Recyclability. Although thermoplastic composites offer the easiest recycling, the supply chain for dry carbon fiber textile waste and carbon fiber-reinforced thermoset composites is growing. Still, in general, recycling of composites is not as straightforward as with metals, wood, or unreinforced plastics. Research continues in this area, and new recyclable thermosets (vitrimers) are now available.

Labor-intensive. Tailoring of properties ordinarily requires exact material placement, either through hand layup or automated processes. Key developments include automated manufacturing and inline inspection processes.

Capital equipment. Ovens, autoclaves, presses, controllers and software are expensive to buy and to operate. Automated machines and programming costs can be a considerable investment.

Damage tolerance. Thin-skinned sandwich panels are particularly susceptible to damage from low-energy impacts. Such impacts can produce delaminations, disbonds, or other structural damage that may not be readily apparent during a visual inspection. Moisture intrusion is a special problem in honeycomb sandwich panels, which is also not apparent by visual inspection. Sometimes water intrusion can occur during manufacturing.

Training requirements. A high degree of knowledge and skill is required to properly fabricate and repair advanced composite structures. Robust initial training as well as recurrent training in this area is mandatory.

Corrosion issues. Carbon fiber reinforcements can cause galvanic corrosion. Metals at the anodic end of the galvanic scale, such as aluminum and magnesium, will corrode if in direct contact with carbon fiber reinforced structures. A conductive path is all that is required for corrosion to begin, so it is important that these metals be electrically insulated from any adjoining or adjacent carbon fiber structure.

Health and safety. Health issues from working with composite materials may include dermatitis from exposure to resins, complications from inhaling respirable fibers, and exposure to suspected carcinogens in some uncured matrix systems. Other safety concerns include transport, storage and disposal of solvents and other materials classified as hazardous.

Composites Development Timeline

1930s

1932   DuPont launches the first polyamide thermoplastic polymer, nylon 6-6.

1936   DuPont is awarded patent for unsaturated polyester resin. Fiberglas® is patented by Owens-Illinois and Corning Glass.

1937   Ray Greene makes the first fiberglass/polyester sailboat.

1938   P. Castan and S. Greenlee first patent epoxy resin chemistry.

1940s

1942   Owens-Corning begins making fiberglass/polyester parts for WWII aircraft.
Ray Greene produces a fiberglass/polyester composite day sailer.

1944   First plane with a GFRP fuselage flown at Wright-Patterson Air Force Base.
Allied forces land at Normandy in ships made of GFRP components.

1945   Over 7 million pounds of fiberglass shipped for primarily military applications.
Hartzite, a proprietary composite material, is developed and used in the construction of propeller blades by the Hartzell Company.

1946   Owens-Corning produces fiberglass reinforced polymer fishing rods, serving trays and pleasure boats.
Ciba commercializes the first epoxy resin.

1947   U.S. Navy issues contract to R.E. Young and M.W. Kellogg for filament winding machinery to manufacture fiberglass-polyester rocket motors and FRP pipe.

1948   Several thousand commercial FRP pleasure boats are produced.
Glastic Corporation develops sheet molding compound (SMC) and bulk molding compound (BMC) type materials for making automotive parts.
FRP pipe is introduced.

1950s

1950   Goldsworthy develops pultrusion composites manufacturing process.
Muskat patents Marco method for liquid resin infusion of FRP parts.
FRP composites are used for equipment in the chemical, pulp and paper, waste treatment, and other industries for corrosion resistance.

1953   The Chevy Corvette is the first automobile with a body made entirely from fiberglass reinforced plastic.

1957   Scientists invent carbon fibers from cotton and rayon.

1958   Saint-Gobain produces fiberglass composite helicopter blades for the Alouette II in their Chambery, France factory.

1959   Smith patents process for vacuum molding large FRP structures.

See "Bibliography and Acknowledgments", page 424, for credit information on these timeline photos.

1960s

1960   Nancy Layman creates the first semi-permanent polymer release agent.

1961   Expanded hexagonal paper cores are used in building panels.
Carbon fibers are first manufactured from polyacrylonitrile (PAN).

1964   H-301 Libelle ("Dragonfly") non-powered glider is the first all-fiberglass
composite aircraft to be issued a type certificate (in U.S. and Germany).
Geringer receives patents for Resin Transfer Molding (RTM) process.

1969   Rigid foams are used as core materials in building panels.
Boron-epoxy rudders are installed on an F-4 jet made by General Dynamics.
Windecker AC-7 is the first composite aircraft to receive FAA certification.

1970s

1971   DuPont releases Kevlar® aramid fiber.
Small amounts of PAN carbon fiber are being sold to
industry.
Hexcel develops the first production ski with a
honeycomb core.

1973   Development work begins on carbon fiber fishing poles
and golf club shafts.
The first fiberglass/foam core water ski is developed.

1974   Lockheed delivers C-130 transporter with boron-epoxy composite wing to
U.S. Air Force.

1975   The first carbon-tubed, metal-lugged bicycle frame is built by Exxon Graftek but
suffers frequent failures.

1976   Modified Vought A-7D Corsair II attack bomber, with graphite, boron, and epoxy
composite wing, flies.

1977   Bill Lear designs the Lear Fan, the first all carbon fiber-epoxy airframe. The aircraft
never made it to certification, however almost every composite aircraft design today
owes something to the technology developed and pioneered at Lear Fan.

1978   Hartzell reintroduces composite propeller blades, the industry's first structural
advanced composite blade for the CASA 212 utility aircraft.

1980s

1981   John Barnard develops, and Hercules Aerospace
builds the first carbon fiber monocoque chassis for
the McLaren Formula-One racing team.

1982   Five Boeing 737-200 commercial airliners are placed
into service with composite horizontal stabilizers.

1985   Carbon fiber bicycle frames are produced in quantity
by Trek.
Airbus A310 is the first commercial aircraft to feature a carbon fiber/epoxy
composite vertical fin (vertical stabilizer) torque box.

1986   Resin film infusion process and apparatus is patented.
World's first highway bridge using composites reinforcing tendons is built in Germany.
Voyager carbon fiber/epoxy composite aircraft built by the Rutan Aircraft Company/
Voyager Aircraft; it made the first nonstop flight around the world without refueling.

1987   350 composite components, mostly secondary structures, entered into commercial airline flight service.

1988   The Beech Starship is the first all-composite engine-powered aircraft to receive FAA certification with a carbon fiber fuselage and wing.
Airbus A320 is the first commercial aircraft to enter service with a complete composite vertical tail.

1989   B-2 bomber flies, featuring all-composite carbon fiber skins and structures, the first aircraft to use composites so extensively.
Goode Skis invents the first carbon fiber snow ski pole.

1990s

1990   Bill Seemann patents Seemann Composites Resin Infusion Molding Process (SCRIMP).

1992   First all-composites pedestrian bridge installed in Aberfeldy, Scotland.

1994   Goode Skis develops the first carbon fiber water ski.

1995   Boeing 777 entered service, boasting the first all-composite empennage on a commercial jet (complete tail assembly including horizontal and vertical stabilizers, elevators, and rudder).

1996   First FRP reinforced concrete bridge deck was built at McKinleyville, WV, followed by the first all-composite vehicular bridge deck in Russell, KS.

2000s

2001   Designs for Airbus A380 finalized, including use of non-crimp fabric and resin film infusion (RFI) for the CFRP rear pressure bulkhead, reducing cost 27% vs. prepreg version used for A340-500/600.

2002   Airbus A340 enters service as the first aircraft to replace aluminum with thermoplastic composites on wing components, using fiberglass-reinforced PPS j-nose leading edge structures made by Stork Fokker AESP.

2003   Completion of the first A380 center wing box section at Airbus Nantes — made using automated tape laying (ATL) and autoclave cure — which will be the first CFRP center wing box to fly on a jet airliner.

2004   Oxeon introduces spread tow fabrics, which via a patented machine using heat and vibration, spreads a 12k carbon fiber tow from 5mm to 25mm widths, achieving an 80% reduction in thickness and 20% cut in weight. These spread tow tapes are then woven into crimpless (higher strength) and oblique angle (+50°/-25°) fabrics, used in sporting goods, aircraft, America's Cup racing boats and more.

2005   Boeing Aerostructures Australia (previously Hawker de Havilland) demonstrates the Controlled Atmospheric Pressure Resin Infusion (CAPRI) process via maximum-span 9.1m long CFRP skins for the Boeing 787's moveable trailing edges, which it will produce out of autoclave.

2005   Airbus parent company (then EADS) selected to build Boeing 787 rear pressure bulkhead using its patented Vacuum Assisted Process (VAP®), which is a type of resin infusion employing semipermeable membranes for VOC removal and low void content.

2007   Zyvex Technologies introduces the world's first commercial carbon nanotube enhanced products in a partnership with Easton Sports.

2008   Assembly and proof testing completed for the X-55A Advanced Composite Cargo Aircraft (ACCA) featuring a 19.8m/65-ft long out of autoclave (OOA) composite fuselage, the first in the industry, made with vacuum-bagged, oven-cured OOA MTM-45-1 carbon fiber/epoxy prepreg.

2008   Advanced Composites Group (ACG, purchased by Umeco, Cytec) qualified MTM44-1 OOA prepreg at Airbus for use on the CFRP outer and mid-section fixed trailing edge panels for the A350.

2009   First flight of the Boeing 787 Dreamliner, featuring a carbon fiber/epoxy fuselage, wing, center wing box and vertical tail. In fact, composites make up more than 50% of the aircraft.

2009   Huntsman Polyurethanes introduces VITROX resin that combines isocyanates, polyols and a unique, proprietary catalyst system that permits processors to "dial in" a desirable gel time and viscosity profile, offering a tailorable/tunable pot life and cure time.

2009   BMW uses fast-cure (snap-cure) epoxy resins to mold carbon fiber roofs for the M3 and M6 models using high-pressure resin transfer molding (HP-RTM).

2010s

2010   Airbus Helicopter (Donauworth, Germany) awarded contract to build all doors for A350 using RTM.

2013   First flight of the Airbus A350 XWB, also with a composite fuselage, wing and vertical tail, and claiming overall 53% composite materials.

2013   BMW begins production of the composites-intensive i3 electric car, including CFRP life module and roof, reaching 100 cars/day by 2014.

2013   After 20 years of research, GE puts ceramic matrix composites (CMCs) into production, building the 125,000 ft^2 Asheville, NC facility to produce the static first stage HP compressor shroud for the LEAP aircraft engine, and GE9X inner and outer combustion liners and stage 1 and 2 nozzles.

2014   First flight of CFM International (GE/Safran joint venture) LEAP engine, featuring 19 turbofan blades and fan case made with 3D carbon fabric preforms and epoxy resin using the resin transfer molding (RTM) process.

2014   Markforged begins delivery of the world's first carbon fiber 3D printer — the Mark One — which uses two print heads to build complex structures from continuous carbon, glass or aramid filaments and nylon resin via a modified fused deposition modeling (FDM) process.

2014   First 3D-printed composite body produced for the Strati electric car at the IMTS show by Local Motors, Oak Ridge National Laboratory (ORNL) and Cincinnati Incorporated's Big Area Additive Manufacturing (BAAM) machine, using a 15% chopped carbon-fiber-reinforced ABS compound.

2014   Nanocomp Technologies receives Edison Award for sheets, tape and yarn using macro-scale carbon nanotube (CNT) fibers to improve strength, flexure fatigue, and thermal and electrical conductivity in composites and other structures; e.g., 100 layers of CNT sheet (the thickness of several business cards) can stop a 9mm bullet, while other products reduce aircraft wiring harness weight by up to 70%.

2015   N12 Technologies launches NanoStitch™, the first commercial product using vertically aligned carbon nanotubes (VACNTs) in a drop-in layer between plies to prevent delamination, increase interlaminar shear strength and double fatigue life in prepreg laminates.

2015   Kreysler & Associates installs over 700 of its FRP panels—the first to pass NFPA 285 fire testing for use as cladding on multi-story buildings—on the SFMOMA expansion.

2017   Dassault Falcon Jet uses 3D printed tooling made with a Stratasys Fortus 900mc and unreinforced Ultem 1010 PEI resin.

2017   Preforming is fully industrialized, with dozens of commercially available automated systems that can produce tailored preforms from dry fiber, prepreg and thermoplastic composite materials for parts molded in minutes.

2017   First flight for Russian MS-21 jetliner, the first commercial aircraft with a CFRP wing and wing box made using out-of-autoclave resin infusion.

2018   Automated inline inspection systems are qualified for AFP composite aerostructures by MTorres and Electroimpact.

2019   Unidirectional carbon fiber thermoplastic prepreg tapes gain applications in consumer electronics, sporting/consumer goods and transportation, as suppliers proliferate and begin the process for commercial aircraft qualification.

2019   3D printing with continuous fiber also expands as multiple companies commercialize systems, including Arevo, CEAD, Continuous Composites, Moi Composites, Orbital Composites, Anisoprint, 9T Labs and more.

2020s

2020   Electroimpact combined an out-of-autoclave, in-situ consolidated, thermoplastic (TP) automated fiber placement process and 3D advanced fused filament fabrication printing process into a unified Scalable Composite Robotic Additive Manufacturing (SCRAM) system, enabling quasi-toolless fabrication of aerospace-quality TP composite structures.

2021   MTorres develops dry carbon fiber unidirectional tape along with an automated fiber placement system for use in post-layup resin infusion processing of large aircraft fuselage and wing components.

2022   Joby Aviation, Inc., announced that it had received Part 135 Air Carrier Certification from the FAA for its composite-intensive eVTOL aircraft slated for use as an air taxi. Joby leads the way for other urban air mobility (UAM) vehicles currently in development.

2023   The National Center for Advanced Materials Performance (NCAMP) qualifies Victrex 250-AS4 unidirectional tape for publication in their allowables database for use in aerospace applications.
Toray announced that they have developed TORAYCA™ T1200 carbon fiber, the world's highest strength at 1,160 kilopounds per square inch (Ksi).

2024   Fraunhofer IWS completes Multifunctional Fuselage Demonstrator (MFFD) using CONTIjoin technology, welding eight-foot-long upper and lower carbon fiber reinforced thermoplastic composite fuselage sections together with a CO_2 laser process.

2

Matrix Technology

CONTENTS

Matrix Systems Overview

A composite matrix acts to bond and/or encapsulate the fiber reinforcement, enabling the transfer of loads from fiber to fiber. It also moderately protects the fibers from degradation due to environmental effects, including moisture, ultraviolet (UV) radiation, chemical attack, minor abrasion and impacts. Matrix materials can be molded, cast, or formed to shape. Types include: polymeric (plastic), metallic, and ceramic. Matrix-dominated structural properties include compression, interlaminar shear, and ultimate service temperature.

Selection of a matrix material has a major influence on the shear properties of a composite laminate, including interlaminar shear and in-plane shear. The interlaminar shear strength is important for structures functioning under bending loads, whereas the in-plane shear strength is important under torsion loads. The matrix also provides resistance to fiber buckling in a laminate under compression loads and therefore is considered a major factor in the compressive strength of a composite.

Thermoset resins are primarily used for highly loaded structures because of their high strength and relative ease of processing. Thermoplastic resins are utilized where toughness or impact resistance is desired, or when high-volume production dictates the need for a fast processing material. Metallic and ceramic matrices such as titanium and carbon are primarily considered for very high temperature applications (> 650°F/343°C.)

Thermosets

With thermoset resins, the molecules are chemically reacted and joined together by **crosslinking**, forming a rigid, three-dimensional network structure. Once these crosslinks are formed during cure, the molecules become locked-in and cannot be melted or reshaped again by the application of heat and pressure. However, when a thermoset has an exceptionally low number of crosslinks, it may still be possible to soften it at elevated temperatures. (*See also* "Vitrimers (Thermoset–Thermoplastics)," page 38.)

Thermoset resins typically require time to fully react or "cure" at temperatures ranging from room temperature to upwards of 650°F (343°C), depending on the chemistry. Examples of thermoset matrix materials include as follows: polyester, vinyl ester, polyurethane, epoxy, phenolic, benzoxazine, cyanate ester, bismaleimide, and polyimide resins.

Many thermoset resins bond well to fibers and to other materials; for this reason, many thermoset resins are also used as structural adhesives, paints, and coatings. Structural properties for some common thermoset matrix systems at 77°F (25°C) are shown in Tables 2.1 and 2.2.

TABLE 2.1 Typical Thermoset Matrix Systems for Composites

Matrix System	Tensile Strength Ksi (MPa)	Tensile Modulus Msi (GPa)	% Elongation to Failure	*Cost
Polyester (UP)	3–11 (20.7–75.8)	0.41–.0.50 (2.8–3.4)	1–5	Low
Vinyl Ester (VE)	10–12 (68.9–82.7)	0.49–0.56 (3.4–3.9)	3–12	Low–Med
Polyurethane (PU)	9–15 (62.1–103.4)	0.35–0.48 (2.4–3.3)	6–14	Low–Med
Epoxy (EP)	7–13 (48.3–89.6)	0.39–0.54 (2.7–3.7)	2–9	Medium
Phenolic (PF)	7–9 (48.2–62.1)	0.43–0.60 (2.9–4.1)	1–2	Low–Med
Benzoxazine (BZ)	7–16 (48.3–110.3)	0.49–0.81 (3.4–5.6)	1–5	Medium
Bismaleimide (BMI)	7–13 (48.3–89.6)	0.48–0.62 (3.3–4.3)	1–3	High
Cyanate Ester (CE)	7–13 (48.3–89.6)	0.40–0.50 (2.8–3.4)	2–4	Very High
Polyimide (PI)	5–17 (34.5–117.2)	0.20–0.70 (1.4–4.8)	1–4	Very High

*Relative cost comparison to polyester.

TABLE 2.2 Initial Cure Temperature vs. Service Temperature for Thermosets

Matrix System	Initial Cure Temperature	Maximum Service Temperature*
Polyester (UP)	R/T–250°F ▪ R/T–121°C	135°–285°F ▪ 58°–140°C
Vinyl Ester (VE)	R/T–200°F ▪ R/T–93°C	120°–320°F ▪ 49°–160°C
Polyurethane (PU)	R/T–390°F ▪ R/T–200°C	140°–355°F ▪ 60°–180°C
Epoxy (EP)	R/T–350°F ▪ R/T–177°C	120°–360°F ▪ 49°–182°C
Phenolic (PF)	140°F–250°C ▪ 60°–121°C	300°–500°F ▪ 148°–260°C
Benzoxazine (BZ)	300°–475°F ▪ 148°–246°C	250°–465°F ▪ 121°–240°C
Bismaleimide (BMI)	375°–550°F ▪ 190°–288°C	400°–540°F ▪ 204°–282°C
Cyanate Ester (CE)	250°–350°F ▪ 121°–177°C	200°–600°F ▪ 93°–316°C
Polyimide (PI)	640°–750°F ▪ 316°–399°C	500°–600°F ▪ 260°–316°C

*Note: The operational service temperature of any thermoset resin will largely depend upon the ultimate Glass Transition Temperature (T_g) of a specified resin chemistry, as well as the cure/post-cure time and temperature that the resin has seen during processing.

❱ GLASS TRANSITION TEMPERATURE (T_g) AND SERVICE TEMPERATURE

The glass transition temperature is the temperature at which increased molecular mobility results in significant changes in the properties of a solid polymeric resin or fiber. In this case, the upper temperature "glass transition" (or T_g, which is pronounced "t-sub-g") refers to the transition in behavior from rigid to "rubbery."

This can be thought of as the temperature above which the mechanical properties of a cured thermoset polymer are diminished. While it is not necessarily harmful for a structure to see temperatures moderately above the T_g, the structure should always be supported above this temperature to prevent laminate distortion.

When curing a thermoset polymer, the "rate of cure" (or rate of chemical reaction) is accomplished faster above the glass transition temperature than below it. Therefore, final cure temperatures are typically engineered to be as near as practical to the final desired T_g in order to minimize the cure time.

During processing it is important to differentiate between the "state-of-cure" (percent of chemical reaction completed) and the glass transition temperature. A common misconception is that a selected cure temperature alone determines the final glass transition temperature of the polymer. In the case of a *partially reacted* thermoset polymer, it will continue to "cure" over time at a given temperature until it is completely reacted, which can actually elevate the T_g. For example, a thermoset polymer that is heated to 350°F (177°C) may initially have a T_g at or below this temperature. However, after several hours at this temperature, it may ultimately attain a much higher T_g. Therefore, just attaining a specific cure temperature for a limited time will not necessarily produce the desired ultimate glass transition temperature of the polymer.

A similar concern is that a material may appear to be properly cured because it has been to a specified temperature and it exhibits the required strength and stiffness properties when tested at room temperature, but it may have not yet achieved a full state of cure nor the ultimate T_g required to carry design loads under "hot/wet" service conditions.

Alternatively, a *fully reacted* polymer with an ultimate T_g of, for example, 275°F (135°C), will not necessarily increase regardless of added temperature or time at that temperature (*see* Figure 2-1 on the next page).

After initially processing a polymer to a full state of cure, the resulting T_g is typically referred to as the "dry" T_g. Ingress of moisture or other fluids into the structure when exposed to a hot-wet service environment reduces the T_g. This is often referred to as the "wet" T_g and it is considered to be the upper temperature limitation of the polymer. As a general rule, the maximum designed service temperature limit of a composite structure is often well below this threshold.

The T_g of a given **neat-resin** or composite sample can be established using one of the following methods of Thermal Analysis (TA): Thermomechanical Analysis (TMA), Dynamic Mechanical Analysis (DMA), or Differential Scanning Calorimetry (DSC).

❱ HEAT DEFLECTION TEMPERATURE, OR HEAT DISTORTION TEMPERATURE (HDT)

Many resin manufacturers will cite the heat deflection temperature (HDT) as a measure of the upper temperature capability of a resin in lieu of the T_g. The HDT is a simpler concept to understand than that of T_g outside of the scientific laboratory, and is used by many part manufacturers as a reliable method of thermal analysis. Both methods are used to define the upper temperature limitations of composite materials within industry.

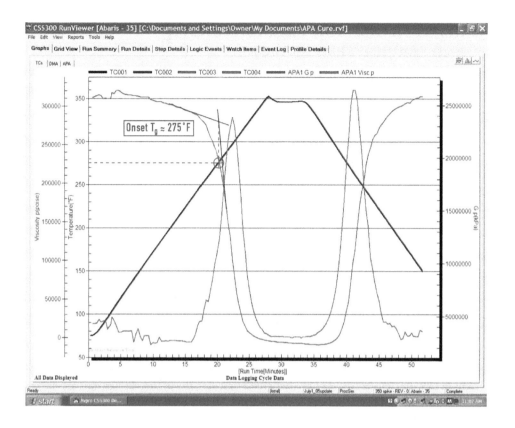

FIGURE 2-1. Example of a post-process DMA to establish a T_g.

Tested on a parallel plate rheometer to approximately 350°F (177°C). The purple trace (Gp) is measuring the modulus (stiffness) of the sample. As the sample is heated, it exceeds its T_g and the modulus drops rapidly. The onset of this slope is approximately 275°F (135°C) according to the graph. *(Rheometer, DSC, process control equipment, software, and all thermal analysis data/graphs courtesy of AvPro, Inc., Norman, OK in collaboration with Abaris Training Resources, Inc. Reno, NV)*

The HDT is measured differently than T_g. The common test is ASTM D648 or ISO 75. This test measures the temperature at which a standard size rectangular bar of cured resin or laminate deflects in a 3-point bending fixture a total distance of 0.010 inches (0.25 mm) under a constant load of 66 psi (0.46 MPa) or 264 psi (1.8 MPa), depending upon the which test method is specified. (Method A or B respectively, per the ISO standard.)

It should be noted that while a given matrix resin would soften at a similar temperature when measured using other methods of thermal analysis, the influence of fiber type (e.g., carbon vs. glass) and fiber axial arrangement in a laminated composite specimen can affect the HDT results. As a result, the fiber type and axial arrangement must be consistent for comparison of one sample set to another when using this test.

❯ COMMON THERMOSET MATRIX SYSTEMS

Polyester

Unsaturated polyester resins typically achieve crosslinking through a multi-phase condensation reaction. They are inexpensive and provide fairly good environmental resistance. Polyesters have lower strength and exhibit higher shrinkage than epoxies during cure. They generally contain between 30–50% styrene in the formula, producing considerable emissions. Polyester resins are primarily used in marine, transportation, and industrial applications, having a large market share in these and related industries.

Vinyl Ester

Vinyl esters have an epoxy backbone with vinyl groups at the end of the molecular chain, connected by ester linkages. Thus, they achieve crosslinking through a multi-phase condensation reaction, similar to polyester, and yet offer better environmental resistance, strength and fiber adhesion than polyester resins due to their epoxy components. For this reason, they are often called "epoxy vinyl ester resins" (EVER). They exhibit less shrinkage than polyester, but more than epoxy. Styrene content and emissions are roughly the same as polyester resins (30–50%).

Polyurethane

Polyurethane resins are formed by reacting isocyanate with polyol monomers. Most are thermosets (PU or PUR) but thermoplastic polyurethanes (TPU) are also widely used. Polyurethanes exhibit high toughness and fracture resistance and have a much higher elongation to yield than many other resins. They have good adhesive properties and are fairly low cost when compared to epoxies. New generation formulas are designed for structural applications and have suitable thermal properties, similar to that of epoxies, depending upon formulation. There are several tunable or snap-cure formulas available.

Epoxy

The workhorse of the *advanced* composites industry, epoxy resins offer better mechanical properties than polyester or vinyl ester resins and also exhibit superior adhesion properties. Epoxies cure by crosslinking reactive polymers at both ambient and elevated temperatures. They make for great adhesives and provide good environmental resistance, but can emit large quantities of toxic smoke when burned unless fire retardant chemicals are added to aid ignition. More expensive than polyester and vinyl ester, epoxies are available with a wide variety of mechanical and thermal properties.

Phenolic

Phenolic resins are obtained by simply reacting phenol and formaldehyde (phenol formaldehyde resin) under either acidic or basic conditions. Acidic conditions lead to solid phenolic resin known as novolak, while basic conditions lead to liquid type resin known as resole. Resole is the preferred phenolic liquid resin for preparing phenolic resin fiber reinforced composites. Resole undergoes cure by thermal, acidic, or basic conditions with the preferred method being thermal. The resulting cured phenolic resin is fairly brittle with good to excellent adhesive characteristics depending on components that are bonded.

   Phenolic resins exhibit good chemical resistance and excellent **static-dissipative** properties. Phenolic resins possess better thermal stability than epoxies at high temperatures as well as excellent flame resistance and low smoke toxicity (FST characteristics). At higher temperatures they undergo carbonization. Thus, phenolics are often used to make carbon-carbon composites that possess exceptional low weight, very high temperature thermal stability, and outstanding ablative properties.

Benzoxazine

Benzoxazine resins are in the same chemical family as phenolic resins and exhibit similar flame resistance and low smoke toxicity but without the health risks associated with phenol formaldehyde. They have good high temperature service properties, can be stored for long periods at room temperature, and have low initial shrinkage and low exothermic potential during cure. The chemistry is such that it can easily be hybridized with epoxy or other formulations for improved overall processing capabilities and high performance properties. Benzoxazine resins are more adaptable to closed molding processes than phenolic.

Cyanate Ester

Cyanate ester resins offer an excellent balance of mechanical performance and toughness. They are very sensitive to moisture uptake prior to processing, but provide high temperature service capabilities after post-cure. They also exhibit minimal microcracking and low moisture absorption after cure. Cyanate ester resins have a low coefficient of thermal expansion (CTE) compared to epoxy. They provide good dielectric properties and are somewhat expensive.

Bismaleimide

Often referred to as BMI, these resins are part of the polyimide family and deliver better thermal stability than epoxies with comparable processing. They typically have very low viscosity when processed and can be brittle, although improved toughness formulations are available. BMI resins exhibit good hot/wet properties, and are typically more expensive than epoxies.

Polyimide

Polyimides exist in both thermoplastic and thermoset formulations. Typically, they require a high temperature cure (640–750°F/316–399°C) and can be difficult to process. They provide good flame resistance and low smoke-toxicity, but are fairly expensive and can be difficult to process.

❱ HYBRID RESINS

Hybrids are blends of resin chemistries that achieve tailored optimization of properties, generally for toughness, impact performance and/or higher heat resistance. Examples include urethane acrylic, urethane ester, and vinyl polyurethane. Hybrids generally have good compatibility with glass fibers and may be enhanced for improved bonding with other reinforcements like carbon, aramid and other high-performance fibers.

TABLE 2.3 Hybrid Resins

Product	Hybrid	Benefits	Applications
Advalite **Company** Reichold	• Vinyl acrylate • Urethane acrylate and other chemistries	• Styrene-free • Monomer-free • Temp resistance to 200°C • Rapid cure • Unsaturated cure chemistry but better properties	• Automotive floorboards, battery trays • Consumer and sporting goods
Crestapol **Company** Scott Bader	• Urethane acrylate	• High toughness • High mechanical properties • Rapid cure • Can be tailored for infusion, fire retardance, pultrusion, etc.	• Boat hulls and decks, car parts
Xycon **Company** Polynt Composites	• Urethane ester	• Low moisture uptake • High toughness • Temp resistance to 120°-176°C • Low viscosity • Room temp or elevated temp cure possible • Rapid dual-cure through urethane reaction and conventional peroxide • Can be made into a prepreg including SMC and BMC	• Manhole covers, utility boxes, poles, oil sucker rods, barrier coats, structural high fiber products

TABLE 2.3 Hybrid Resins *(continued)*

Product	Hybrid	Benefits	Applications
P-100 **Company** Plexinate	• Vinyl polyurethane	• Styrenated, one component processing • High toughness • Priced between PE and VE • Can be modified for pultrusion	• Hulls and decks of Elliott sailboats • Pultrusion
CSRVE **Company** Polynt Composites	• Hybrid VE*	• High toughness • Low viscosity • High fracture toughness • High elongation • Reduced shrinkage vs. conventional VE • Compatible with multiple reinforcing fibers (glass, aramid, carbon)	• Tanks, pipes, transportation, marine, filament winding, hand lay- up, applications where high impact damage is expected. Can be blended with conventional UPR and VEs
781-7300 **Company** Polynt Composites	• VE acrylate	• Styrene-free • High mechanical properties • Compatible with multiple reinforcing fibers	• CIPP (cured-in-place pipe)

*VE=Vinyl Ester

❭ BIO-RESINS

Bio-based resins and polymers may be developed in a variety of ways. One is to simply replace part or all of the traditional petrochemical building blocks with ones derived from plants or other biomass. For example, glycerol—a byproduct of rapeseed oil used in biodiesel production—can be used to make epichlorohydrin (ECH), which is used in production of epoxy resin. With abundant feedstocks, due to increased biodiesel production, glycerol-to-ECH is not only cheaper, it also avoids solvents and results in less chlorinated waste. Producers include Olin Epoxy (previously Dow Epoxy). Glycerol may also be used to synthesize acrylic acid or propylene glycol, used to produce acrylic and unsaturated polyester resins, respectively.

In addition to these "drop-in" solutions, it is also possible to make new building blocks for resins from biomass. This could involve, for example, epoxidized vegetable oils for epoxy resins, phenols from lignin for phenol formaldehyde resins, or sugar derivatives such as polyol in polyurethane resins.[1] *See* Table 2.4.

Current, first generation bio feedstocks are mainly sugars and starches from corn, cane, potatoes, beets and grains. Second generation lingo-cellulose feedstocks are 1,000 times more abundant, and include forestry and agricultural waste and grasses. Future, third generation feedstocks include macro and micro algae and bacteria, and are even more abundant.

(The text below is edited with permission from several articles available at CompositesWorld.com.)

Ashland Performance Materials commercialized Envirez unsaturated polyester resin (UPR), made from corn-based alcohols and soybean oil, in 2002. It was used initially in a program to make tractor body panels from SMC at John Deere. Envirez has since expanded into nearly all markets and manufacturing processes as a "drop-in" replacement for petrochemical UPRs, particularly for building and construction companies involved in the U.S. Green Building Council's popular LEED (Leadership in Energy and Environmental Design) building program.

---

1 Wageningen University & Research (Wageningen, The Netherlands) https://www.wageningenur.nl/en/show/Biobased-resins.htm

TABLE 2.4 Bio-based Resins

Product Name	Product Type	Supplier	Bio Source	Bio Content
Envirez	Unsaturated polyester resin (UPR)	Ashland	corn-based alcohols, soybean oil	8–20%
Envirolite	UPR	Reichhold	soybean oil	up to 25%
EkoTek	UPR	AOC	soy, corn	up to 30%
Maleinated Acrylated Epoxidized Soybean Oil (MAESO)	Similar to UPR	Dixie Chemical	soy	up to 65% 85% with MFA
Maleinated Acrylated Epoxidized Linseed Oil (MAELO)	Similar to UPR	Dixie Chemical	Linseed	up to 65% 85% with MFA
Methacrylated Fatty Acid (MFA)	Partial or total styrene replacement (UPR, VE)	Dixie Chemical	Palm kernel or coconut oil	60%
Super Sap	Epoxy resin	Entropy Resins	Pine oil plus pulp & paper and/or bio-diesel waste	20–40%
Epicerol	Substitute for propylene-based ECH in epoxy resin	Solvay	Glycerol (bio-diesel and rapeseed oil byproduct)	30% in epoxy resin
GreenPoxy	Epoxy resin	Sicomin	Plant matter	28–51%
JEFFADD™ B650	Polyol for polyurethane spray & rigid foam, coatings and adhesives	Huntsman	Vegetable Oil	65%
Bio-based epoxy	Epoxy	Huntsman	Plant matter	50–85%
Ingeo	Polylactic Acid (PLA)	NatureWorks	Plant sugars (corn starch, sugar cane sucrose)	32–80%
Bio PP	Polypropylene (PP)	NatureWorks	Plant sugars (corn starch, sugar cane sucrose)	32–80%
Bio PET	Polyethylene terephthalate (PET)	NatureWorks	Plant sugars (corn starch, sugar cane sucrose)	32–80%
Bio PE	Polyethylene (PE)	NatureWorks	Plant sugars (corn starch, sugar cane sucrose)	32–80%
Rilsan	Polyamide 11 (PA 11)	Arkema	Castor oil	100% bio
Rilsan S	PA 6/PA 6.6	Arkema	Castor oil	Up to 60% renewable
ExaPhen Novocards	Novolacs for phenolic epoxy liquid resins	Elmira Limited	Cashew nut shell liquid (CNSL)	95%
ExaPhen Polycards	Polyols	Elmira Limited	Cashew nut shell liquid (CNSL)	73–95%
SuperCoral	Epoxy Prepregs	Elmira Limited	Cashew nut shell liquid (CNSL)	16–71%

Canada's Campion Marine was one of the first volume boat builders to adopt an eco-resin, using Ashland's specially formulated Envirez L 86300 laminating series across all models in 2009, after field tests verified that the bio-resin had strength equal to, and elongation and elasticity superior to, previously used petroleum-based polyesters.

Entropy Resins, which produces Super Sap epoxy, also claims better elongation properties than petroleum-based epoxies, as well as excellent adhesion to fibers. Dixie Chemical formulates bio-based unsaturated polyesters (ortho-, iso-, and terephthalic, DCPD-modified and bisphenol A fumerate) and vinyl esters using a methacrylated fatty acid (MFA) as its reactive diluent, which can serve as a partial or total styrene replacement. It reduces emissions and odor, increases toughness while minimizing shrinkage, is less sensitive to moisture than other diluents, and can eliminate volatile organic compounds (VOCs) and hazardous air pollutants (HAPs).

SABIC produces a bio-based version of the resin diluent and clean-up solvent acetone, which has a bio-content of 18% with no loss of properties versus normal acetone. Another "green" acetone is Bio-Solv, produced by Bio Brands LLC, which is 100% bio-based, made from corn. As a Class 2 combustible liquid, it requires no special storage, yet has been demonstrated to provide five times stronger ability to clean resins and adhesives from tools and guns. With a low vapor pressure and evaporation rate, it lasts much longer than petroleum-based solvents. It leaves no residue and is easily recycled and reused.

❱ PRINCIPLES OF CURING AND CROSSLINKING

Initiated Systems

Polyester monomers want to "polymerize" together into long chains. Because of this, the resin manufacturers add "inhibitors" which prevent polymerization. Later, an initiator is used to consume these inhibitors, thus allowing the chain-polymerization to proceed. These molecular chains are then crosslinked together with a functional monomer (usually styrene) in a second phase of reaction during the cure.

As mentioned, most of these systems contain up to 50% styrene (or methacrylate or vinyl toluene) as a functional monomer. This typically leads to a strong styrene odor during mixing and curing, which may be objectionable. In addition, there are often limits imposed by regulatory authorities on airborne styrene concentrations. "Closed molding" processes are now typically prescribed for use with these systems to effectively reduce the airborne styrene emissions.

Other items, such as promoters and accelerators, are added to these resin mixes to influence the rate of cure. Cobalt Napthenate (CoNap) is a common *promoter*, and Dimethylaniline (DMA) is an *accelerator*.

Technically not a catalyst, but often referred to as such, the *initiator* starts the reaction that allows the chemistry to evolve to a cured solid. The most commonly used materials for this purpose are:

- Methyl Ethyl Ketone Peroxide (MEKP)
- Benzoyl Peroxide (BPO)

One must be particularly careful not to mix MEKP or BPO and CoNap directly together, because they will violently react with each other and catch on fire or explode. Because of this fact, the CoNap is diluted into the resin first by the manufacturer.

Normally, 0.5–2.0% by volume of the initiator is used to facilitate the reaction, depending on the ambient temperature. Within limits, one can control the speed of polymerization of these systems by varying the amount of initiator. Mixing "hot" (extra initiator) causes the reaction to happen faster and adding less ("cold") slows it down.

Condensation Reaction or Polymerization

A condensation polymerization is simply a chemical reaction wherein bifunctional monomers react to form a long chain polymer molecule and smaller molecules (e.g., water and alcohol) are released as a byproduct of the reaction. Also known as a step reaction, many thermoset resin chemistries such as polyesters, phenolics, and imides polymerize in this manner prior to maturation into a fully crosslinked molecule. This can sometimes present problems when curing at temperatures above the boiling point of water, whereas the gas (steam) that is produced must be removed or converted back into solution with pressure before the resin ultimately cures.

Epoxy Systems

Epoxies are cured by mixing the base resin (part A) with a "hardener" (part B) rather than an "initiator." Most epoxy base resins are derived from Bisphenol A or Bisphenol F materials. Bisphenol A or F is reacted with epichlorohydrin to form "diglycidyl ether of bisphenol A or F" (DGEBA or DGEBF). This is the base resin in a typical two-part epoxy system.

It might be noted that bisphenol A is the most common prepolymer used in epoxy formulations. Bisphenol F provides for lower viscosity and additional chemical resistance in an epoxy formulation. Epoxy phenol novolacs (EPN) and epoxy cresol novolacs (ECN) can also be used with epichlorohydrin in formulating a more highly crosslinked, higher modulus epoxy that has both high chemical and temperature resistance.

Several amine and anhydride curing agents are used to crosslink with epoxy resins depending on the desired end-properties:

- **Amines** are basically ammonia with one or more hydrogen atoms replaced by organic groups. Aliphatic amines and polyamines are common curing agents for epoxy resins.
- **Cycloaliphatic amines** (carbon-ring structured aliphatics). These hardeners provide better moisture and UV resistance and are often prescribed when moisture is an issue in handling or processing the epoxy resin.
- **Amides and polyamides.** Amides are basically ammonia with a hydrogen atom replaced by a carbon/oxygen and organic group.
- **Anhydride curing agents** such as Dodecenyl Succinic Anhydride (DDSA), Methyl Hexahydro Phthalic Anhydride (MHHPA), Nadic Methyl Anhydride (NMA), and Hexahydrophthalic Anhydride (HHPA), provide for good electrical and structural properties and exhibit a long working time. Anhydrides require an elevated temperature cure.

Amine-based curing agents are more durable and chemical resistant than amide-based curing agents but most have a tendency to react with water or moisture in the air. Amides, on the other hand, are more tolerant and less affected by moisture. Aliphatic amines generally have a short reaction time (short pot life) and aromatic amines a longer reaction time (longer pot life). In these systems, the base resin molecules attach to the hardener molecules when mixed. These polymers crosslink, or form strong chemical links to each other.

Since the resin and hardener molecules are crosslinked to each other to make one large molecular structure, it is of utmost importance that the two materials be mixed at the proper ratio. Unlike polyesters and vinyl esters, one cannot mix these systems "hot" or "cold." They may never obtain the desired structural or thermal properties if mixed incorrectly. These systems typically go through a change from a liquid to a gel to a solid, as illustrated in Figure 2-2.

The time it takes for a thermoset resin to gel at a given temperature is an important marker in processing these resins. All flow and compaction steps must be accomplished prior to this point. Figure 2-2 describes the various stages of a thermoset resin as it crosslinks.

FIGURE 2-2. Crosslinking: Polymerization —> Gelation —> Fully reacted.

Tunable, Snap-cure Thermoset Resins

Epoxy, polyurethane and many other thermoset resin chemistries can be formulated for greatly accelerated and tunable curing.[2] Changing the "protonation," or catalytic amine reaction at the terminal carbon atom sites, with proprietary additives results in a "thermolatent" system—that is, one that maintains very low viscosity for a specified length of time, delaying the onset of cure and allowing easy resin flow to fully wet out fiber reinforcements. This is coupled with fast reaction with the hardener when the process temperature reaches a defined target and subsequent rapid crosslinking, with cure times as low as 30 seconds. Snap-cure resins can be used with liquid or prepreg molding processes and can be formulated to react at several different temperatures or at a specific time at an isothermal temperature. Thus, they are considered to be tunable and also offer rapid molding cycles for high-rate production.

One characteristic parts manufacturers are seeking in snap-cure resins is good "hot-in/hot-out" processing. This minimizes time in the mold while preserving structural properties, and depends on how rapidly the resin can: (1) be reacted, (2) achieve its T_g, and (3) sustain sufficient modulus to maintain shape during demolding.

TABLE 2.5 Selected Snap-cure/Tunable Thermoset Resin Systems

Company	Product Name	Resin Type	*Tg(°C)	Cure Time	Cure Temp.
Cytec	XMTR 750	Epoxy	120°C	3–5 min	120–130°C
	XMTM710	Epoxy prepeg	145°C	3 min	130–150°C
Dow Automotive Systems	VORAFORCE 5330	Epoxy	120°C	30 sec–3 min	120°C
	Applications: Automotive				
	VORAFORCE 7500	Epoxy	>200°C		
Hexcel	HexPly M77	Epoxy prepeg	130°C	2 min	150°C
	Applications: Automotive, sporting goods				
Toray	G83C	Prepeg	160°C	20 min @ 149°C	85–149°C

(continued)

2 This section includes information taken from the article "Automotive composites: Thermosets for the fast zone" by Sara Black, CompositesWorld magazine, September 2015; reproduced by permission of CompositesWorld magazine, copyright 2015, Gardner Business Media, Cincinnati, Ohio, USA.

TABLE 2.5 Selected Snap-cure/Tunable Thermoset Resin Systems *(continued)*

Company	Product Name	Resin Type	*Tg(°C)	Cure Time	Cure Temp.
Henkel	Loctite MAX 2	Polyurethane	115°C	45 sec–3 min	80–130°C
	Applications: Auto suspension (leaf & coil spring, stabilizer bar), body and exteriors (backwall, roof, underbody), CFRP wheels, etc.				
	Loctite MAX 3	Polyurethane	125°C	45 sec–3 min	80–130°C
	Loctite MAX 5	Epoxy	230–270°C	8–20 min	100–150°C
Hexion/ Momentive	Epikote TRAC 06170/ Epikure TRAC 06170	Epoxy	120–135°C	45 sec–3 min	115–145°C
	Epikote TRAC 05475/ Epikure TRAC 05500	Epoxy	120–135°C	90 sec	110–140°C
	Epikote TRAC 06425/ Epikure TRAC 06465/ Epikure TRAC 06825	Epoxy	135–160°C	90 sec	135–150°C
	Epikote TRAC 06465/ Epikure TRAC 06465/ Epikure TRAC 06865	Epoxy	165–185°C	90 sec	150–170°C
Huntsman	Araldite LY3585/ Aradur 3475	Epoxy	105–115°C	2 min (RTM) 1 min (LCM)	115–140°C
	Applications: Automotive, Industrial				
	Araldite FST 40002/40003 i	Epoxy	185–260°C	5 min @150°C 1hr @100°C	100–150°C
	Applications: Aero interiors				
	Araldite LY 3031/3032	Epoxy	95–120°C	30 sec	120–150°C
	Applications: Automotive, Industrial				
	VITROX	Polyurethane	200–250°C	5–13 min	70–80°C
	Applications: Automotive, CIPP				

* Glass transition temp

Thermoplastics

Thermoplastic matrix materials undergo a physical change from a solid to a liquid when heated, and then re-solidify upon cooling. This is because thermoplastic polymers have linear or branch-chain molecular structures, with no chemical links between them. They are instead held in place by weak intermolecular bonds (called **Van der Waals force**).

Because thermoplastic molecules are not crosslinked, the intermolecular bonds in a thermoplastic polymer can be broken when heated, allowing the molecules to flow and move around. Upon cooling, the molecules freeze in their new position, restoring the weak bonds between them, thus returning to a solid form. Theoretically, a thermoplastic polymer can be heated, softened, melted, and reshaped as many times as desired. Practically, this may cause aging of the matrix or a change in crystallinity and physical properties.

Depending on the chemistry, thermoplastics typically require temperatures in excess of 300°F (149°C) before they start to melt or flow. Some high performance thermoplastics require temperatures in excess of 650°F (343°C) to achieve flow.

❱ THE IMPORTANCE OF CRYSTALLINITY IN PLASTICS PERFORMANCE

(Included and edited with permission from Jeffrey Jansen, The Madison Group, TMG News, November 2014.)

One of the fundamental characteristics of polymeric materials is the organization of their molecular structure. Broadly, plastics can be categorized as being semi-crystalline or amorphous. Understanding the implications of the structure, and specifically the crystallinity, is important as it affects material selection, part design, processing, and the ultimate anticipated service properties.

Most non-polymeric materials form crystals when they are cooled from elevated temperatures to the point of solidification. This is well demonstrated with water. As water is cooled, crystals begin to form at 0°C as it transitions from liquid to solid. Crystals represent the regular, ordered arrangements of molecules, and produce a distinctive geometric pattern within the material. With small molecules such as water, this order repeats itself and consumes a relatively large area relative to the size of the molecules, and the crystals organize over a relatively short time period. However, because of the relatively large size of polymer molecules and the corresponding elevated viscosity, crystallization is inherently limited, and in some cases, not possible.

Polymers in which crystallization does occur still contain a relatively high proportion of non-crystallized structure. For this reason, those polymers are commonly referred to as semi-crystalline. Polymers that, because of their structure, cannot crystallize substantially are designated as amorphous. As illustrated in Figure 2-3, amorphous polymers have an unorganized loose structure. Semi-crystalline polymers have locations of regular patterned structure bounded by unorganized

FIGURE 2-3. Amorphous vs. semi-crystalline thermoplastics are illustrated by the loose tangled amorphous region depicted on the left side, as compared to the aligned semi-crystalline regions within the structure depicted on the right.

amorphous regions. While some modification can be made through the use of additives, the extent to which polymers are semi-crystalline or amorphous is determined by their chemical structure, including polymer chain length and functional groups.

The ordered arrangement of the molecular structure associated with crystallinity results in melting when a sufficient temperature is reached. Because of this, semi-crystalline polymers such as polyethylene, polyacetal and nylon will undergo a distinct melting transition, and have a melting point (T_m, "T-sub m."). Amorphous polymers, including polystyrene, polycarbonate, and polyphenylsulfone (PPSU), will not truly melt, but will soften as they are heated above their glass transition temperature (T_g). This is represented by the differential scanning calorimetry thermograms shown in Figure 2-4.

The difference between semi-crystalline and amorphous molecular arrangement also has an implication on the mechanical properties of the material, particularly as they relate to temperature dependency. In general, amorphous plastics will exhibit a relatively consistent modulus over a temperature range. However, as the temperature approaches the glass transition temperature of the material, a sharp decline will be observed. In contrast, semi-crystalline plastics will exhibit modulus stability below the glass transition temperature, which is often sub-ambient, but show a steady decline between the glass transition temperature and the melting point. This is shown in Figure 2-5.

The importance of controlling the cooling rate during processing is very important to the extent of crystallinity achieved in the thermoplastic polymer and thus must be carefully controlled in processing fiber-reinforced thermoplastic (FRTP) composites. Typically, the goal is to achieve 35%–45% crystallinity in the matrix, depending upon the polymer chemistry and desired end use properties.

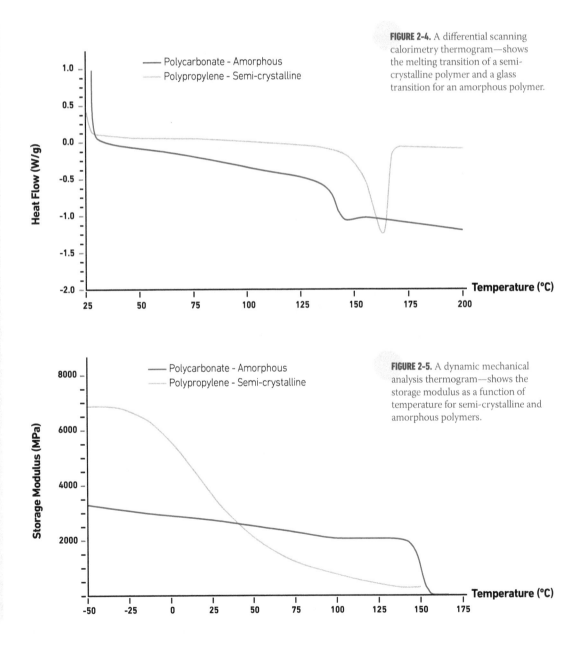

FIGURE 2-4. A differential scanning calorimetry thermogram—shows the melting transition of a semi-crystalline polymer and a glass transition for an amorphous polymer.

FIGURE 2-5. A dynamic mechanical analysis thermogram—shows the storage modulus as a function of temperature for semi-crystalline and amorphous polymers.

Other key properties of thermoplastic polymeric materials are determined by their semi-crystalline/amorphous structure, as partially summarized in Figure 2-6.

❱ TYPES OF THERMOPLASTICS

Examples of thermoplastics include polycarbonate (PC), Polyamide (PA), polyethylene (PE), polyethylene terephthalate (PET), polyetheretherketone (PEEK), polyetherketoneketone (PEKK), polyaryletherketone (PAEK), polysulfone (PSU), polyamideimide (PAI), polyphenylene sulfide (PPS), and many others.

Commodity and engineering thermoplastics tend to be used more in consumer goods, industrial, automotive and sporting goods applications, with some implementation in smaller unmanned aerial vehicles (UAVs). Aerospace, oil and gas, and the more demanding medical applications—as well as end-uses that demand higher temperature and moisture resistance—favor high-performance thermoplastics. (*See* Tables 2.6, 2.7, and 2.8)

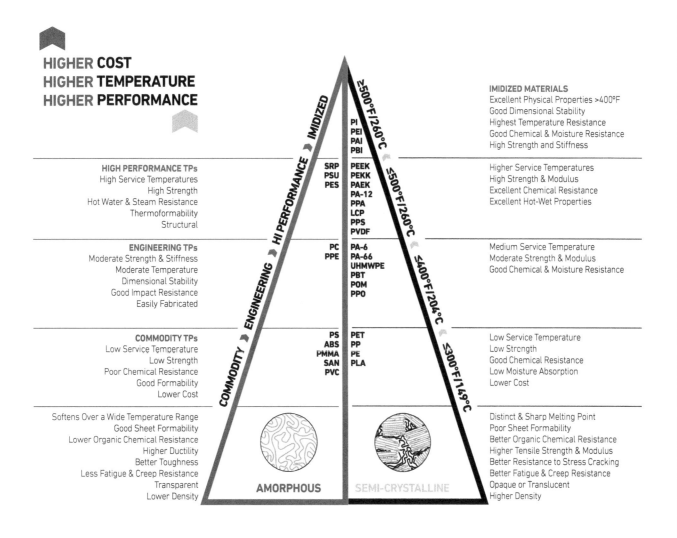

FIGURE 2-6. Amorphous vs. semi-crystalline thermoplastics.

TABLE 2.6 Commodity Thermoplastic Matrix Systems for Composites

Commodity Thermoplastics	Process Temperature	Maximum Service Temperature
Polyethylene terephthalate (PET)	509–536°F ▪ 265–280°C	284°F ▪ 140°C
Polypropylene (PP)	428–536°F ▪ 220–280°C	266°F ▪ 130°C
Polyethylene (PE)	239–275°F ▪ 115–135°C	230°F ▪ 110°C
Polylactid Acid (PLA)	320–352°F ▪ 160–178°C	230°F ▪ 110°C
Styrene-acrylonitrile (SAN)	392–518°F ▪ 200–270°C	230°F ▪ 110°C
Polymethyl methacrylate (PMMA or Acrylic)	460–536°F ▪ 240–280°C	217°F ▪ 103°C
Polystyrene (PS)	387–482°F ▪ 197–250°C	194°F ▪ 90°C
Acrylonitrile butadiene styrene (ABS)	392–536°F ▪ 200–280°C	192°F ▪ 89°C
Polyvinyl chloride (PVC)	320–428°F ▪ 160–220°C	140°F ▪ 60°C

TABLE 2.7 Engineering Thermoplastic Matrix Systems for Composites

Engineering Thermoplastics	Process Temperature	Maximum Service Temperature
Polybutylene terephthalate (PBT)	428–536°F ▪ 220–280°C	284°F ▪ 140°C
Polycarbonate (PC)	500–644°F ▪ 260–340°C	282°F ▪ 125°C
Polyamide (PA or Nylon)	446–536°F ▪ 230–280°C	266°F ▪ 130°C
Polyphenylene oxide (PPO)	480–580°F ▪ 249–304°C	248°F ▪ 120°C
Polyphenylene Ether (PPE)	446–608°F ▪ 240–320°C	230°F ▪ 110°C
Polyacetal (POM)	356–446°F ▪ 180–230°C	221°F ▪ 105°C
Ultra-high-molecular-weight polyethylene (UHMWPE, UHMW)	392–482°F ▪ 200–250°C	212°F ▪ 100°C

TABLE 2.8 High Performance Thermoplastic Matrix Systems for Composites

High Performance and Imidized Thermoplastics	Process Temperature	Maximum Service Temperature
Polybenzimidazole (PBI)	797–932°F ▪ 425–500°C	788°F ▪ 420°C
Polyimide (PI)	540–750°F ▪ 280–399°C	680°F ▪ 360°C
Polyetherimide (PEI)	625–675°F ▪ 329–357°C	572°F ▪ 300°C
Polyamide-imide (PAI)	540–650°F ▪ 280–343°C	536°F ▪ 280°C
Polyetherketoneketone (PEKK)	620–680°F ▪ 327–360°C	500°F ▪ 260°C
Polyaryletherketone (PAEK)	689–734°F ▪ 365–390°C	500°F ▪ 260°C
Polyetheretherketone (PEEK)	725–755°F ▪ 385–413°C	500°F ▪ 260°C

TABLE 2.8 High Performance Thermoplastic Matrix Systems for Composites *(continued)*

High Performance and Imidized Thermoplastics	Process Temperature	Maximum Service Temperature
Liquid Crystal Polymer (LCP)	540–563°F ▪ 280–295°C	469°F ▪ 243°C
Polyphenylene Sulfide (PPS)	610–650°F ▪ 321–343°C	428°F ▪ 220°C
Polyarylsulphones (PSU, PPSU)	425–475°F ▪ 321–390°C	410°F ▪ 210°C
Polyethersulfone (PES, PESU)	610–735°F ▪ 321–390°C	365°F ▪ 185°C
Self-reinforcing Polymer (SRP)	649–710°F ▪ 343–376°C	340°F ▪ 171°C

❱ ADVANTAGES

- Faster processing—cycle times of minutes instead of hours.
- Can be welded in assembly.
- Enables **overmolding** using injection molding processes.
- Re-flow capability—possible recycling.
- Room temperature storage.
- Minimal out-gassing.
- Greater damage tolerance with some systems.
- Possibility of fast, easy temporary repairs.
- High resistance to delamination.
- Typically, low moisture uptake and low microcracking.

❱ DISADVANTAGES

- Higher performance systems require high to very-high temperature processing and often high pressure.
- Semi-crystalline systems require controlled cool-down to achieve sufficient crystallinity to support structural properties.
- Lower-end systems may have lower HDT, moisture pickup issues and lack sufficient mechanical properties for structural applications.
- Complex shapes can be difficult due to lack of **tack** and **drape** in prepregs.
- Higher temperature process requires more expensive tooling.
- Design database available to engineers may be more limited compared to thermosets.
- Woven fabric prepregs of higher-end, more viscous systems are typically expensive due to difficulty of impregnation—alternatives may include biaxial or multi-axial stitched fabric forms, **organosheets** made with monomers and new, lower viscosity products.

❱ VITRIMERS (THERMOSET–THERMOPLASTICS)

More recently, polymers have been developed that blur the line between thermosets and thermoplastics. These resins are classified as vitrimers, and they include products like those listed in Table 2.9, which include thermosets that can be "switched" to thermoplastics during or after cure, or thermoplastics that process and have properties like thermosets. These new chemistries are helping to improve process cycle time, part cost and recyclability of both thermoset and thermoplastic composites. (*See also* Figures 2-7, 2-8)

Vitrimers were first discovered (and named) in 2011 by Ludwik Leiber, a French physicist and researcher known for his work in polymer self-assembly and dynamics. His work has enabled the industry to design and develop these resins from epoxies, aromatic polyesters, polyhydroxy urethanes, and more.

A dynamic chemistry, vitrimers are derived from thermoset polymers comprised of molecular, covalent polymer chain (crosslinked) networks. Sometimes referred to as "reversible" or "self-healing" resins, these polymers allow covalent bonds to be exchanged and rearranged at elevated temperatures and still exhibit thermoset properties, especially below the glass transition temperature (T_g). Therefore, these resins are reprocessable, reformable, repairable, recyclable, and weldable.

Initial vitrimeric formulas exhibited a high level of creep, even below the T_g. Newer formulas have emerged that have much better properties. One way this is achieved is through formulations that incorporate permanent crosslinks along with rearrangeable dynamic bonds. These formulas are made up of tetraglycidyl methylene dianiline (TGMDA) and diglycidyl ether of bisphenol F (DGBF) epoxy resins crosslinked with dynamic disulfide-amine hardener. With increased crosslink densities, these newer vitrimer resins have increased creep resistance and a higher T_g because partial crosslinks are maintained at temperature.

TABLE 2.9 Matrix Systems that Transition between Thermoset and Thermoplastic

Co./Product	Resin Type	Applications	Description
Arkema			
Elium®	Acrylic		• 100 cps at RT enables infusion and RTM w/o heat • 20-120 min cure at RT but 30-120 sec at 100°C • Recyclable, weldable • Burns readily • Mech. props near epoxy but 50% higher toughness • Low water pickup (0.5%) vs. nylon (5-10%) • Thermoformable at 200-220°C
L&L Products			
L-F601 Reformable Epoxy Adhesive Film	Thermoplastic Epoxy (TPER)	Automotive	• 40% tensile elongation • Can be de-bonded and re-formed • Transparent • Long shelf life • Tough • Dry to touch • RT storage • Tg 185-195°C

TABLE 2.9 Matrix Systems that Transition between Thermoset and Thermoplastic *(continued)*

Co./Product	Resin Type	Applications	Description
Connora Technologies			
Recyclamine® hardener	Epoxy hardener	Solid surfaces, sporting goods, anywhere epoxy is used	• Applicable to all types of epoxies • Does not diminish processability or properties • Makes epoxy recyclable • Allows designed-in cleavage of crosslinks (e.g., at high temp or low pH)
Cornerstone Research Group			
MG Resins	Shape memory		• High temp properties like polyimide • Cures to TP below 110°C and to TS above 180°C • Ultra-low flammability
Evonik			
Thermoreversible Crosslinkable Thermoplast-Thermoset Hybrid	Acrylate copolymer which undergoes a Diehls-Alder reaction	Automotive (e.g. B pillars) Transportation	• Compression moldable prepregs stables at RT for >2 yrs • Crosslinked TS below 100°C • Can be heated and cooled many times • Better properties and lower water uptake vs. nylon • Processes like a TP above 180°C

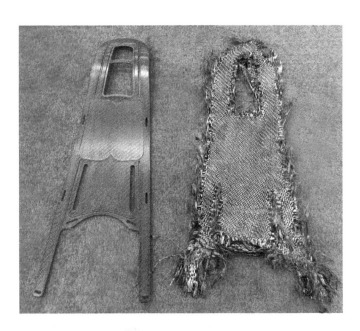

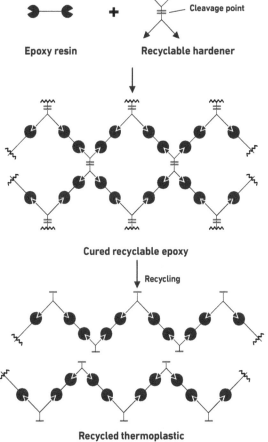

FIGURE 2-7. Cleavage points designed into crosslinks via Recyclamine® hardener enable cured thermoset epoxy resins to be converted into thermoplastic epoxy for easy recyclability. For example, both resin and fiber can be recycled using heat and acetic acid for the carbon fiber epoxy backpack frame shown here. *(Photo courtesy of Connora Technologies)*

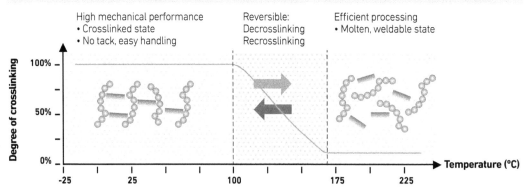

FIGURE 2-8. Evonik has developed a thermoplastic-thermoset hybrid which reversibly crosslinks according to temperature. It offers high mechanical properties in its crosslinked state below 100°C, yet processes like a thermoplastic when all crosslinks are opened above 170°C and can be repeatedly heated and cooled without loss of properties. *(Diagram info. courtesy of Connora Technologies)*

Other Matrix Materials

❱ METAL MATRIX

Metal matrix composites (MMC) have a metallic, rather than an organic, matrix system. Metals such as aluminum, titanium, copper, and magnesium are used, with aluminum and titanium being the most common. Fiber reinforcements include carbon, silicon carbide, boron, aluminum oxide, tungsten, and other metallic and ceramic fibers.

Metal matrix composites can be very strong, and exhibit excellent high-temperature properties. However, they tend to be quite expensive and difficult to process. They are largely found in aircraft and spacecraft applications. Examples of metal matrices include:

- aluminum and AL alloys
- titanium alloys
- magnesium alloys
- copper-based alloys
- nickel-based alloys
- stainless steel

❱ CERAMIC MATRIX[3]

Ceramic matrix composites (CMCs) comprise ceramic fibers in a ceramic matrix. CMCs are used for very high-temperature applications (up to 2,000°F/1,093°C). Examples include:

- aluminum oxide
- carbon
- silicon carbide
- silicon nitride

---

3   Some text in this section, and Table 2.10, reprinted with permission from "Ceramic Matrix Composites—an Alternative for Challenging Construction Tasks," Ceramic Applications 2013 by Friedrich Raether, Fraunhofer Center for High Temperature Materials and Design HTL.

Two major forms of fiber reinforcements are used:

- Short lengths of reinforcement fibers randomly distributed within the ceramic matrix;
- Continuous fiber reinforcements with high fiber content.

The former has considerably greater fracture toughness than unreinforced ceramics, while the latter gives a higher temperature capability than do metal matrix or organic matrix systems. (Figure 2-9)

The preceramic polymers used typically contain carbon (C) and/or silicon (Si), but may also contain boron (B), aluminum (al), titanium (ti) and/or other elements. For silicon carbide (SiC) matrices, these polymers include polycarbosilanes, siloxanes and silazanes. For carbon matrices, thermosetting resins (e.g., high-temperature epoxy, phenolic, and benzoxazine) and thermoplastic matrices including pitch and coal tar can be used.

The leading types of CMCs include SiC fibers reinforcing a SiC matrix (SiC/SiC), carbon/carbon (C/C), C/SiC and Ox/Ox where the oxide is typically alumina ($Al2O3$). SiC/SiC components used in oxidation-causing environments must be protected using environmental barrier coatings (EBCs), and even the SiC fibers must be coated to prevent attack from oxygen molecules diffusing through the porous matrix. Because they do not require a carbon coating on the fibers or EBCs, Ox/Ox composites offer lower cost. However, they lag in thermomechanical properties compared to SiC/SiC. (*See* Tables 2.10, 2.11)

FIGURE 2-9. CMC applications.

[a, b] SGL manufactures both carbon-carbon materials and high-performance C-C brakes for aircraft *(bottom)* and automobiles *(top). (Photos courtesy of SGL Carbon and Brembo SGL Carbon Ceramic Brakes)*

[c] This Ox/Ox CMC mixer cuts 20 kg from the GE Passport 20 engine, used to power Bombardier 7000 and 8000 ultra-long range business jets. *(Photo courtesy of Composite Horizons)*

TABLE 2.10 Properties of CMC Materials

Property	SiC/SiC	C/SiC	C/C	Ox/Ox
Fiber content (vol %)	40–60	10–70	40–60	30–50
Porosity (vol %)	10–15	1–20	8–23	10–40
Density (g/cm3)	2.3–2.9	1.8–2.8	1.4–1.7	2.1–2.8
Tensile strength (MPa)	150–360	80–540	14–1100	70–280
Bending strength (MPa)	280–550	80–700	120–1200	80–630
Strain-to-failure (%)	0.1–0.7	0.5–1.1	0.1–0.8	0.12–0.4
Young's modulus (GPa)	70–270	30–150	10–480	50–210
Fracture toughness (MPa·m$^{1/2}$)	25–32	25–30	5.7–.3	58–69
Thermal conductivity (W/m·K)	6–20	10–130	10–70	1–4
CTE* (ppm/K)	2.8–5.2	0–7	0.6–8.4	2–7.5
Max. service temp. (°C)	1100–1600	1350–2100	2000–2100	1000–1100

*Coefficient of thermal expansion

TABLE 2.11 Properties of C/SiC by Manufacturing Method

Manufacturing Method	C/SiC Chemical Vapor Infiltration* (CVI)	C/SiC Liquid Polymer Infiltration* (LPI)	C/SiC Liquid Silicon Infiltration* (LSI)
Porosity (%)	12	12	3
Density (g/cm3)	2.1	1.9	1.9
Tensile Strength (MPa)	310	250	190
Elongation (%)	0.75	0.5	0.35
Young's Modulus (GPa)	95	65	60
Flexural Strength (MPa)	475	500	300

*Infiltration or Infusion

CMC Manufacturing

CMCs are produced using ceramic fibers with a diameter of 3 to 20 μm, which enables flexibility for textile processing. Fibers are usually produced as yarns with hundreds up to several ten thousands of filaments.

CMC preforms are made by cutting the yarns and forming short fiber bundles or by weaving, knitting and braiding continuous fibers. Non-woven structures like uniaxial or multiaxial fabrics, fleeces and felts are also used. The preforms have a one-, two- or three-dimensional fiber structure, tailored to handle the anisotropic loads on the CMC component. The ceramic matrix is then introduced into the preform as either a gas or liquid.

Liquid Silicon Infiltration (LSI)

Also referred to as reactive melt infiltration (RMI), liquid silicon—which melts at 1,414°C—is infiltrated as a ceramic particle slurry, as a polymer or as a molten metal. The preform is infiltrated by capillary forces, and reacts with carbon in the preform to form silicon carbide. This process is called liquid silicon infiltration (LSI).

Liquid aluminum may also be used, in which case infiltration is in an oxidizing atmosphere and this RMI process forms an alumina-aluminum (Al_2O_3-Al) matrix. RMI is fast and relatively cost-effective, producing low-porosity CMCs that can have high thermal and electrical conductivity.

Polymer Infiltration and Pyrolysis Process (PIP)

Polymers may also be introduced as organometallic compounds either in dissolved or molten state, where infiltration of the preform is aided by external pressure. Alternatively, preceramic polymers may be formed into composite parts using traditional thermoset polymer forming processes. The polymers are pyrolyzed during a subsequent heat treatment and the final structure of carbon or silicon carbide is formed. Since the volume of the polymers decreases during pyrolysis, this polymer infiltration and pyrolysis process (PIP) has to be repeated typically 3 to 10 times to achieve densities required for the final CMC.

PIP uses relatively low temperatures, and thus fiber damage is minimized. Net shape parts may be fabricated with good control of the matrix composition and microstructure. However, the fabrication time is longer due to repeated infiltration-pyrolysis cycles, meaning higher production cost, and there is typically residual porosity that can lower mechanical properties. (Figure 2-10)

Ceramic Slurry Infiltration (CSI)

CSI (also called slurry infiltration process, SIP) is another liquid process where ceramic particle slurries are infiltrated into preforms at ambient temperature. This impregnation allows a quasi-ductile manufacturing of complex shapes analogous to the processing of sheet metals. After

FIGURE 2-10. An example of PIP.

During PIP, a ceramic fiber preform is infiltrated with a pre-ceramic polymer and then heated to pyrolize the polymer. However, because the volume of the pyrolized polymer decreases, this process may be repeated up to 10 times to achieve the final densified ceramic fiber/ceramic matrix composite.

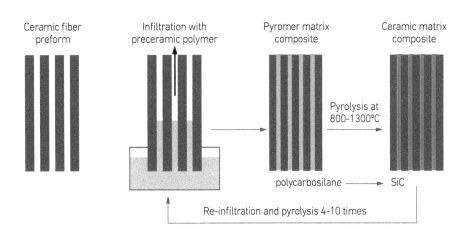

the preforms are dried, a heat treatment sintering process is required to increase the strength of the ceramic matrix. Usually sintering is interrupted in the initial stage because excessive shrinkage of the matrix would lead to cracks in the CMC structure. CSI leads to matrices with an open porosity of 20%–50%.

Chemical Vapor Infiltration (CVI)

CVI is similar to the well-known chemical vapor deposition (CVD) process. It is performed in a controlled atmosphere of the reactive gases at temperatures above 800°C and needs infiltration times of hours to days to achieve sufficient densities. An advantage of the CVI process is its capability to apply coatings on the fibers before introducing the matrix. Using the liquid infiltration routes the coatings are applied either via an additional CVD process or by cheaper wet chemical routes.

Chemical Vapor Deposition (CVD)

CVD begins with a **preform** gas, such as methane, acetylene or benzene, and then under high heat and pressure, the gas decomposes and deposits a layer of carbon onto the preform fibers. The gas must diffuse through the entire preform for the matrix to be uniform; thus, the process is very slow, often requiring several weeks and several processing steps to make a single part.

Pyrolization

Another method is to use a thermosetting resin such as epoxy or phenolic, and apply it under pressure to the preform, which is then pyrolized into carbon at high temperature. A preform can also be made from resin-impregnated carbon reinforcements, and then cured and pyrolized. Shrinkage in the resin during carbonization results in tiny cracks in the matrix and reduced density, so that the part must then be re-injected and pyrolized up to a dozen times in order to fill in the small cracks and achieve the desired density. CVD may also be used to complete densification.

❱ CARBON MATRIX

Carbon used as a matrix material is **amorphous carbon**. It is usually reinforced with carbon fiber, and is referred to as "carbon-carbon." Typical applications include brakes, clutches, and high-temperature aerodynamic surfaces, such as the nose cone and wing leading edges of the space shuttle. Carbon-carbon (C-C) brake disks are considerably lighter than steel disks, and due to the material's high coefficient of friction and low wear at high temperatures, the hotter the brakes get, the better they work.

Carbon-carbon composites have very low thermal expansion coefficients and maintain excellent structural properties at temperatures in excess of 3,650°F (2,000°C) in non-oxidizing atmospheres. They also have high thermal conductivity but resist thermal shock, meaning they won't fracture even when subjected to rapid and extreme temperature changes. Their mechanical properties vary depending on the fiber type, fiber fraction, textile weave and matrix precursor properties. Modulus can be very high, from 15–20 **GPa** for composites made with a 3D fiber felt perform, to 150–200 GPa for those made with high modulus carbon fibers.

Carbon-carbon composites are actually low in strength and the matrix is very brittle. They oxidize readily at temperatures between 1,112–1,292°F (600–700°C) in the presence of oxygen. To prevent this, a protective coating (usually silicon carbide) must be applied, adding an extra manufacturing step and cost to the production process, which is already very expensive. Because of this, and the complexity of their manufacture, carbon-carbon composites have been mostly limited to aerospace, defense and some Formula 1 racing applications.

Carbon-carbon composites are made by building up a carbon matrix around carbon fibers. There are two common ways to do this: chemical vapor deposition and resin pyrolization (*see* above).

Liquid Resins

❱ INTRODUCTION TO LAMINATING RESINS

Liquid laminating resins typically have a fairly low viscosity at room temperature (< 60 poise/6,000 centipoise at 77°F/25°C). This allows for easy fiber **wet-out** in a **layup** or an **infusion** operation. In general, the lower the viscosity, the better the flow and fiber permeability.

Structural laminating resins are usually unfilled or lightly filled with select fiber, mineral, or metallic fillers that aid in the retention of the matrix between the fibers and/or provide resiliency. A variety of modifiers may also be used in laminating resins to improve toughness properties. These modifiers range from reactive liquid elastomers (butadiene-acrylonitrile) to particulate thermoplastics.

❱ POT LIFE, WORKING TIME, AND OPEN TIME

Mixed reactive resins and adhesives have a designated amount of time in which to use the material before it eventually becomes gelatinous and unusable in the cup. The amount of time-to-gel in the cup is usually listed on the material's technical data sheet and is specified as the "pot life." The pot life refers to the time that a specific amount of mixed resin takes to gel at room temperature (77°F/25°C) in a standardized container.

The "working time" of a resin may be quite different than its pot life. The working time is measured after the resin is mixed and distributed into a thinner cross-section (smaller mass) within the laminate or along the bondline, thus allowing more time for vacuum bagging or other processing before the resin gels. Note: The working time can be greatly affected by the amount of time that the mixed resin is left in the cup or bucket prior to use. It is always suggested that the bucket time be minimized and the resin distributed quickly to allow for maximum working time.

"Open time" refers to the time that a mixed resin or adhesive is open and exposed to the air. Open time can be an issue with certain amine-cured epoxy systems that take on CO_2 and H_2O from the air and form a carbonate layer (hydrates of amine carbonate). This layer can greatly affect the bond strength of the resin or adhesive at the interface and steps should be taken to minimize such exposure.

❱ MIX RATIOS

Often when a technician is required to use a two-part resin or adhesive system, he or she will refer to the mixing instructions and determine if the mix is to be parts by weight (P.B.W.), or parts by volume (P.B.V.).

For example, at a mix ratio of 100:10 P.B.W., technicians could simply mix 100 grams of resin to 10 grams of hardener, for a total of 110 grams of mixed resin. (Or, 50 grams to 5 grams for a total of 55 grams.) They would then use what is needed and dispose of the rest. Although this method generates considerable waste, it is easy to figure out, and maintains the proper ratio of resin to hardener, thus ensuring good resin properties.

When a larger quantity is desired, it may be necessary to calculate the total amount of mixed resin that will be required. For example, say that the technician needs a total amount of 340 grams of mixed resin. For this calculation, it may be convenient to convert the ratio to a percentage and do the math to find out exactly how many grams of resin and hardener will be weighed to mix the required amount at the correct ratio.

In order to proceed, it is important to understand the difference between a percentage and a ratio. A percentage is by definition, a portion of one hundred. A ratio is used to express the relationship of two quantities.

The following example may be used as a guideline to properly convert a ratio to a percentage and then figure out the net amount of resin to hardener that will be needed to make a 340 gram total mixed batch:

Mix ratio.	100 : 10 P.B.W
Add the two amounts to get the sum	100 + 10 = 110
Divide each side of ratio by the sum.	100 ÷ 110 = .91
	and 10 ÷ 110 = .09

(Result is percent resin and percent hardener; 91% and 9% respectively.)

Now, plug in the total amount required:.	340 x .91 = 309.4
	and 340 x .09 = 30.6

The final weight of each material would be 309.4 grams of resin plus 30.6 grams of hardener to equal the 340 gram total mixed weight required.

Attention to detail in mixing is critical to the performance of the resin and subsequently, the performance of the laminate or adhesive bonded joint. It is extremely important to attain the proper mix ratio with any given resin system so as to achieve the desired mechanical and thermal properties of the final mix. In addition to proper mix ratio, it is also important to thoroughly mix the components so that there is sufficient distribution of each component to promote a complete chemical reaction.

❱ UNDERSTANDING VISCOSITY

Poise and Centipoise Measurements

The common unit for expressing absolute viscosity is "centipoise" (1/100 Poise, 1/1,000 of a Poiseuille.) The name Poiseuille (pronounced pwäz-'wē) and the shortened form, *Poise*, come from the French physician, Jean Louis Poiseuille, who performed numerous tests on the flow resistance of liquids through capillary tubes and published a paper on his research in 1846.

Poise (P) is a measure of resistance to flow and is expressed in units of one dyne-second per cm^2. Industry uses the centipoise (cP) or one-hundredth of a Poise because water has a viscosity of 1.002 cP (which is very close to 1 cP), making it easy to compare other viscosities.

TABLE 2.12 Viscosity Comparisons of Some Common Materials

Material (at 68°F/20°C)	Viscosity (cP)
Acetone	0.3
Water	1.002
Ethylene Glycol	19.9
Huntsman RenInfusion® 8601/8602	325
Huntsman Epocast® 50A/946 Lam. Resin	2,400
Pancake Syrup	2,500
Huntsman Epocast® 52 A/B Repair Resin	5,500
Chocolate Syrup	25,000
Ketchup	50,000
Loctite Hysol EA 9394 Paste Adhesive	60,000
Peanut Butter	250,000
Tar or Pitch	3×10^{10}

The lower the Poise or centipoise value, the more easily a material will flow. If a fluid has a high viscosity it strongly resists flow. If a fluid has a low viscosity, it offers less resistance to flow. The viscosities of liquids are greatly affected by temperature. Fluids exhibit a lower viscosity when heated, while the viscosity of gas rises with temperature.

❱ SHELF LIFE CONSIDERATIONS

Base resins typically store well at room temperature, usually one to three years depending on the product. Some systems may require cold storage (typically < 40°F/4.4°C). However, they must be kept moisture-free. Because of this, many manufacturers state that the shelf life is assured as long as the product is unopened. "Pre-promoted" polyesters and vinyl esters may have shorter shelf lives, on the order of three to six months. Catalysts, hardeners, accelerators, etc. also may have short shelf lives at room temperature, on the order of several weeks to a year.

❱ CURING CONSIDERATIONS

It is misleading to think that a resin has fully cured just because it appears to be hard. Most "room-temperature-curing" resins and adhesives will develop a partial cure (polymerized, partially reacted, or "gelled") within a few hours. Complete reaction and full cured properties may require several days or weeks at room temperature (77°F/25°C). An elevated temperature post-cure is sometimes prescribed for these resins to shorten the cure time and/or to raise the T_g.

Other thermoset chemistries require heat in order to facilitate active molecular crosslinking within a reasonable time, and thus will require an elevated temperature cycle to make this happen. These materials are generally specified for more high-performance applications and are usually processed in a similar manner to prepregs. (*See* "Curing Thermoset Prepregs," page 53.)

Prepregs

Many of the disadvantages encountered with liquid resin processes can be overcome with the use of **prepregs**. A "prepreg" is a reinforcement material that is pre-impregnated with resin. The reinforcement can be made from any type of fiber, in a unidirectional, stitched, braided, woven, or non-woven form. With thermosets, the resin is already mixed prior to application to the reinforcement. The amount of resin in the prepreg is strictly controlled to achieve the desired resin content.

Prepregs can be purchased from numerous sources. Companies that produce prepregs are commonly called "prepreggers." These companies specialize in the resin formulation and pre-impregnating processes used to produce prepregs and film adhesives.

❱ PREPREG MATERIAL CONSIDERATIONS

Prepregs require different handling procedures than wet lay-up materials. Because prepregs are made with resins that are already mixed, they must be stored in cold lockers (< 40°F/4.4°C) or in frozen storage (< 0°F/-17.8°C), to slow the reaction of the resin. (*See* "Prepreg Storage and Handling," page 51.) Some prepreg resins have been developed that can be stored at room temperature for up to one year. These unique formulations require an elevated temperature cure. Common cure temperatures for thermoset-matrix prepregs range from room tempera-ture to > 600°F, depending on the matrix.

❱ PREPREG MANUFACTURING METHODS

Several common manufacturing methods are used to impregnate the resin into the fiber form. In the "resin-bath" or "solvent bath" system, solvent is added to the mixed resin to dilute it and to permit rapid saturation of the dry fiber form. This prepreg is then taken through a heating tower to evaporate most of the solvent, raising the viscosity of the resin, and creating a usable material. These materials have some cost advantages but lack quality in relation to resin content control, aesthetic appearance, handling and structural properties. (Figure 2-11)

The resin bath method is most applicable to woven forms as it is difficult to do with uni-tape, stitched forms or fabrics. Unidirectional prepreg manufactured via solvent coating methodology is preferred using a drum winding machine. These types of machines limit the length of material sheets produced to the diameter of the winding drum. This is also a lengthy process compared to **hot melt prepregs**.

There are two methods of hot-melt coating used in the prepreg industry: (1) hot-melt direct impregnation coating route, and (2) hot-melt film route. In these techniques, resin is not diluted with solvent, but instead is quickly heated to reduce the viscosity to a suitable value for quick distribution into a film and/or saturation into a dry cloth. The resin may be chilled to raise the viscosity and minimize crosslinking prior to rolling it up on the roll. The majority of hot-melt systems are resin-coated on both sides of the cloth or unidirectional tape. One side coating (or a semi-impregnation process) is used to coat one side of the reinforcement and is useful for molding applications of thick laminate structures. Hot-melt systems typically have a low volatile content as a result of this process.

Hot melt process flexibility has come a long way with the invention of new types of prepreg machines. Modern prepreg machines manufactured by Century Design, Inc., for instance, give the manufacturer the ability to set process parameters and program into the machine Human Machine Interface (HMI) recipes for materials and store them for easy retrieval and comparison for process improvements. (Figures 2-12, 2-13)

❱ SEMIPREG AND THERMOPLASTIC PREPREG/ORGANOSHEET

Thermoset (TS) coated **semipreg** is a form of prepreg that has resin film on one side, of which the resin is typically impregnated only 10 to 30% into the fabric, resulting in a mostly dry fabric layer with sufficient resin content (42-45% by volume) to allow for good wetting during processing.

One-side coated TS semipregs are commonly prescribed for out-of-autoclave (OoA) or vacuum-bag-only (VBO) processes, where the dry fabric layer acts as an internal channel layer for air and gas to escape via edge breathing during processing, resulting in a lower void volume panel. Another advantage is the film coating has a very low or zero volatile content that also contributes to low void volume in the end product.

Thermoplastic (TP) coated semipreg fabrics typically have both sides coated with a TP film that is partially impregnated into the fabric layer with a heat pressing process. For example, Fokker Aerostructures uses a fiberglass/polyphenylene sulfide (PPS) material used to form the J-nose fixed leading edge for Airbus' A380 aircraft wings. In this case, the PPS film is applied to both sides of a (re-sized, eight-harness satin-weave) fiberglass fabric in a heated laminating process, to form a semipreg product that is later fully consolidated during forming of the leading edge components. Thermoplastic semipreg fabrics may also be referred to as organic sheets or *organosheets*.

Other semipreg fabrics are made using a TP powder coating and oven melt/crystallization processes, or a solvent dip-coating process where the fabric (or other fiber form) is run through a liquid TP resin. This results in partially impregnated (or encapsulated) fiber bundles that will later wet-out in a final process where the material is heated and molded in a press or vacuum bag-autoclave process.

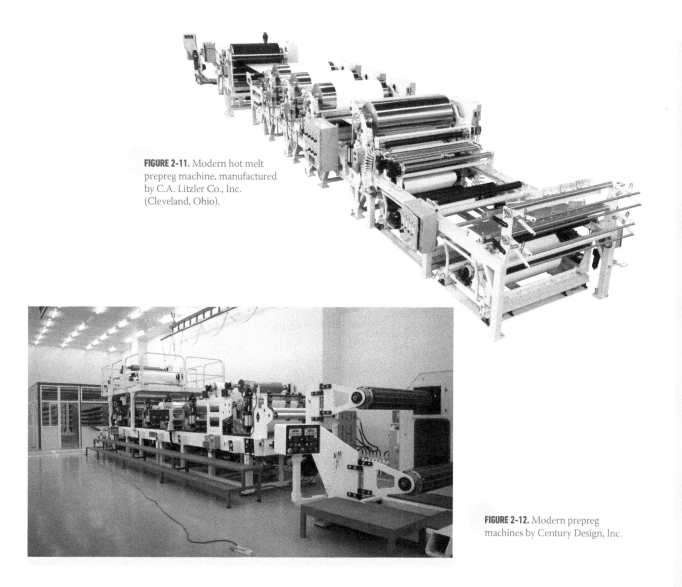

FIGURE 2-11. Modern hot melt prepreg machine, manufactured by C.A. Litzler Co., Inc. (Cleveland, Ohio).

FIGURE 2-12. Modern prepreg machines by Century Design, Inc.

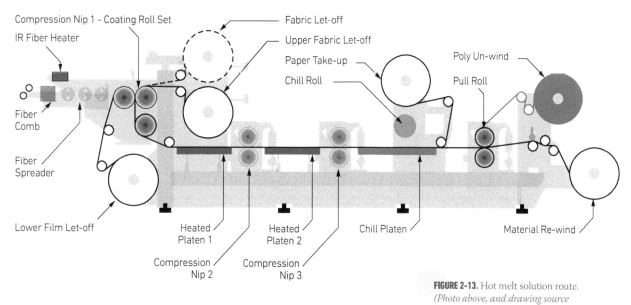

Compression Nip 1 - Coating Roll Set

IR Fiber Heater

Fabric Let-off

Upper Fabric Let-off

Paper Take-up

Poly Un-wind

Chill Roll

Pull Roll

Fiber Comb

Fiber Spreader

Lower Film Let-off

Heated Platen 1

Compression Nip 2

Heated Platen 2

Compression Nip 3

Chill Platen

Material Re-wind

FIGURE 2-13. Hot melt solution route. *(Photo above, and drawing source courtesy of Century Design, Inc.)*

TABLE 2.13 Thermoplastic Prepreg and OrganoSheet — Continuous Fiber

Company	Product	Polymers	Fibers
BASF	Ultralaminate™ Ultratape™	PA	Glass, carbon
BGF	PolyPreg®	PP	Glass
Bond-Laminates, div. of Lanxess	Tepex® Dynalite	PPS, PP, PC, PA, TPU	Glass, carbon
Chomarat	Tpreg R™	PP, PA, PET	Twintex®, Comfil®
Covestro	Maezio™ CFRTP	PC	Glass, carbon
DSM		PA	Glass, carbon
DuPont	Vizilon™	PA66	Glass
Evonik	VESTAPE®	PA	Glass, carbon
Fibrtec	FibrFlex™	PEEK, PEI, PPS, PP, PA, FEP (Tefon)	Glass, carbon, aramid
Frenzelit	HICOTEC	PEEK, PEI, PPS	Glass, carbon, aramid
Porcher	PiPreg®	PEEK, PEI, PPS, PC, PA, TPU	Glass, carbon, aramid
SABIC	UDMAX™	PE, PP	Glass, carbon
Solvay	APC-2 APC Evolite®	PEEK PEKK PA66, PPA, PPS	Carbon Carbon Glass
Suprem	Suprem™ T	PA12, PEEK, PES, PPS, TPI	Glass, carbon, aramid
PolyOne	Polystrand™, X-Ply™ 0°/90°, multi-ply	PA6, PA66, PC, PE, PETG	Glass
SGL Carbon	Sigrafil tapes, organosheet	PA, PP	Glass, carbon
Teijin	TPUD TPWF (woven fabric)	PEEK	Carbon
Toray	Cetex®	Acrylic, HDPE, PA, PC, PE, PEEK PEI, PEKK, PET, PMMA, PP, PPS	Aramid, carbon, glass
Vector Systems	Flexile™	PA, PBT, PC, PE, PEEK, PEI, PEKK, PES, PET, PP, PPS	Aramid, carbon, glass, PBO, steel, alum, ti

TABLE 2.14 OrganoSheet — Discontinuous Fiber

Company	Product	Polymers	Fibers	Fiber Length
Frenzelit	HICOTEC	PEEK, PEI, PPS	Glass, carbon, aramid	6–10 mm
Lanxess/Bond-Laminates	Tepex® Flowcore	PA	Glass	30–50 mm

❯ STAGES OF A RESIN SYSTEM

For thermosetting resins, there is an initial A-stage, a secondary B-stage, and a final C-stage, which describes its status regarding chemical reaction and cure.

A-Stage

A-stage describes the first reaction stage for resins. This stage can be considered "as mixed," in a liquid stage and ready for use in a variety of liquid molding or wet-layup processes. Prepreg manufacturers may also use resins in this stage to easily impregnate cloth.

B-Stage

Prepreg resins are advanced to this intermediate B-stage, where the resin is somewhat viscous and tacky, but not flowing. This facilitates resin filming prior to prepregging. B-staged resin in the prepreg allows the resin to stay put during prepreg storage, handling, and layup of the material. When heat is later applied during the cure process, the resin viscosity will drop and flow and the chemical reaction (crosslinking) will proceed until it reaches gelation.

C-Stage

This is the final reaction stage for thermoset resins, in which the resin has partially crosslinked to the point where it is insoluble and infusible without applying heat. Usually prepreg in this stage is not used to make parts, due to very poor handling characteristics and bonding between plies. However, by design there are some prepregs in which the resin is intentionally C-staged (or vitrified) and this allows for the material to be wrapped around columns or pipe prior to being cured at elevated temperatures.

❯ PREPREG STORAGE AND HANDLING

Date of Manufacture

The rolls of material are marked with the date of manufacture (DOM), which refers to the date the prepreg was originally manufactured, for the purpose of tracking the total combined storage life including time spent in inventory at the prepreg manufacturer or supplier. The supplier may have the material in frozen storage (≤ 0°F/-18°C) inventory for up to 2 years prior to shipping to the purchaser, dependent upon the resin type.

Shelf life usually starts on the date received by the user and is defined as the time the materials can be stored and still processed to the desired condition. These materials almost always require frozen storage. Usually, the shelf life of a prepreg is between six months to one year. Some resin systems such as benzoxazine chemistries will allow up to one year of storage at room temperature (≤ 77°F/25°C).

Prepreg Handling and Out-Time/Mechanical Life Limits

The total out-time (or mechanical life) for prepreg materials is defined as the amount of time out of frozen storage (> 0°F/-18°C), with specific upper temperature limits (≤ 77°F/25°C), during which prepreg can still be used to make good parts. A subgroup of total out-time is the handling time, which is typically limited to a period of 2–30 days. This is a period in which the prepreg can be out and exposed to a controlled environment after which it must be vacuum bagged until cured or otherwise molded. (Figure 2-14)

In order to control prepreg out-time, a record is kept with the roll or with pre-cut kits of the material in the form of an out-time record (see Table 2.15). This record is usually kept with each roll or kit of material in order to track its out time.

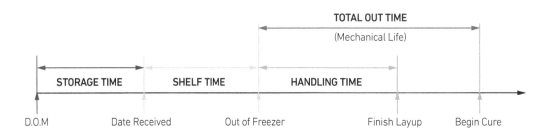

FIGURE 2-14. Storage, handling, and out-time of prepreg materials.

TABLE 2.15 Out-Time Tracking Record Sheet

OUT-TIME / RECORD SHEET									
Batch No.	Roll No.	Time Out	Date Out	Freeze Temp.	Time In	Room Temp.	Date In	Add Up Hours	Total Out-Time

Tack and Drape

Tack refers to the stickiness of prepreg at room temperature. Prepreg manufactures may have several different tack levels available if specified.

Drape describes the ability of materials to conform to compound curves without wrinkling or buckling. A very drapeable prepreg is required for structures with intricate geometries, like ducting. Drape varies greatly, dependent on both fiber/fabric form and the viscosity and tack of the matrix resin.

Moisture Contamination

The risk of excessive moisture uptake is a big problem with frozen prepregs. This can happen by opening the frozen prepreg bag and allowing condensation to form on the exposed material. For this reason, prepreg materials should always be allowed to fully thaw to room temperature prior to removal from the moisture barrier.

With epoxy prepregs, excessive moisture can greatly affect the crosslink chemistry and weaken the matrix resin's structure. Minor moisture in an epoxy matrix may be effectively removed with an extra-long vacuum debulk step prior to processing.

With cyanate ester resins, moisture can accelerate the resin and the resultant carbamate reaction creates CO_2 as a byproduct. The CO_2 acts as a blowing agent and will embrittle the matrix. The presence of moisture in cyanate ester resin can also cause transesterification of the matrix (i.e., the matrix polymerizes into a linear-chained thermoplastic, instead of a crosslinked thermoset structure).

Recertification

Prepreg materials that have exceeded their shelf life or out-time can usually be tested and requalified for a shorter period of time. Typically, this requires that the material be tested in the lab for flow and gel properties and compared to the original certification data. This is usually done in accordance with ASTM D3531 or other international standards (ISO).

Kit-Cutting Prepreg Materials

The technique of kitting materials into single-use kits is especially advantageous when small quantities of material are needed periodically, such as in a repair station or small-scale fabrication environment. This best preserves the out-time of the roll.

Kit-cutting for production is normally done using 2-D automated table cutters that are programmed to cut specific ply patterns from the roll materials in a "just in time" (JIT) manufacturing environment where the materials are cut and placed either in a pre-layup staging area, or directly in the mold or on the mandrel.

❯ CURING THERMOSET PREPREGS

Prepreg materials often require an elevated temperature process to facilitate the cure reaction. This is usually done under pressure in a heated press or within a vacuum bag in the oven or autoclave. Historically, a time/temperature recipe was prescribed with specific heating rates and **isothermal** holds to produce the desired state of cure (legacy cure cycle).

Processing prepreg parts per these recipes can be tricky. This is especially true when the measured process conditions go outside of the specified boundaries of the recipe, bringing to question the true cure-state of the material. Often, when such "out-of-spec" conditions happen the prepreg parts may be rejected, scrapped, or otherwise subjected to further testing because of the unknown properties of the material.

To better understand the nature of how a thermosetting prepreg material cures, you must examine the change in viscosity and modulus of the resin as it cures during the elevated temperature cycle. The term "viscoelastic" is used to describe both the viscous (liquid) initial phase of the resin and the final elastic (hardening) stage of the resin/laminate during completion of the cure reaction.

Figure 2-15 outlines how the viscoelastic properties of a thermoset resin change during a thermal cycle. The annotations on the graph in Figure 2-17 point to significant changes in the resin during the cure of a typical epoxy/carbon fiber prepreg, as follows:

- The viscosity drops as a function of temperature; this is measured with a torsion-plate rheometer (refer to the blue line on the graph, "APA1 Visc p").
- The chemical reaction is measured by the heat that is released during the cure. The maximum heat evolution corresponds to the rapid rise in viscosity and modulus. This is measured via differential scanning calorimeter (DSC); refer to the green line ("DSC Heat Flow").
- Both the viscosity (flowing liquid) and rise in modulus (springiness of the solid)—see the red line, "APA1 Gp"—exhibits a similar change at the beginning of the cure cycle. Because the sample is a prepreg, both fibers and resin are present and as the resin viscosity rises and falls it causes more or less bending of the fibers and therefore the elastic response.
- Once the resin solidifies and the flow of the resin becomes immeasurable, the apparent viscosity (blue line) will drop on the graph. This is due to a lack of molecular friction in the solid and not an actual drop in viscosity.

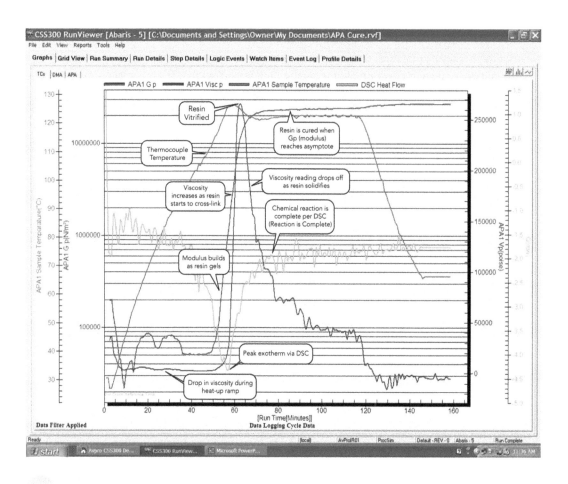

FIGURE 2-15. Example of viscosity and modulus vs. time/temperature.

Flow and Gel

The definition of the "flow" phase is when the resin is at its lowest viscosity and maintains this viscosity for a period of time. The measured drop in viscosity and the time that the resin maintains this low-level of viscosity is highly dependent upon the temperature ramp-rate. (*See also* "Too-Fast/Too-Slow Ramp Rate Problems," page 58)

The interpretation of "gel" varies since it is actually the transition phase from a liquid to a solid state. Prepreg resins start out as a true liquid and end as a true solid. The gel-point is technically measured at the point where the modulus is equal to the viscosity (where the red and blue lines cross on the graph as the viscosity and modulus build). However, the most easily observed and typically referenced gel-point is at the point where the viscosity is at its peak.

Optimal Compaction Point

During the latter part of the flow phase, where the resin viscosity is starting to rise, is the optimal time to add pressure and compact a laminate (at a point where the resin viscosity is about equal to the original B-stage viscosity of the prepreg when it was at room temperature). Compacting at this point ensures that all of the air and gas has had sufficient time to escape and that the resin will not re-soften or be "blown out" of the laminate with expanding gases. Lower void content can be achieved in laminates processed using this "optimized compaction" method.

Unfortunately, it is a common practice to apply pressure at the beginning of the cure cycle because of a lack of actual resin flow/gel information on all of the parts being processed in a production load. Under these conditions it is critical to remove as much of the trapped gas as possible prior to starting the cure cycle. This can be achieved by placing the laminate under vacuum for a long period of time before processing (vacuum debulking).

Vitrification and Cure

Eventually, **vitrification** takes place (the resin becomes a glassy solid) at a given temperature. If the temperature is raised further before full cure time is achieved, this glassy solid will re-soften. When the T_g is above a specified temperature value, the resin is considered to be "cured" and is structurally functional below this temperature.

Cure Cycles

Different "cure cycles" are specified for different prepreg resins, depending on their chemistries. For example, epoxy resins may be processed with a single or dual (heat) ramp/soak cycle with simple vacuum and pressure steps to get the best results. Bismaleimide and polyimide resins, on the other hand, may require a multi-step ramp/soak, high temperature cycle, with several different intermittent pressurization steps to do the job.

Most legacy cure cycles are simply recipes that define time/temperature and vacuum and pressure requirements at specified time intervals. Originally these cure cycles were derived from the viscoelastic properties of the different neat resins as measured in the laboratory and then mapped-out as a standard time/temperature recipe with upper and lower time and temperature boundaries.

Figures 2-16 and 2-17 show the basic recipe for a 350°F carbon epoxy prepreg cure cycle, as interpreted after the viscoelastic properties were previously tested and analyzed in the laboratory.

Predictable Viscoelastic Change

As long as the laminate sample is actually heated at the prescribed rate and meets the target temperature for the specified time, the viscoelastic properties change as predicted. In fact, the graphs indicate that the modulus (the "APA1 Gp" blue line) becomes **asymptotic** about midway through the soak in the single ramp profile and even earlier in the two-step process, indicating that the resin is probably fully cured at that point.

The information gleaned from these graphs seem to indicate that the soak time is longer than necessary, and in the latter case, much longer than needed to cure this resin. This extra time at temperature may or may not be necessary to achieve full properties but since it is often specified that the time/temperature requirements be documented in accordance with the legacy recipe, it has become common practice to run material for the entire duration to "ensure" a full cure.

Practical Considerations

In theory this recipe should work for this resin within the target parameters given, and with a bit of a safety factor added to accommodate slight variations in the process (including shelf life and out-time factors). In practice it is often more difficult to meet the prescribed ramp rate and soak time/temperature requirements for every part in a large load of parts in an oven or autoclave, as there can be many factors affecting the ability to heat everything uniformly. (*See* Chapter 9, "Introduction to Tooling" for more information.)

Because of these variables, actual thermal sensor measurements in large size loads of real prepreg parts often demonstrate a high degree of variability and many can be out of the boundaries of the recipe or specification. This often results in parts that see too slow or too fast of a heat rate, or too long at the soak temperature in some parts to accommodate the lagging thermocouple readings in other parts.

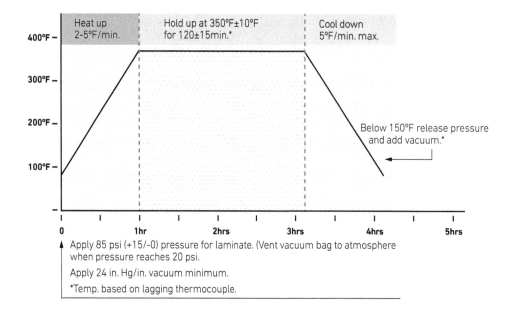

Apply 85 psi (+15/-0) pressure for laminate. (Vent vacuum bag to atmosphere when pressure reaches 20 psi.

Apply 24 in. Hg/in. vacuum minimum.

*Temp. based on lagging thermocouple.

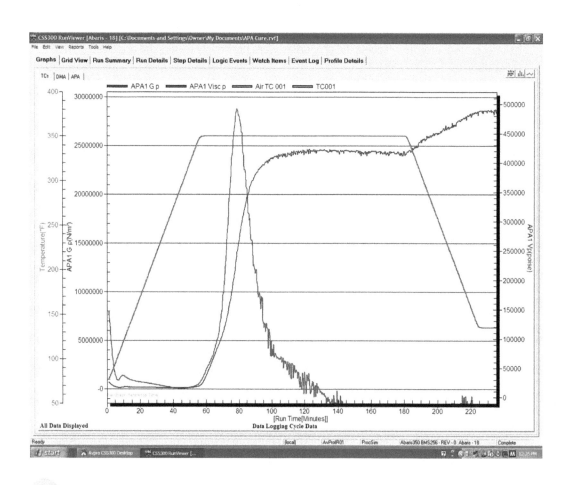

FIGURE 2-16. Single-step 350°F (177°C) legacy cure profile.

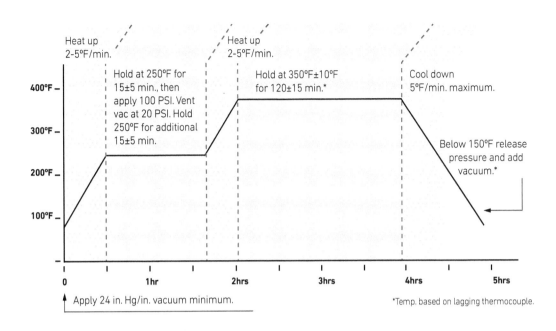

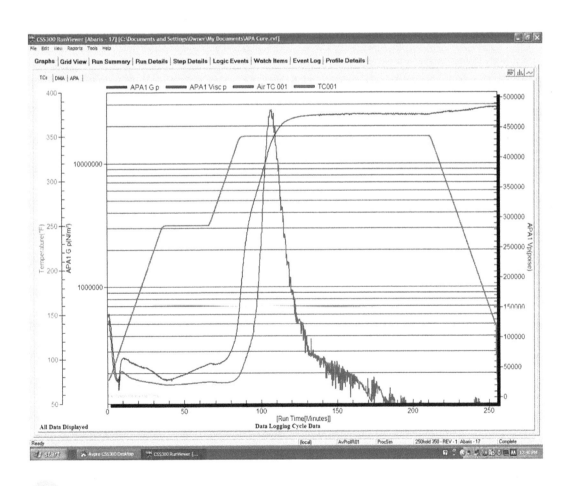

FIGURE 2-17. Two-step 350°F (177°C) legacy cure profile.

This makes it difficult to run large loads of various sized parts and be sure that all of the parts have fully cured and/or are not held for unnecessarily long periods of time per the legacy cure cycles. Future process specifications most likely will include options to the recipe-type of cure cycles that use actual or statistically-predicted viscoelastic properties based on actual thermal sensor feedback from multiple parts using new generation computer process control capabilities. (*See also* "Computer Controlled Processing," on the next page)

Too-Fast/Too-Slow Ramp Rate Problems

The issue of a *too-fast* or *too-slow* heat ramp can be somewhat confusing. Unfortunately, the complex relationship among viscosity, pressure, cure rate, gas devolution, etc. is a fact of life that must be managed. If the resin is heated too quickly, the flow time is shortened and the viscosity will not stay low for a long enough period of time. (Figure 2-18)

This can diminish the exchange of volatiles out of the laminate and interrupt the movement of resin into low-pressure sites. Rapid heating will generally cause the resin viscosity to drop to a lower value than slow heating. If air or moisture is present, a rapid rise in temperature will cause the gas to rapidly expand and the trapped moisture to boil, thus creating resin voids and potentially displacing resin out of the laminate. In addition, too much heat, too fast, can generate an uncontrolled exothermic reaction in thicker structures that can greatly affect the resin properties.

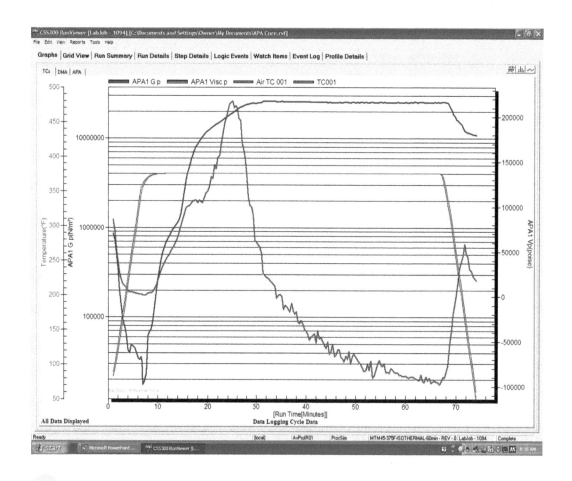

FIGURE 2-18. "Too-fast" ramp rate.

If the prepreg is heated too slowly (Figure 2-19), the viscosity may never get low enough to allow sufficient resin movement to take place. As a consequence, trapped air and gas in the laminate may not adequately move through the viscous liquid and may result in a higher than normal void content in the laminate.

This can be a problem with a honeycomb sandwich structure as the heat-rate directly affects the flow of the resin or adhesive to the edge of the core cells. If good filleting of these cells is not achieved, a poor bond is the result. (Slow heating molds made of high-mass materials can contribute to this effect.)

Too Long at Soak Problems

Many prepreg thermoset resin systems are cured at temperatures that push or exceed their own service temperature limits. Long soaks at these upper temperatures may actually begin to degrade, rather than improve, the matrix properties. For this reason an upper soak-time tolerance is usually specified in the cure cycle. Because of this issue, care must be taken not to exceed the maximum time boundary just to facilitate lagging thermocouple readings in the process. This may be contrary to the process specification and must be addressed by a material review board (MRB) or other engineering group to identify the root cause.

Computer Controlled Processing

New computer-aided processing technology allows us to monitor and control to actual material state conditions, optimizing the thermal-process cycle, and ensuring that a full cure is achieved

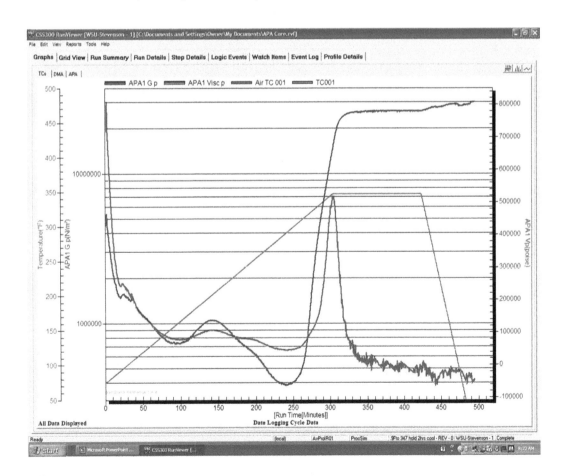

FIGURE 2-19. "Too-slow" ramp rate.

in the minimum amount of time possible. Vacuum and pressure targets during the process can be adjusted by the computer based on actual viscosity changes in the curing prepreg when coupled with real-time analytical equipment (i.e., DSC and/or torsion-plate rheometer).

In order to facilitate such capabilities, specifications can be written that allow options to time/temperature legacy profiles for curing prepreg composite parts. Such specifications model the viscosity and modulus change of the material and the completion of the reaction instead of relying on the legacy recipe cure cycle approach.

The value of processing to **material state conditions** using computer process controls is found in the inherent quality assurance of such a system. Not only does this allow for a more accurately controlled cure of the composite materials, it also provides a myriad of sensor feedback on large production loads and the data storage capabilities relevant to each cycle. This data can be used by both quality assurance and materials engineering for a multitude of purposes from quality control and statistical analysis to lean production improvement in processing.

Another advantage of processing with material state feedback is the ability to adjust the time/temperature cycle as needed to build large parts with minimal effect from thermal expansion of the tooling. Figure 2-20 depicts a range of time/temperature options for a selected prepreg material using the storage modulus (Gp) feedback as an indication of when the T_g exceeds the cure temperature. The resin will remain solid as long as the T_g is not exceeded at any time during an elevated temperature post cure or in-service.

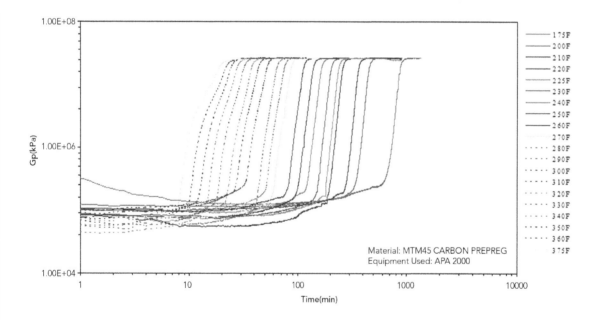

FIGURE 2-20. Isothermal cycles at different temperatures vs. time.

3

Fiber Reinforcements

CONTENTS

Introduction and Overview

This chapter discusses fiber types, fiber properties, and the different fiber forms that are used in the fabrication of advanced composite structures. Table 3.1 on the next page lists the most common fiber types used in composites. These fibers (e.g., glass, carbon, ceramic and aramid) are detailed in the "Fiber Types and Properties" section beginning on page 65, which examines how they are manufactured, their mechanical properties and how they are prepared for use in composites.

The reinforcement used in composites is also an important consideration. Structural composites typically use a long or continuous form of fiber reinforcement. Non-structural composites may use chopped fibers, which are typically ¼- to 1-inch in length. Fibers that are shorter than this—for example, milled fibers, which are approximately 1/32-inch (0.8 mm) in length—have a low aspect ratio (length/diameter) that does not provide much strength. Thus, these are usually not considered as reinforcement, but as fillers offering improved matrix properties such as dimensional stability. We'll discuss the "Forms of Reinforcement" in more detail starting on page 90. (Figure 3-1)

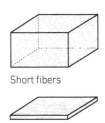

Short fibers

Random short fibers

Oriented short fibers

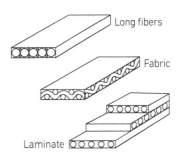

Long fibers

Fabric

Laminate

2D fabrics offer planar properties (X and Y axes) while laminates extend this into the Z-axis by stacking plies in different orientations.

3D fabrics orient fibers in the X, Y and Z axes to provide improved out-of-plane properties.

FIGURE 3-1. Forms of reinforcement in composites.

TABLE 3.1 Common Composite Reinforcement Fibers and Properties

FIBER - Type	Density oz/in3 (g/cm3)	Tensile strength Ksi (GPa)	Tensile modulus Msi (GPa)	Elongation to break %	Melt or decomposition temp.
CARBON					
- PAN-SM*	1.04 (1.8)	500–700 (3.4–4.8)	32–35 (221–241)	1.5–2.2	
- PAN-IM*	1.04 (1.8)	600–1200 (4.1–6.2)	42–47 (290–297)	1.3–2.0	
- PAN-HM*	1.09 (1.9)	600–800 (4.1–5.5)	50–65 (345–448)	0.7–1.0	6,332°F 3,500°C
- Pitch-LM*	1.09 (1.9)	200–450 (1.4–3.1)	25–35 (172–241)	0.9	
- Pitch-HM*	1.2 (2.0)	275–400 (1.9–2.8)	55–90 (379–621)	0.5	
- Pitch-UHM*	1.24 (2.15)	350 (2.4)	100–140 (690–965)	0.3–0.4	
BASALT					
- Standard	1.56 (2.7)	400–695 (2.8–4.8)	12.5–13.0 (86.2–89.6)	3.0–3.5	2,654°F 1,450°C
GLASS					
- E-glass	1.47 (2.54)	450–551 (3.0–3.8)	11.0–11.7 (75.8–80.7)	4.5–4.9	1,346°F 730°C
- S-2 glass	1.43 (2.48)	635–666 (4.3–4.6)	12.8–13.2 (88.3–91.0)	5.4–5.8	1,562°F 850°C
ARAMID					
- Twaron® 1000	0.84 (1.45)	450 (3.1)	17.6 (121.3)	3.7	
- Kevlar® 29	0.83 (1.44)	525 (3.6)	12.0 (82.7)	4.0	850°F 455°C
- Kevlar® 49	0.83 (1.44)	525 (3.6)	18.9 (130.3)	2.8	
POLYETHYLENE (PE)					
- Spectra® 900	0.56 (0.97)	380 (2.6)	11.5 (79.3)	3.6	
- Spectra® 1000	0.56 (0.97)	447 (3.1)	14.6 (100.7)	3.3	302°F 150°C
- Spectra® 2000	0.56 (0.97)	485 (3.3)	18.0 (124.1)	2.8	
POLYPROPYLENE (PP)					
- PPF	0.55 (0.95)	98.6 (6.8)	0.54 (3.7)	14.0	302°F 150°C
HIGH MOLECULAR WEIGHT POLYPROPYLENE (HMPP)					
- Innegra® S	0.49 (0.84)	96.7 (0.7)	2.5 (14.8)	9.5	302°F 150°C

*Carbon fiber is classified by the stiffness or modulus designation:
  SM = Standard Modulus   HM = High Modulus   LM = Low Modulus
  UHM = Ultra High Modulus fiber   IM = Intermediate Modulus

TABLE 3.1 Common Composite Reinforcement Fibers and Properties *(continued)*

FIBER - Type	Density oz/in³ (g/cm³)	Tensile strength Ksi (GPa)	Tensile modulus Msi (GPa)	Elongation to break %	Melt or decomposition temp.
POLYPHENYLENE BENZOBISOXAZOLE (PBO)					
- Zylon® AS	0.89 (1.54)	841 (5.8)	26.1 (180.0)	3.5	1,202°F 650°C
- Zylon® HM	0.90 (1.56)	841 (5.8)	39.2 (270.3)	2.5	
OTHER CERAMIC FIBERS					
- Boron	1.49 (2.58)	522 (3.6)	58 (400)	--	3,769°F 2,076°C
- Silicon Carbide (SiC)	1.45–1.73 (2.5–3.0)	360–900 (2.5–6.2)	27–62 (186–428)	0.7–2.2	4,172°F 2,300°C
- Alumina-silica	1.76 (3.05)	250–290 (1.7–2.0)	22–28 (152–193)	0.8–2.0	1,400°F 760°C
- α-alumina	1.97–2.37 (3.4–4.1)	260–510 (1.8–3.5)	38–67 (262–462)	--	1,800°F 982°C
- Quartz	1.25 (2.17)	493 (3.4)	10.0 (68.9)	5.0	1,920°F 1,050°C

❯ REINFORCEMENT TERMINOLOGY

Figure 3-2 is an illustrated glossary of common terms used in referring to fiber reinforcements.

[a] **Filament**—The very smallest form of the fiber as extruded or spun from the raw materials. Filaments are then combined to make various fiber bundles as described below.

[b] **Strand**—A bundle of filaments twisted into a "mini yarn" for subsequent weaving operations. A term specific to glass fiber, strands of glass can be anywhere from 51 to 1,624 filaments in size.

[c] **Strands to yarn**—Glass yarns are made of a single strand or multiple twisted strands depending on the weight and diameter requirements for subsequent weaving operations and/or processing. (*See* "Typical Glass Fiber/Fabric Forms and Styles" starting on page 70.)

[d] **Filaments to yarn**—In carbon fiber, common yarn sizes range from 1,000 to 50,000 filaments (1K to 50K). In aramid fiber, yarns can contain from 25 to 5,000 filaments.

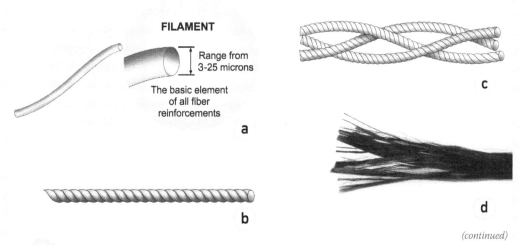

FILAMENT

Range from 3-25 microns

The basic element of all fiber reinforcements

a

b

c

d

(continued)

FIGURE 3-2. Fiber reinforcement terminology.

[e] Yarn—A small, continuous bundle of filaments, tows or combined strands that are gently twisted to enhance bundle integrity. This is important for keeping the yarn intact during subsequent weaving operations.

[f] Commingled yarn—Intimately blending thermoplastic fibers with reinforcing fibers at the filament level. Examples include PEEK, PPS, PE, and PP thermoplastic fibers combined with carbon, glass, aramid, or other reinforcing fibers.

[g] Tow—Basically the same as a yarn, but not twisted, and specific to carbon fiber. The number of filaments comprising a carbon tow is designated in units of 1,000 (e.g. 1K, 12K, 50K). Unidirectional tape is made from multiple tows laid side by side. *Smaller tow sizes* (e.g., 1K, 3K, 6K) are normally used in weaving, winding and braiding applications, and also for very thin unidirectional tapes. *Larger tow sizes* (e.g., 12K, 24K, 50K) are used in unidirectional tapes, stitched multi-axial fabrics, and automated fiber placement processes, and are often spread into thinner spread tows. In general, larger tow sizes are cheaper than smaller tow sizes for the same fiber type (e.g., compare a 50K tow carbon fiber at $5–10/lb to a 12K at $15–20/lb).

[h] Spread tow—A tow that has been spread into a wider, thinner form, offering a thinner laminate. The woven fabric form results in straighter fibers and fewer crimps when compared to conventional fabric forms.

[i] Roving—a number of strands, yarns, or tows collected into a parallel bundle with little or no twist. It is a term usually used to describe fiberglass and sometimes aramid reinforcements. **Single-end roving** is made from continuous fiber filament or strand and is typically prescribed for filament winding and pultrusion processes. **Multi-end roving** is made up of long but discontinuous filament bundles that are leafed together into lengthy fiber bundles. Rovings are most commonly used for constructing heavy fabrics called **woven roving** and may also be used for chopper gun/spray-up processes.

[j] Tape—The common name for "unidirectional tape"—all flat parallel tows, available in both dry and prepreg forms. Unidirectional tape is considered to be the most efficient fiber form, as it has no crimps in its construction and each layer is specific to one axis of orientation in layup. Unidirectional tape provides the highest strength-to-weight ratio and has the most efficient on-axis load transfer compared to other forms.

[k] Textile—Item manufactured from natural fiber or man-made filaments, including yarns, tows, threads, roving, cords, ropes, braids, nets and fabrics. Techniques include weaving, knitting, braiding, felting, bonding and tufting.

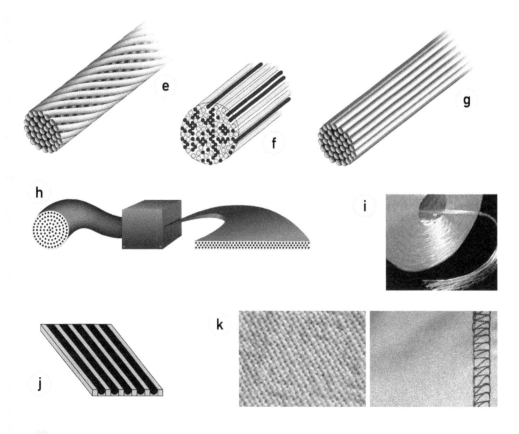

FIGURE 3-2. Fiber reinforcement terminology. *(continued)*

[l] **Mat**—Sheets of chopped or continuous fiber held together with a small amount of adhesive binder.

[m] **Veil**—Similar to mat, but thinner and lighter weight (lower fiber areal weight). Used for surface cosmetics, electrical properties (e.g., electromagnetic interference shielding, lightning strike protection) and/or fire resistance properties.

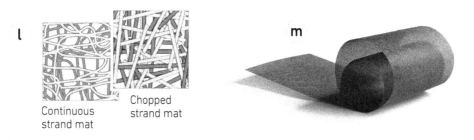

FIGURE 3-2. Fiber reinforcement terminology. *(continued)*

Fiber Types and Properties

Figure 3-3 lays out the different types of fibers used in composites, under two main categories: natural and synthetic.

❯ GLASS FIBER

NOTE: Text from this section is adapted from "The Making of Glass Fiber" by Ginger Gardiner, Composites Technology Magazine, April 2009; reproduced by permission of Composites Technology Magazine, © 2009 Gardner Publications, Inc., Cincinnati, Ohio

Fiberglass yarns are available in different formulations. "E" glass (electrical grade) is the most common all-purpose glass fiber, while "S-2" (high strength) glass is used for special applications.

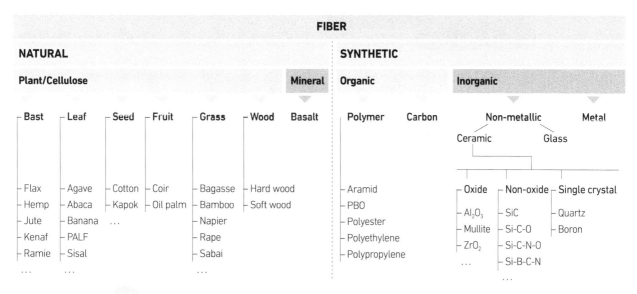

FIGURE 3-3. Types of fibers used in composites.
Note: Modified from (1) "Fibers for Ceramix Matrix Composites" Clauss, B., in *Ceramic Matrix Composites*, pp.1-22, Walter Krenkel, Ed., © 2008 Wiley-VCH Verlag GmbH & Co. KGaA, Weinheim; and (2) "Cellulosic/synthetic fibre reinforced polymer hybrid composites: a review" Jawaid, M.; HPS Abdul Khalil, in *Carbohydrate. Polymers* 2011 (Vol. 86, pp. 1 – 18).

Manufacture

Textile-grade glass fibers are made from silica (SiO_2) sand which melts at 1,720°C/3,128°F. Though made from the same basic element as quartz, glass is amorphous (random atomic structure) and contains 80% or less SiO_2, while quartz is crystalline (rigid, highly-ordered atomic structure) and is 99% or more SiO_2. Molten at roughly 1,700°C/3,092°F, SiO_2 will not form an ordered, crystalline structure if cooled quickly, but will instead remain amorphous—i.e., glass. Although a viable commercial glass fiber can be made from silica alone, other ingredients are added to reduce the working temperature and impart other properties useful in specific applications. (*See* Table 3.2, "Comparison of E and S-2 glass fibers.")

For example, E-glass, originally aimed at electrical applications, with a composition including SiO_2, Al_2O_3 (aluminum oxide or alumina), CaO (calcium oxide or lime), and MgO (magnesium oxide or magnesia), was developed as a more alkali-resistant alternative to the original soda lime

TABLE 3.2 Comparison of E and S-2 glass fibers

Composition	E-glass	S-2 Glass®*
Silicon Dioxide	52–56%	64–66%
Calcium Oxide	16–25%	
Aluminum Oxide	12–16%	24–26%
Boron Oxide	8–13%	
Sodium & Potassium Oxide	0–1%	
Magnesium Oxide	0–6%	9–11%

*S-2 GLASS® is a registered trademark of AGY.

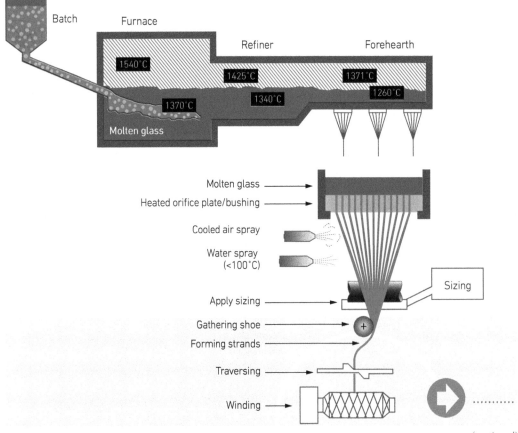

(continued)

FIGURE 3-4. Glass fiber manufacturing process.

glass. S-glass fibers (i.e., "S" for high strength) contain higher percentages of SiO_2 for applications in which tensile strength is the most important property.

Glass fiber manufacturing begins by carefully weighing exact quantities and thoroughly mixing (batching) the component ingredients. The batch is then melted in a high temperature (~1,400°C/2,552°F) natural gas-fired furnace. Beneath the furnace, a series of four to seven bushings are used to extrude the molten glass into fibers. Each bushing contains from 200 to as many as 8,000 very fine orifices. As the extruded streams of molten glass emerge from the bushing orifices, a high-speed winder catches them and, because it revolves very fast (~2 miles/3 km per minute—which is much faster than the speed the molten glass exits the bushings), tension is applied and this draws the glass streams into thin filaments (i.e., fibrous elements ranging from 4–34 µm in diameter, or 1/10 that of a human hair). (Figure 3-4)

Fiber Diameter and Yield

The bushings' orifice or nozzle diameter determines the diameter of the glass filament; nozzle quantity equals the number of ends. A 4,000-nozzle bushing may be used to produce a single roving product with 4,000 ends, or the process can be configured to make four rovings with 1,000 ends each. The bushing also controls the fiber yield or yards of fiber per pound of glass.

The metric unit, tex, measures fiber linear density: 1 tex = 1 g/km; while yield is the inverse, yd/lb. A fiber with a yield of 1,800 yd/lb (275 tex) would have a smaller diameter to a 56 yd/lb (8,890 tex) fiber, and an 800-nozzle bushing produces a smaller yield than a 4,000-nozzle bushing.

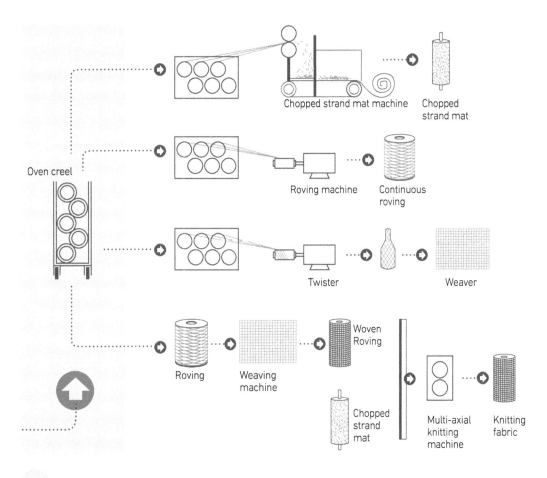

FIGURE 3-4. Glass fiber manufacturing process. *(continued)*

For example, OCV Reinforcements (Toledo, Ohio) commonly uses a 4,000-nozzle bushing for optimizing production flexibility, while AGY uses 800-orifice bushings because it is a smaller company that produces finer filaments and smaller-run niche products. The range of glass fiber diameter, or micronage, has become more varied as composite reinforcement applications have become more specialized. OCV observes 17 μm and 24 μm as the most popular glass fiber diameters for composite reinforcements, although its reinforcement products vary from 4 μm to 32 μm. AGY's products typically fall into the 4 μm to 9 μm range.

Size vs. Finish

In the final stage of glass fiber manufacturing, a chemical coating, or size, is applied. (Although the terms binder, size, and sizing often are used interchangeably in the industry, *size* is the correct term for the coating applied and *sizing* is the process used to apply it.) Size is typically added at 0.5 to 2.0 % by weight and may include lubricants, binders, and/or coupling agents. The lubricants help to protect the filaments from abrading and breaking as they are collected and wound into forming packages and, later, when they are processed by weavers or other converters into fabrics or other reinforcement formats.

Coupling agents cause the fiber to have an affinity for a particular resin chemistry, improving resin wet-out and strengthening the adhesive bond at the fiber-matrix interface. Some size chemistry is compatible only with polyester resin and some only with epoxy, while others may be used with a variety of resins.

Glass fiber manufacturers agree that size chemistry is crucial to glass fiber performance and each company considers its size chemistry proprietary. Some believe that in many composite applications, performance can be achieved via size chemistry as effectively as with glass batch chemistry. One example is PPG's 2016 size chemistry with HYBON products for wind blades, which reportedly achieved an order of magnitude improvement in blade fatigue life by significantly improving the fiber's ability to be wet-out by and adhere to all resin types.

After textile glass fibers are processed by weaving, the completed fabric is heat cleaned to remove any remaining size, and then a finish is applied. Finishes are applied to glass fabrics to improve resin wet-out and enable a strong bond between the resin and the glass fibers. Some finishes are just for polyester, some are just for epoxy, and others are good for many resins. An incorrectly selected finish can greatly reduce the ability of the matrix to wet and bond to the fiber, which subsequently diminishes the overall laminate properties.

Volan (chromium methacrylate) and silane (silicon tetrahydride) are common finishes. Volan is much older and was originally developed to promote adhesion to polyester resins, although it will bond to epoxies. Silane is more common for composites used in aircraft because it forms a stronger bond with epoxy resins. Volan has traditionally been a softer finish than silane yielding a more pliable fabric, but recent developments have produced some excellent soft-silane finishes.

What is the difference between size and finish? Size is applied to the fiber, finish is applied to fabric. As the composites industry moves more toward unidirectional tape (unitape) materials, size chemistry continues to be an area of development, including formulations specifically for thermoplastic matrix materials. Several fiber and unitape producers are already promoting such products for improved performance in thermoplastic composites.

Glass Fiber Products History and Development

E-glass, ECR-glass and ADVANTEX

The industry norm for composite reinforcement worldwide is E-glass; S-glass is the standard for higher-performance applications such as aerospace and ballistics. A major evolution for E-glass has been the removal of boron. Although at one time boron facilitated fiberization, it is expensive and produces undesirable emissions. Its removal has produced multiple benefits, including lower cost and a more environmentally-friendly glass fiber.

TABLE 3.3 Types of Glass Fiber

S	Structural: ½ fiber diameter of E-glass; stronger	A	High-Alkali
S-2*	Commercial grade S-glass, has larger filaments	E	Electrical
Advantex®	Produced as a replacement for E-glass and E-CR Glass	D	Dielectric grade
E-CR	High chemical resistance	M	High-modulus (boron-rich) glass
Lead	Radioactivity resistance	C	Chemical resistance
Lithium oxide	X-Ray transparency	R	High strength-high modulus

*S-2 GLASS® is a registered trademark of AGY.

OCV's boron-free product, ADVANTEX, is actually its second-generation ECR-glass. Its first iteration in the 1980s was a response to a market need for even higher corrosion resistance coupled with good electrical performance. However, this original patented ECR-glass was difficult to make and therefore more expensive to end-users. Thus, OCV developed ADVANTEX, which is more cost-effective due to a lower-cost, boron-free batch composition and the elimination of scrubbers and other environmental equipment previously required to capture boron emissions. ADVANTEX is also made using higher temperatures, which produces higher properties.

OCV is converting all of its global reinforcements manufacturing to ADVANTEX, including the 19 Saint-Gobain Vetrotex reinforcements plants it acquired in 2007. Its patent on ECR-glass has recently expired, enabling companies like Fiberex (Leduc, Alberta, Canada) and CPIC (Chongqing, China) to emerge with their own versions.

S-2 and S-3 GLASS, T-glass and HS2/HS4

S-glass has also evolved. Driven by U.S. military missile development in the 1960s and its need for high-strength, lightweight glass fiber for rocket motor casings, Owens Corning pioneered the production of S-glass, and subsequently developed an improved form trademarked as S-2 GLASS, featuring a 40% tensile strength and 20% higher tensile modulus than E-glass. These properties are derived from its composition, though the manufacturing process helps to maintain that performance, as does using the correct size for the polymer matrix in the final composite structure.

As a business, Owens Corning's fine glass yarns and S-2 GLASS fiber products was spun off in 1998, then reorganized and emerged in 2004 as AGY. S-3 GLASS is described as "designer glass," in that AGY can make small amounts (e.g.,100 tons) of it customized to meet one customer's precise specifications. One example is AGY's HPB biocompatible glass fiber, developed for long-term medical implants (over 30 days) such as orthodontics and orthopedics.

Other higher-performance products include T-glass manufactured by Nittobo (Tokyo, Japan) and Sinoma Science & Technology's (Nanjing, China) HS2 and HS4 products, distributed in Europe and North America exclusively by PPG.

TABLE 3.4 Comparison of Mechanical Properties for Various Glass Fiber Products

	E-glass	R-glass	HS2/HS4	T-glass	S-2 GLASS
Tensile Strength (GPa)	1.9–2.5	3.1–3.4	3.1–4.0	not available	4.3–4.6
Tensile Modulus (GPa)	69–80	89–91	82–90	84	88–91

S-1 and R-glass

Higher-performance glass fiber types have traditionally been harder to produce, requiring higher melt temperatures and paramelt processing using smaller paramelter furnaces with low throughput, all of which increases cost. Both AGY and OCV saw a need for higher performance glass fiber at a lower cost. AGY developed S-1 GLASS, situated between E-glass and S-2 GLASS in performance and cost, and targeted for composite wind blades where its higher properties reduce the amount of glass fiber required as blade lengths are extended.

OCV developed its similarly situated high-performance glass (HPG) process in 2006, which produces higher-performance glass fibers on a larger scale. This process uses furnaces smaller than those used to make all-purpose E-glass, but 50 times larger than a paramelter, resulting in lower cost than S-glass. The resulting OCV products are all R-glass by composition—FliteStrand, WindStrand, XStrand and ShieldStrand—and feature different fiber diameters (12–24 μm), mechanical properties, and size chemistry, specific to the performance required for each end-use. For example, ShieldStrand requires a lower micronage of 12.5 μm, while WindStrand, aimed at very large wind blade structures, requires a greater micronage of 17 μm. WindStrand features a tensile modulus not quite as high as S-glass, but higher than E-glass at an affordable cost.

The size applied to ShieldStrand enables progressive delamination of the composite armor. It does this by allowing separation at the fiber-matrix interface upon impact while maintaining static mechanical properties of the composite.

Typical Glass Fiber/Fabric Forms and Styles

(The following was provided by Hexcel-Schwebel, *Technical Fabrics Handbook*.)

The wide variety of fiberglass yarns produced requires a special system of nomenclature for identification. This nomenclature consists of two parts: one alphabetical and one numerical.

Example: ECG 150-1/2 (U.S. System)

First Letter (E)—Characterizes the glass composition (E-glass)

Second Letter (C)—Indicates the yarn is composed of continuous filaments. S indicates staple filament. T indicates texturized continuous filaments.

Third Letter—Denotes the individual filament diameter, in this case the G indicates the diameter is 0.00036 inches (9 microns) in diameter. (Commercial glass filaments range in size from 0.00017–.00051 inches (4–13 microns) in diameter).

First Number (150)—Represents 1/100 the normal bare glass yardage in one pound of the basic strand. In the above example, multiply 150 by 100, which results in 15,000 yards in one pound.

Second Number (1/2)—Represents the number of basic strands in the yarn. The first digit represents the original number of twisted strands. The second digit separated by the diagonal represents the number of strands plied (or twisted) together. To find the total number of strands used in a yarn, multiply the first digit by the second digit (a zero is always multiplied as 1).

Note that literally hundreds of styles are available from various manufacturers; these examples are only a few of the many available. Unidirectional forms are also available.

TABLE 3.5 Common Fiberglass Fabric Weave Styles

Hexcel-Schwebel Style Number	Weave Style	Yarn Count–Yarns per inch Warp/Fill	Fabric Weight oz/sq. yd (g/m^2)	Fabric Thickness inches (mm)	Common U.S. Fiber nomenclature
120	4 HS	60/58	3.16 (107)	0.0035 (0.09)	ECG 450 1/2
1581	8 HS	57/54	8.90 (301)	0.008 (0.20)	ECG 150 1/2
6781	8 HS	57/54	8.90 (301)	0.009 (0.23)	S-2CG 75 1/0
7500	Plain	16/14	9.60 (325)	0.011 (0.28)	ECG 37 1/2
7544	2-End Plain	28/14	18.20 (617)	0.020 (0.51)	ECG 37 1/2 ECG 37 1/4
7597	Dbl. Satin	30/30	38.60 (1309)	0.041 (1.03)	ECG 37 1/4
7725	2x2 Twill	54/18	8.80 (298.0)	0.0093 (0.24)	ECG 75 1/0 ECH 25 1/0
7781	8 HS	57/54	8.95 (303)	0.009 (0.23)	ECDE 75 1/0

❱ BASALT FIBER

In the 1960s, both the U.S. and former Soviet Union explored basalt fiber applications, particularly in missiles. Continued development was pursued primarily in former Soviet countries, but basalt fiber is now also produced and/or sold in Belgium, China, Ireland and the U.S. Basalt is rock formed by rapidly cooling lava and the most common type of rock in the Earth's crust. Though the rock's properties vary by geographical location, quality fibers are made from basalt deposits that have a consistent and uniform mineral/chemical makeup.

Basalt rock is crushed, melted and then formed into continuous fibers in a process similar to that for glass fibers. However, pulverized basalt rock is the only raw material required for manufacturing the fiber. Continuous filaments are extruded from molten basalt at 2,700°F (1,500°C) and receive a silane-based size to facilitate fiber processing and weaving. Though early weaving techniques produced many broken fibers, today basalt fiber can be braided and woven to achieve desired performance.

Basalt fibers are not translucent but are naturally resistant to ultraviolet (UV) and high-energy electromagnetic radiation. They have good alkali, chemical, and fire resistance properties, reportedly better than that of E-glass and when compared to carbon and aramid fiber, basalt fiber exhibits a wider application temperature range (-452°F to 1,200°F or -269°C to 650°C), high oxidative and radiation resistance, and higher compression and shear strength properties. Basalt fiber has been used as an alternative for asbestos.

Basalt fiber is a good intermediary reinforcement for lower cost applications. In testing performed by the Fraunhofer Project Center (London, Ontario, Canada), basalt/epoxy panels made using high-pressure RTM (HP-RTM) showed higher tensile modulus and strength compared to E-glass panels, and an interlaminar shear strength midway between glass and carbon fiber comparison panels. The basalt composite showed a 40% higher specific strength and 20% higher specific stiffness versus the E-glass composite. Meanwhile, its pricing is roughly the same as E-glass fiber.[1]

---

1 Taken from CompositesWorld blog "Can basalt fiber bridge the gap between glass and carbon?" by Sara Black, www.compositesworld.com.

Basalt fiber has been used in composite snowboards, skis, surfboards, paddles, tennis rackets, bicycle frames, blast protection, vertical wind turbines, pipes, rebar, canoes, boats, utility poles, storage tanks, car and bus parts, prosthetics, bridge repair/rehabilitation and 3D printed continuous fiber structures. (Figure 3-5)

❱ CARBON FIBER

Carbon vs. Graphite Fiber

Carbon fiber has historically been manufactured using a carbonization/graphitization furnace process and is often called "graphite" fiber. The truth is that carbon fiber is considerably different in form than true graphite, which is very soft and brittle (like pencil "lead"). (Figures 3-6, 3-7)

Unlike graphite, which has a layered or flake-like construction, carbon filament (fiber) has an aligned crystalline structure making it stronger and more durable. Since carbon fiber is atomically the same as graphite, it has historically been called graphite fiber and this terminology is widely accepted in the composite materials industry. Because the term carbon fiber is technically correct, it is the term that will be used throughout this text to describe these fibers.

Types and Precursors

Carbon fiber has a higher modulus (stiffness) and compressive strength, is lighter and is generally more brittle than fiberglass. The properties of carbon fiber vary, depending upon the type of fiber and the **precursor** material used to make it. Most aerospace grade carbon fiber is made

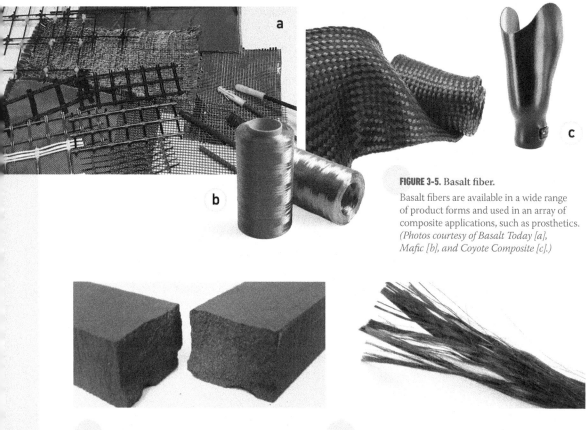

FIGURE 3-5. Basalt fiber.
Basalt fibers are available in a wide range of product forms and used in an array of composite applications, such as prosthetics. *(Photos courtesy of Basalt Today [a], Mafic [b], and Coyote Composite [c].)*

FIGURE 3-6. Monolithic graphite.

FIGURE 3-7. Carbon tow.

from either polyacrylonitrile (PAN), an acrylic fiber material, or oil or coal pitch. PAN is often used to make Standard Modulus (SM) and Intermediate Modulus (IM) carbon fiber with notable tensile strength. Pitch is often used to make High Modulus (HM) and Ultra High Modulus (UHM) carbon fibers that are very stiff and brittle and typically have a lower tensile strength than that made with PAN. (Figure 3-8)

Newer precursor materials such as polyolefin and lignin (a wood byproduct) are being considered in the development of lower cost carbon fiber for industry. These precursor materials provide a high yield of carbon fiber with significantly lower stiffness and strength properties when compared to standard modulus PAN precursor materials.

Research into improving the properties of precursor materials through manipulation of process conditions, hybridization of precursors (PAN and lignin for example), and experiments with various other sustainable and low-cost materials continues in an effort to make lower cost, high performance carbon fiber in the future.

Carbon Fiber Manufacture

Carbon fiber filaments are typically made by first oxidizing the precursor materials at temperatures between 392°–572°F (200°–300°C), then carbonizing and graphitizing the materials at progressively higher temperatures, ranging from 932° to 2,732°F (500°–1,500°C) in an inert gas environment (*see* Figure 3-9). The filaments are constantly pulled in tension, aligning the crystalline structure of the carbon filament as the material is processed to give the fiber high tensile and modulus properties.

After processing, the carbon fiber is again oxidized by immersing the fibers in air, carbon dioxide, or ozone. This can also be accomplished with an immersion bath in sodium hypochlorite or nitric acid. The fiber is then treated with the appropriate finish for bonding to the matrix resin.

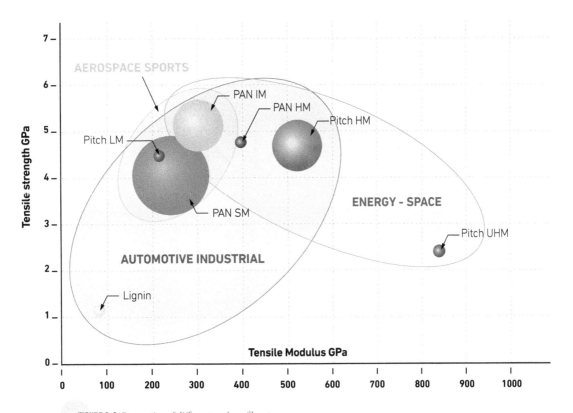

FIGURE 3-8. Properties of different carbon fiber types.

FIGURE 3-9. Carbon fiber manufacturing.
(Photos courtesy of C. A. Litzler Co., Inc.)

Oxidation and carbonization ovens

Treatment and sizing lines

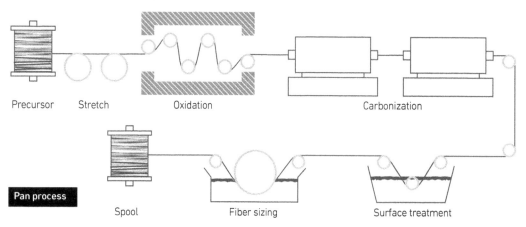

Precursor   Stretch   Oxidation   Carbonization

Pan process

Spool   Fiber sizing   Surface treatment

Newer manufacturing processes have been developed by carbon fiber manufacturers and equipment suppliers that reduce both the energy and time required to produce carbon fiber. 4M Industrial Oxidation LLC, a subsidiary of RMX Technologies (Knoxville, TN), has commercialized a plasma oxidation oven that operates three times faster and uses 25% less energy per pound of fiber than current commercial technology.

Toho Tenax Company (Tokyo, Japan) is using microwave ovens to carbonize PAN precursor, eliminating the need for high-temperature furnaces, resulting in improved fiber properties at faster production rates, while using less energy than traditional processes.

In a process developed by UHT Unitech Co. Ltd. (Zhongli, Taiwan), T-700 grade PAN-based carbon fibers can be reprocessed using a microwave graphitization oven that increases fiber modulus by 20–30% while adding 2–5% strength (T-800 and T-1000 grade). In the past, this type of reprocessing has decreased strength. The throughput is 50% faster than conventional graphitization furnace heating processes.

Carbon Fiber Sizing

Sizing is the process of applying a size or coating to fiber. The functions of size formulations for carbon fiber are similar to those for glass fiber: to protect the fiber during handling and to provide good adhesion between the fiber and matrix resin, keeping the molding process in mind. During handling and weaving, size applied to carbon fiber prevents individual filaments (5–7 μm in diameter) from breaking and creating issues with equipment as well as damage to the fiber. During molding, the size must be compatible with the type of matrix (e.g., epoxy resin or thermoplastic polymer) and be able to withstand and perform under the heat and pressure of the parts fabrication process.

Formulations for carbon fiber sizing have not been as complex as those for glass fiber because the latter require a silane component to build the bond between the glass and polymer. This is not an issue for carbon fiber with epoxy resins, which wet the organic carbon fiber quite well. However, other thermoset resins and thermoplastic matrices are different. Specialized size formulations have been developed for use of carbon fiber with thermoplastic polymers and also with PMR-15 polyimide resin. More recently, carbon nanotubes and nanosilica particles have been used to enhance sizing of both glass and carbon fibers with epoxy and thermoplastic matrix resins.

Carbon Fiber/Fabric Forms and Styles

Unidirectional tape is one of the most common forms of carbon fiber reinforcement. Uni-tape is typically available in widths of 6, 12 and 24 inches, with cured ply thickness ranging from 0.003 to 0.009 inches (0.08 to 0.23 mm). *See* Table 3.6.

Table 3.7 on the next page shows some of the many weave patterns and fabric styles available for woven carbon fiber reinforcements.

TABLE 3.6 Common Grades of Carbon Fiber Unidirectional Tape

Grade	Areal Weight oz/yd^2 (g/m^2)	Thickness Inches (mm)
95	2.8 (95)	0.0045 (0.11)
145	4.3 (145)	0.006 (0.15)
180	5.3 (180)	0.0075 (0.19)
190	5.6 (190)	0.008 (0.20)

TABLE 3.7 Carbon Fiber Fabric Weave Styles

Hexcel-Schwebel Style Number	Yarn Size	Fabric Weave Style	Yarn Count–Yarns per inch Warp/Fill	Fabric Weight oz/sq. yd. (g/m2)	Fabric Thickness Inches (mm)
130	1 K	Plain	24 /24	3.7 (125)	0.0056 (0.14)
282	3 K	Plain	12.5/12.5	5.8 (197)	0.0087 (0.22)
284	3K	2x2 Twill	12.5/12.5	5.8 (197)	0.0087 (0.22)
433	3 K	5-HS	18/18	8.4 (285)	0.0126 (0.32)
444	3 K	2x2 Twill	18/18	8.4 (285)	0.0126 (0.32)
584	3 K	8-HS	24/24	11.0 (373)	0.0165 (0.42)
613	6 K	5-HS	12/12	11.1 (376)	0.0166 (0.42)
670	12 K	2x2 Twill	10.7/10.7	19.8 (671)	0.0297 (0.75)
690	12 K	2x2 Basket	10/10	18.7 (634)	0.0280 (0.71)

❯ CERAMIC FIBERS

Boron Fiber

Boron fiber was the first high strength, high modulus, low density reinforcement developed for advanced composite aerospace applications and has been in commercial production for over forty-five years. Boron is an ultra-high modulus fiber, very stiff and strong, with **thermal expansion** characteristics similar to that of steel. Unlike carbon fiber, boron exhibits no galvanic corrosion potential with metals.

The sole manufacturer is Specialty Materials, Inc. (Lowell, Massachusetts), who uses chemical vapor deposition (CVD) to make their boron fibers. This CVD process heats a fine tungsten wire substrate in a furnace pressurized with boron trichloride gas, which decomposes and deposits boron onto the tungsten wire.

Boron is available in 3.0 mil (76μ), 4.0 mil (102μ), 5.6 mil (142μ), and 8.0 mil (203μ) diameter filaments—4.0 mil diameter is the most common size available. Boron is available in unidirectional prepreg tape only. It cannot be woven into cloth since the filaments cannot be crimped without snapping. Hy-Bor®, a boron-carbon hybrid, is also available.

Use of boron fiber in composites began in the 1970s for new part production in aerospace components. Now, it has largely been replaced by high-modulus carbon fibers in those applications. However, it has become very popular for repairs to metallic structures. Boron is also very popular for sporting goods applications, where it offsets the drawbacks of high modulus carbon fiber by offering improved damage tolerance and a compressive strength that is twice its tensile strength, resulting in a superior all-around composite structure.

Quartz Fiber

Quartz fiber can be used at much higher temperatures than either E-glass or S-glass fiber, up to 1,050°C (1,922°F) and Saint Gobain's Quartzel® 4 can be used at temperatures of up to 1,200°C (2,192°F). Quartz fiber has the best dielectric constant and loss tangent factor among all mineral fibers. For this reason, along with its low density (2.2 g/cm3), zero moisture absorption, and high mechanical properties, quartz fiber has been used frequently in high-performance aerospace and defense radomes. It also has almost zero coefficient of thermal

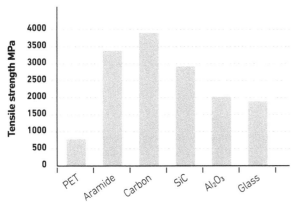

Fiber types

Typical tensile strengths (averages) of different fiber types.

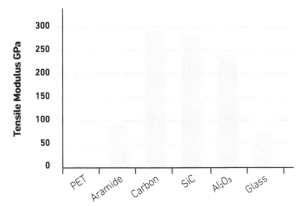

Fiber types

Typical tensile moduli (averages) of different fiber types.

FIGURE 3-10. Comparison of ceramic fiber tensile strength and modulus. *(Source: Ceramic Matrix Composites, Walter Krenkel, Ed.; © 2008 WILEY-VCH Verlag GmbH & Co. KGaA, Weinheim)*

FIGURE 3-11. Ceramic fibers used in ceramic matrix composites.

[a] Nextel™ fibers are available as continuous fiber spools, chopped fibers, braided sleeves and woven fabrics. *(Photo courtesy of 3M)*

[b] COI Ceramics' Nicalon™ fibers are available in a wide variety of formats including continuous fiber spools, woven fabrics and mats. *(Photo courtesy of COI Ceramics)*

expansion in both radial and axial directions, giving it excellent dimensional stability under thermal cycling and resistance to thermal shock.

Quartz fiber is produced from quartz crystal, which is the purest form of silica. The crystals, which are mined mainly in Brazil, are ground and purified to enhance chemical clarity. The quartz powder is then fused into silica rods, which are drawn into fibers and coated with size for protection during further processing. This fiber can be supplied as single and plied yarn, roving, chopped strand, wool, low density felt, needle-punched felt, sewing thread, braid, veil, fabric and tape. Saint-Gobain offers quartz yarn and roving in a wide range of tex (linear density) and with three different size coatings. Note that with quartz fiber, the size is maintained through textile processing and is said to improve resin wet-out and fiber compatibility with the resin matrix.

Saint-Gobain produces Quartzel® fiber products, JPS Composite Materials supplies Astroquartz® products, and CNBM International Corporation is a supplier in China.

Fibers used in Ceramic Matrix Composites

Fibers used as reinforcements in ceramic matrix composites (CMCs) fall into two broad categories: (1) non-oxide fibers, such as silicon carbide (SiC), boron nitride (BN), and silicon-boron-nitride-carbide (Si-B-N-C); and (2) oxide fibers, such as aluminum oxide (Al_2O_3, including single-crystal alumina), alumina zirconia mixtures ($Al_2O_3 + ZrO_2$), YAG (yttria-alumina-garnet), and mullite ($3Al_2O_3 - 2SiO_2$).

Non-oxide fibers have superior strength and good creep resistance but are susceptible to degradation by oxidation. Oxide fibers are inherently resistant to oxidation but have limited creep resistance at high temperatures. Non-oxide fibers are generally derived from polymer precursors. These often require complex processing before they can be **pyrolyzed** into ceramic fibers. (Figure 3-10)

Non-oxide polymer-derived SiC-based fibers are the strongest ceramic fibers. Early versions of these fibers, typified by Nicalon, had low elastic modulus. Later versions had very low oxygen content and were pyrolyzed at higher temperatures, increasing modulus, high-temperature strength and creep resistance. Most commercial fibers are SiC-based fibers.[2] (Figure 3-11)

2   *Ceramic Fibers and Coatings: Advanced Materials for the Twenty-First Century*, 1998, The National Academies Press, National Academy of Sciences, 500 Fifth St., NW, Washington, DC 20001.

Fiber reinforcement in CMCs may be in the form of short fibers, continuous fibers, textiles, unidirectional prepreg, felts or mats. Ceramic fiber manufacturers are listed in Table 3.8 below. For more details on how ceramic matrix composites are formed, see the "Ceramic Matrix" section in Chapter 2.

TABLE 3.8 Ceramic Fiber Manufacturers

Product	Fiber Type	Denier	Filaments per Tow	Filament Diameter (μm)	Tensile Strength (GPa)	Tensile Modulus (GPa)	Thermal Stability
Company: 3M							
Nextel™ 312	$Al_2O_3/$ SiO_2/B_2O_3	600 to 3,600	400 to 1,375	8–12	1.6	150	1,800 C°
Nextel™ 440	$Al_2O_3/$ SiO_2/B_2O_3	1,000 2,000	400 750	10–12	1.8	190	1,800 C°
Nextel™ 610	Al_2O_3	1,500 to 20,000	400 to 5,100	11–13	2.8	370	2,000 C°
Nextel™ 720	Al_2O_3/SiO_2	1,500 3,000 10,000	400 750 2,550	12–14	1.9	250	1,800 C°
Company: COI Ceramics Inc. (an affiliate of ATK)							
Nicalon™	Si-C-O	1,800	500	14	3.0	200	1,473 K
Hi-Nicalon™	Si-C-O	1,800	500	14	2.8	270	2,073 K
Hi-Nicalon™ Type S	Si-C-O	1,800		12	2.6	420	2,073 K
Company: NGS Advanced Fibers Co., Ltd.*							
Nicalon CG					3.2	200	
Nicalon HVR					3.1	170	
Nicalon LVR					3.1	190	
Hi-Nicalon	Si-C-O				3.2	270	
Hi-Nicalon Type S	Si-C				3.1	380	
Company: UBE Industries Ltd.							
Tyranno Fiber® S	Si-C-O-Ti	1,980	1,600	8.5	3.3	170	
Tyranno Fiber® ZMI	Si-C-O	1,800	800	11	3.4	195	
Tyranno Fiber® SA	Si-C-O	1,530 1,710	800 1,600	10 7.5	2.4	380	1,800 C°

* Joint venture between GE, Safran and Nippon Carbon Co., Ltd.

TABLE 3.9 Ceramic Fiber Fabric Weave Styles

Fiber	Denier	Fabric Weave Style	Yarn Count–Yarns per inch Warp/Fill	Fabric Weight (g/m²)
Company: COI Ceramics				
Hi-Nicalon™	1,800	Plain	16/16	250
		5-HS	16/16	250
		8-HS	22/22	380
Company: 3M				
Nextel™ 312	600	10 Mesh Leno	20/10	81
	600	5-HS	46/46	240
	900	Plain	20/17	300
	1,200	5-HS	25/25	270
	1,800	Plain Double Layer	40/20	980
	1,800	5-HS	32/20	810
	1,800	4-HS	19/18	610
	3,600	4-HS	17.5/17.5	540
Nextel™ 440	2,000	5-HS	30/26	510
	2,000	4-HS	21/20	680
	2,000	5-HS	32/20	880
Nextel™ 610	1,500	4-HS	18.5/18.5	250
	1,500	8-HS	27.5/27.5	370
	3,000	8-HS	23.5/23.5	610
Nextel™ 720	1,500	8-HS	27.5/27.5	370
	3,000	8-HS	23.5/23.5	610
	10,000 warp 1,500 fill	Plain semi-UD	16/5	750

❱ SYNTHETIC POLYMER FIBERS

Aramid Fiber

Aramid fibers are different from carbon and glass fibers in that they are textile fibers, spun from an organic polymer versus fibers created by drawing out an inorganic material. The properties of aramid fiber vary according to type. For example, Kevlar® 29 is used for ballistic protection and has a lower modulus and greater elongation, while Kevlar® 49 is used for structures and has a higher modulus and lower elongation. Aramid fibers are 43% lighter than, and twice as strong as E-glass, and ten times stronger than aluminum on a specific tensile strength basis.

Aramid fibers have excellent resistance to high and low temperatures. They excel in tensile strength but have somewhat poor laminate-compressive properties. Their main disadvantage is that they are hydroscopic (absorb water) due to their inherent aramid chemistry. Aramid fibers have much better impact resistance than carbon fiber and are often used in "crack-stop" layers between plies of carbon. Aramid fiber is cheaper than carbon fiber but more expensive than fiberglass.

Manufacturers

Kevlar® is a **para-aramid** fiber that was developed by DuPont and introduced in 1973. It is currently produced and supplied by DuPont. Technora is also a para-aramid fiber, but uses a slightly different chemistry in its manufacturing process. It was introduced in 1976 by the Japanese company Teijin, who also makes Teijinconex, a **meta-aramid** fiber that competes with DuPont Nomex®. Twaron® is a para-aramid fiber basically the same as Kevlar®, originally developed in the early 1970s by the Dutch company AKU, which later became Enka (Glanzstoff) and then Akzo Nobel, but since 2000 has been owned by the Teijin Group. Thus, Teijin now produces and supplies Twaron®, and the Technora® brand name is no longer supported.

Typical Fiber/Fabric Forms and Styles

Aramid tows and yarns are defined by weight using the **Denier** or Decitex system. Denier weight is listed in grams/9,000 meters. For example: 1,140 denier = 1,140 grams/9,000 meters. The Decitex (dtex) system measures grams/10,000 meters.

TABLE 3.10 Kevlar® 49 Fabric Weave Styles

Hexcel-Schwebel Style number	348	350	352	353	354	386
Yarn Type Denier	380	195	1,140	1,140	1,420	2,130
Fabric Weave Style	8-HS	Plain	Plain	4-HS	Plain	4x4 basket
Yarn Count Yarns per inch Warp/Fill	50/50	34/34	17/17	17/17	13/13	27/22
Fabric Weight oz/yd² (g/m²)	4.9 (166)	1.7 (58)	5.3 (180)	5.3 (180)	4.8 (163)	13.6 (461)
Fabric Thickness Inches (mm)	0.008 (0.20)	0.003 (0.08)	0.010 (0.25)	0.009 (0.23)	0.010 (0.25)	0.025 (0.63)

TABLE 3.11 Twaron® T-1055, T-2000, T-2200, T-2040 Fabric Weave Styles

Hexcel-Schwebel Style number	5348	5352	5353	5704	5722
Yarn Type Denier	405	1,210	1,210	930	1,680
Fabric Weave Style	8HS	Plain	4HS	Plain	Plain
Yarn Count Yarns per inch Warp/Fill	50/50	17/17	17/17	31/31	22/22
Fabric Weight oz/yd^2 (g/m^2)	4.9 (166)	5.0 (170)	5.0 (170)	6.5 (198)	8.25 (280)
Fabric Thickness Inches (mm)	0.008 (0.20)	0.010 (0.25)	0.009 (0.23)	0.012 (0.30)	0.015 (0.38)

Spectra® and Dyneema®

Spectra® and Dyneema® are ultra-lightweight, high-strength thermoplastic fibers made from ultra-high molecular weight polyethylene (UHMWPE), also known as high modulus polyethylene (HMPE) or high performance polyethylene (HPPE). Dyneema® was developed by the Dutch company DSM in 1979, and Spectra® was developed by Honeywell International Corporation. Though the two products are chemically identical, they have slightly different production details, resulting in mechanical properties that are not identical, but very comparable. HPPE features high tensile strength, very light weight, low moisture absorption, and excellent chemical resistance. Its high specific modulus, high energy-to-break and damage tolerance, non-conductivity and good UV resistance make it a good alternative to aramid fiber. It has limited temperature service (< 220°F/104°C) and is used in underwater, ballistic, and body armor applications. Types include Spectra® 900, Spectra® 1000, Dyneema® SK76 for ballistic armor and Dyneema® UD-HB for hard ballistic armor applications. Fabrics made with Spectra® fiber are typically used in ballistic and high impact composite applications. Table 3.12 shows common fabric styles for Spectra®. Unidirectional forms are also available.

TABLE 3.12 Spectra® Fabric Weave Styles

Yarn Type Denier	Hexcel-Schwebel Style Number	Fabric Weave Style	Fabric Weight oz/yd^2 (g/m^2)	Fabric Thickness Inches (mm)
215	945	Plain	2.6 (88)	0.006 (0.15)
650	951	Plain	2.9 (108)	0.011 (0.28)
650	985	8-HS	5.5 (186)	0.014 (0.36)

Innegra®

Innegra S® fibers are manufactured by Innegra Technologies (Greenville, SC), and are extruded from high molecular weight polypropylene (HMPP) into aligned, semi-crystalline filaments. Much like other fibers in the polyolefin (olefin) family, they are low-cost, lightweight, tough, ductile, impact resistant and energy dissipative. They are also hydrophobic in nature and do not hydrate like aramid fibers. With mechanical properties that are somewhat lower than those of other noted olefins such as Spectra® and Dyneema®, they are often used as a stand-alone impact/fracture resistant layer in a hybrid laminate structure, or they are intermingled with stronger and stiffer filaments such as glass, basalt, and carbon to create multi-filament

(Innegra H®) tows or yarns that are adaptable to most textile reinforcement and composite manufacturing processes.

HMPP fibers tend to have a low melt point temperature (T_m) and therefore must be used with matrix resins that process well below the T_m. Innegra® fibers are used in a wide range of applications, from ballistics to sporting goods and more recently, use in the aerospace and automotive industries.

Zylon®

Zylon® fibers consist of a rigid chain of molecules of PBO (poly p-phenylene-2, 6-benzobisoxazole). This fiber from Toyobo has excellent tensile strength and modulus. Fabrics made from Zylon® are found in both ballistic and structural applications (*see* Table 3.13).

TABLE 3.13 Zylon® Fabric Weave Styles

Yarn Type Denier	Hexcel-Schwebel Style Number	Fabric Weave Style	Fabric Weight oz/yd^2 (g/m^2)	Fabric Thickness Inches (mm)
246	550-HM	Plain	1.8 (61)	0.003 (0.08)
500	530-AS	Plain	3.8 (129)	0.008 (0.20)
500	545-AS	Plain	6.25 (212)	0.013 (0.33)
1000	1030-AS	Plain	8.0 (271)	0.016 (0.41)
1000	1035-AS	Plain	8.85 (300)	0.017 (0.43)

❱ NATURAL FIBERS

Though natural fibers technically include those made from mineral (e.g., basalt, asbestos), animal (e.g., silk, wool) or plant matter, only the latter is commonly used in natural fiber composites (NFC). (*See* "Basalt Fiber" section on page 71). This section covers plant- and cellulose-based fibers and their use in composites.[3]

Wood is perhaps the original natural composite, comprised of cellulose fibers held together by lignin as a matrix or glue. Cellulose is a natural polymer with high strength and stiffness per weight. The most abundant biological material on Earth, cellulose forms long fibrous cells that help make up the stems, leaves and seeds of plants. In general, the highest performing natural fibers are those with the highest cellulose content and also those where the fiber filaments (microfibrils) are more aligned, such as in bast fibers, where this alignment provides strength and stiffness in the plant stem.

Plant and cellulose-based fibers have become an attractive substitute for glass and other synthetic fibers in composite materials due to their potential for weight reduction, lower raw material price, recycling and sustainability, vibration damping and aesthetic appeal. However, they also have issues including water absorption, variability and fire/high temperature performance. (Figure 3-12; Table 3.14)

---

3   Key references for this section include https://doi.org/10.1016/j.compositesa.2015.08.038 and "Biocomposites reinforced with natural fibers: 2000-2010" by Omar Faruk, Andrzej K. Bledzki, Hans-Peter Fink, and Mohini Sain, from *Progress in Polymer Science 37* (pp. 1552-1596), 2012 Elsevier.

TABLE 3.14 Properties of Glass and Natural Fibers

Fiber	Diameter (µm)	Density (g/cm³)	Tensile strength (MPa)	Elastic modulus (GPa)	Specific modulus	Elong. at failure (%)	Moisture absorption (%)	$/kg
E-glass	<17	2.5–2.6	2400–3500	70–76	29	1.8–4.8	–	1.3–3.3
Abaca	–	1.5	400–980	6.2–20	9	1.0–10		1.6–2.0
Bamboo	25–40	0.6–1.1	140–800	11–32	25	2.5–3.7		0.3–0.5
Banana	12–30	1.35	500	12	9	1.5–9		
Coir	10–460	1.15–1.46	95–230	2.8–6	4	15–51.4	10	0.3–0.5
Flax	12–600	1.4–1.5	343–2000	27.6–103	45	1.2–3.3	7	0.3–1.5
Hemp	25–600	1.4–1.5	270–900	23.5–90	40	1–3.5	8	0.3–1.7
Jute	20–200	1.3–1.49	320–800	30	30	1–1.8	12	.3
Kenaf	–	1.4	223–930	14.5–53	24	1.5–2.7		0.3–0.7
Ramie	20–80	1.0–1.55	400–1000	24.5–128	60	1.2–4.0	12–17	1.6–2.0
Sisal	8–200	1.33–1.5	363–700	9.0–38	17	2.0–7.0	11	0.4–0.7

* Modulus/density
SOURCES: "Flax fibre and its composites—A review," Yan, Libo, Chouw, Nawawi and Jayaraman, Krishnan, in *Composites Part B: Engineering, An International Journal* (Vol. 56, pp. 296–317); Elsevier 2013 (10.1016/j.compositesb.2013.08.014), also https://www.researchgate.net/publication/256761992). *Natural Fibre Composites in Structural Components: Alternative Applications for Sisal?* W.D. Brouwer, Delft University, The Netherlands.

Bast	Leaf	Seed	Fruit	Grass	Wood
Flax	Abaca	Cotton	Coir	Bagasse	Hard wood
Hemp	Banana	Kapok	Oil palm	Bamboo	Soft wood
Jute	PALF	...		Napier	
Kenaf	Sisal			Rape	
Ramie	...			Sabai	
...			...		

Bast a / Sattar Jute
Leaf b / Celtex-Abaca
Seed c / Nimbus Company
Fruit d / Palm Plantations of Australia
Grass e / iStockphoto BIHAIBO - Sugar cane
Wood f / Kentucky Division of Forestry

FIGURE 3-12. Classification of plant- and cellulose-based fibers.

Bast Fibers

Bast fibers include flax, hemp, jute, kenaf, ramie and harakeke, a type of flax grown in New Zealand. They usually consist of a woody core surrounded by a stem. The stem is composed of bast fiber bundles, which contain individual cellulose fibers bonded together by lignin or pectin. The pectin is removed during processing of the fibers after harvest. (Figure 3-13)

Bast Fiber Production

Flax may be grown for linseed oil, linen clothing or composite reinforcement. After a 100–120 day growing cycle, the crop is cut at its base and either left in the field to achieve *dew retting* or collected and processed through water/vet retting or chemical/enzyme retting. This removes the pectin that holds the cellulose fibers together, enabling separation of the fiber bundles from the rest of the stem via *breaking and scutching*. (Figure 3-14) These process steps are achieved using mechanized equipment.

Fiber filaments are then spun into strands or yarns for converting into unidirectional, woven and stitched fabrics, as well as braids and nonwoven mats. Chopped fibers are used in plastic molding compounds. According to the European Confederation of Linen and Hemp (CELC, Paris, France), one hectare of cultivated flax produces 0.65 metric tons of yarn and 3,750 m² of fabric. Over 300,000 metric tons of flax fiber were produced in 2015.

Leaf, Seed, Fruit and Grass Fibers

The most common leaf fibers used in composites are sisal (from the agave plant), banana, abaca (also called Manila hemp and a close relative of the banana), palm (made from the dwarf palm in southern Europe and northern Africa) and pineapple leaf fiber (PALF) made from waste after harvesting the fruit. They are generally coarser than bast fibers.

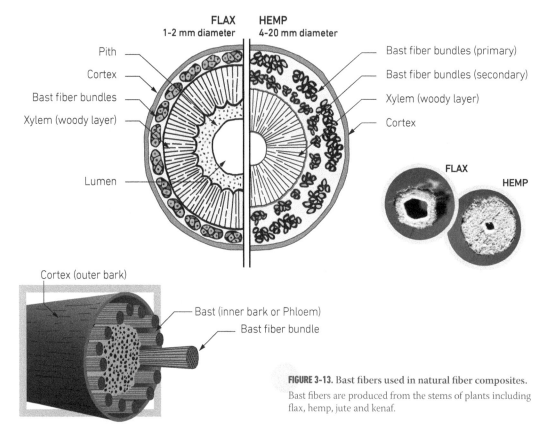

FIGURE 3-13. Bast fibers used in natural fiber composites.
Bast fibers are produced from the stems of plants including flax, hemp, jute and kenaf.

Seed fibers include cotton and kapok, though neither is currently used heavily in composites. Fruit fibers include coir taken from coconut husks and oil palm fiber (OPF), extracted from the empty fruit bunches after the palm fruit has been pressed to make oil.

Fibers are also extracted from a wide variety of grasses, including rape, which is typically grown for rapeseed or canola oil, bagasse (the remains of sugar cane after juice extraction) and bamboo.

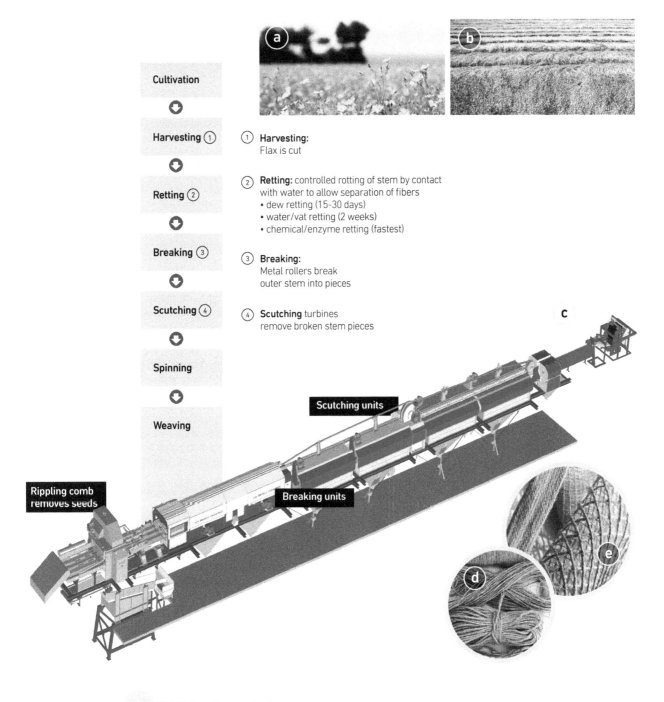

Cultivation

Harvesting ①

① **Harvesting:**
Flax is cut

Retting ②

② **Retting:** controlled rotting of stem by contact with water to allow separation of fibers
• dew retting (15-30 days)
• water/vat retting (2 weeks)
• chemical/enzyme retting (fastest)

Breaking ③

③ **Breaking:**
Metal rollers break
outer stem into pieces

Scutching ④

④ **Scutching** turbines
remove broken stem pieces

Spinning

Weaving

Rippling comb removes seeds

Breaking units

Scutching units

FIGURE 3-14. Flax fiber production process.

Flax fiber production is global and industrialized. *(Images courtesy of [a, b] CELC, [c] Cretes, [d, e] Bcomp, and Van Dommele Engineering)*

TABLE 3.15 Natural Fiber Reinforcement, Overview of Matrices, Experiments and Results

Fiber	Matrices	Test Results
Abaca/Banana	Cement, epoxy, Polyester, PE, PP, urethane	Injection molding provided higher properties vs. compression molding. Modulus highest with 40% fiber loading in polyester composites. Properties higher with acetylation. Enzyme treatment improved composite tensile strength 5–45%.
Bagasse	Cement, HDPE, polyester, PP, PVC	Injection molding with vacuum worked better than compression molding using PP, mechanical properties require improved adhesion to matrix.
Bamboo	Epoxy, PP, rubber	Highest mechanical properties via steam vs. mechanical extraction of fibers.
Coir	Cement, epoxy, PP, rubber	Acetylated coir fiber/polyester composites showed higher bio-resistance, less tensile strength loss vs. silane-treated fibers. Silane-treated fibers in polyester performed better vs. untreated fibers.
Flax	Epoxy, polyester, PP	Plasma treatment improved properties with polyester.
Hemp	Epoxy, PP	Recycled hemp/PP composites maintain properties even through numerous cycles. Treatments to improve fiber-to-matrix adhesion significantly improve composite properties. Corona-treated fibers increased tensile strength with PP.
Jute	VE, PA, polyester, PP	Vinyl ester composites with alkali-treated fibers had higher properties vs. untreated fibers. 6mm long fibers with maleated PP (MAPP) showed 72% higher flexural strength, improved fiber-matrix adhesion and fatigue performance. Corona-treated jute fibers performed better in epoxy composites. Oxygen plasma treatment increased properties with HDPE.
Kenaf	Polyester, PP	Success with thermoforming sheets of chopped fibers with PP powder, 30-40 wt% fiber. Kenaf/PS composites improved with silane coupling agent.
Oil Palm		Acetylated oil palm fiber/polyester composites showed higher bio-resistance, less tensile strength loss vs. silane-treated fibers. Silane-treated fibers in polyester performed better vs. untreated fibers.
PALF	Epoxy, phenolic, PC, polyester, PP, PVC	Highest properties with silane sizing and 30 wt% fiber in composite. Alkaline treatment also shown to improve composite properties.
Ramie	Epoxy, polyester, PP, PLA	Highest properties with silane coupling agent, 5–6mm length fibers and 45 wt% fiber in composite. Level II bulletproof panels made with epoxy. Biodegradable composites made with PLA. Alkaline-treated fibers showed 4–18% higher tensile strength than untreated fibers.
Sisal	Bio-polyurethane, cement, epoxy, phenolic, polyester, PE, PP, rubber	Composites with alkali-treated fibers had higher properties vs. untreated fibers. Plasma treatment improver properties with HDPE.

Wood Fibers

Cellulose fiber may also be extracted from wood. It is processed as pulp first and then made into uniform-sized filaments using methods similar to those for manufacturing glass and other synthetic fibers. Wood fiber is reportedly 40% lighter than glass fiber. It became popular in the 1990s as a reinforcement for plastic extrusion of polypropylene in products such as tubes and profiles. This is now a sub-segment of a very large market for wood plastic composites.

A wood plastic composite (WPC) is made of wood fiber and/or wood flour mixed with a thermoplastic polymer such as polypropylene (PP), polyethylene (PE), polyvinyl chloride (PVC), and many others. Extrusion is the most popular process but injection molding is used as well. Thermoset resins may also be used, including polyurethane and phenolic and other phenol formulations. Wood flour is wood fiber that has been processed into a powder with a specified particle size. Its manufacture often begins with sawdust. (Figure 3-15)

WPCs are used in residential decking, park benches and outdoor furniture, interior/exterior molding and trim, window and door frames, indoor furniture, house cladding, fencing and landscape timbers. It is also used in a wide range of injection-molded consumer goods. WPCs offer resistance to rot, decay and insect attack, though they do absorb water. A main advantage is the ability to mold them into almost any shape—as well as the elimination of paint, because color can be added to the plastic mixture before processing. Cellulose content can be up to 70%, though 50% is more common. WPCs are touted as sustainable and eco-friendly because they can be made from trees that are regrown relatively quickly, and also often use recycled and/or bio-based plastics such as thermoplastic starch (TPS) and polylactic acid (PLA).

Advantages of Natural Fiber Composites

Beyond their ability to improve sustainability, natural fibers offer lower density. For example, the 1.5 g/cm^3 density of flax fibers compared to glass fibers at 2.6 g/cm^3 yields a net weight savings of nearly 40%. The cost per kilogram of natural fibers also produces savings in the range of 20–100%.[4]

Another key benefit is the ability of natural fibers to provide vibration damping. One study found flax fiber composites had 51% higher vibration damping vs. glass fiber composites.[5] Natural fibers also impart a natural, wood-like appearance. These two characteristics have been combined successfully in all types of sporting goods (e.g., boards, bikes, skis, tennis rackets) and also in musical instruments offering the look and tone of wood but with less weight, higher performance and better durability and sustainability.

FIGURE 3-15. Wood plastic composites.
Wood plastic composites are used in a wide range of applications, from extruded planks and profiles for decks, windows, doors, fencing, etc. to injection-molded cases, gears, car interiors, toys, consumer goods and much more. *(Images courtesy of JELUPLAST® of JELU-Werk)*

4 From www.compositesworld.com/articles/natural-fiber-composites-market-share-one-part-at-a-time.
5 *See* https://doi.org/10.1016/j.proeng.2014.12.285.

Disadvantages of Natural Fiber Composites

Natural fiber composites present several challenges. The range in fiber properties is greater than for glass or carbon fiber, for example. This is because the different types of fibers have different structures based on the plants they are processed from and also because the source plants vary based on environmental conditions during growth. This property variation can be minimized by using hybrids of different types of fibers and also hybrids with glass, carbon and other fibers.

Natural fibers are also hydrophilic—i.e., they will absorb moisture from their environment. This again varies by type of plant and may be minimized by very good encapsulation of each fiber with a moisture-resistant matrix. However, just as glass, carbon and other fibers require a sizing or coupling agent to achieve good fiber to resin bonding, natural fibers also require some type of surface treatment. Significant improvement has been documented with the following treatments:

- Silane-based sizing and coupling agents
- Alkaline treatment (increases surface roughness)
- Acetylation (adding acetyl functional groups)
- Corona and plasma treatment
- Maleated coupling (maleic anhydride used to modify fiber surface)
- Enzyme treatment (environmentally friendly)

Using nanofibers and nanoparticles to improve fiber-to-matrix bonding of natural fibers is being investigated as well.

Higher temperatures may also present issues for natural fibers, with most of the types used in composites being thermally unstable above 200°C/392°F. This must be taken into account when choosing matrix resins and molding methods and also noted when defining in-service conditions.

Applications, Manufacturers and Product Forms

The largest markets for natural fiber composites are construction and automotive. The global market for wood plastic composites (WPCs) was estimated between 4 and 5 million metric tons in 2016 and projected to grow at more than 10% annually. Composite decking and profiles make up 80% of this market. Estimates of the natural fiber composites market range from

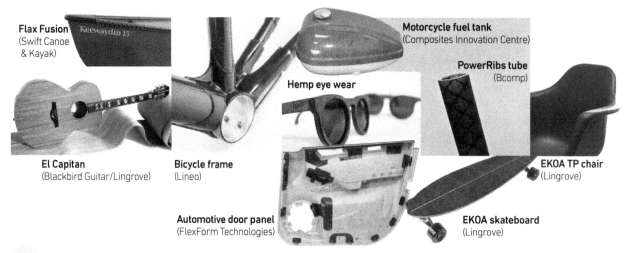

Flax Fusion (Swift Canoe & Kayak)
Keewaydin 15
Motorcycle fuel tank (Composites Innovation Centre)
PowerRibs tube (Bcomp)
Hemp eye wear
El Capitan (Blackbird Guitar/Lingrove)
Bicycle frame (Lineo)
EKOA TP chair (Lingrove)
Automotive door panel (FlexForm Technologies)
EKOA skateboard (Lingrove)

FIGURE 3-16. Examples of natural fiber-reinforced composites.

hundreds of millions to billions of dollars, with leading applications in automotive interiors and sporting goods, as well as projected growth in aircraft and building interiors. (Figure 3-16)

Table 3-16 below shows applications of natural fiber composites in the automotive industry. Natural fiber products for use in composites are available in a variety of forms, listed in Table 3-17.

TABLE 3.16 Natural Fiber Composites Used in Automotive Applications

Body panel	Door panel	Head liner panel	Instrument panel	Pillar cover panel	Rear parcel panel	Seat back	Spare tire well*	Spoiler	Trunk liner	Other
MANUFACTURER: Audi				**MODEL:** A2, A3, A4, A4 Avant, A6, A6 Avant, A8, Roadster, Coupe						
■					■	■	■		■	
MANUFACTURER: BMW				**MODEL:** 3, 5, 7 series and others						
	■	■				■	■		■	Noise insulation panels
MANUFACTURER: Citroen				**MODEL:** C5						
		■								
MANUFACTURER: Daimler				**MODEL:** A, C, E and S-Class, EvoBus (exterior)						
	■		■	■						
MANUFACTURER: Ford				**MODEL:** Modeo CD 162, Focus						
	■								■	B-pillar
MANUFACTURER: Lotus				**MODEL:** Eco Elise						
■						■		■		
MANUFACTURER: Mercedes-Benz				**MODEL:** Trucks						
		■								Sun visor, engine cover, bumper
MANUFACTURER: Opel GM				**MODEL:** Astra, Vectra, Zafira						
	■	■	■	■						
MANUFACTURER: Peugeot				**MODEL:** New model 406						
					■	■				
MANUFACTURER: Renault				**MODEL:** Clio, Twingo						
					■					
MANUFACTURER: Saab										
	■				■					
MANUFACTURER: SEAT										
■						■		■		
MANUFACTURER: Toyota				**MODEL:** Brevis, Harrier, Celsior, RAUM						
	■					■	■			
MANUFACTURER: Volkswagen				**MODEL:** Golf, Passat, Bora, Fox, Polo						
	■					■			■	Trunk lid finish panel
MANUFACTURER: Volvo				**MODEL:** C70, V70						
										Cargo floor tray

* or wheel well.

TABLE 3.17 Natural Fiber Product Forms for Use in Composites

Company	Product names	Form	Fiber
Bcomp	ampliTex ampliTex Braids ampliTex Fusion ampliTex PowerRibs	Unidirectional (UD), noncrimp fabric (NCF) and woven fabrics Braid Hybrid with carbon Grid creates stiffening rib structure	Flax
Composites Evolution	Biotex Biotex Flax/PP, PLA	UD and woven dry fabrics Fiber commingled with TP fibers	Jute, Flax Flax
Flaxcomposites		UD, NCF, woven fabrics and braids Prepreg with epoxy Nonwoven mat	Flax
FlexForm Technologies	FlexForm MT FlexForm LD FlexForm HD	Nonwoven mat Low density board High density board	Bast fibers
GreenCore Composites	NCell	Fiber/PP or /PE pellets for injection molding and extrusion	Cellulose microfibers
Lineo	FlaxTape FlaxPreg FlaxPly	Dry UD fiber tape UD or woven fabrics with epoxy Fabrics	Flax
Lingrove	Lineo flax tape Bcomp Amplitex EKOA tape	See Lineo above See Bcomp above UD bio-epoxy prepreg tape	Flax
Sunstrand		Bulk fibers Chopped strand mat Continuous fiber mat	Bamboo, Flax, Hemp, Kenaf and other fibers

Forms of Reinforcement

This section discusses the different forms of fiber reinforcement used in composites. We will explore important terminology and definitions, show examples of materials and applications, and examine some of the key considerations when choosing fiber reinforcement.

In a composite, reinforcement provides the strength and stiffness. Particles or fibers may be used as reinforcement. Particles are smaller and have roughly the same dimensions in all directions, while a fiber has a higher aspect ratio which is the ratio of length-to-diameter (l/d). Strength increases with aspect ratio, thus particle-reinforced composites are lowest in strength. Particle and short-fiber composites also contain less reinforcement, described as fiber volume, typically capped at 40 to 50% due to processing difficulties, and in the case of particles, brittleness. (Figure 3-17)

Composite strength, modulus and fiber volume all increase with the length of reinforcement, as shown in Figure 3-18. Longer, straighter fibers can carry more load, compared to composites with discontinuous fibers where the load from one fiber end to the next must be carried by the matrix in shear.

Historically, the practical limit for fiber volume, as shown below, has been about 70%. However, the maximum *ideal* fiber volume, calculated for perfectly round fibers (cross section)

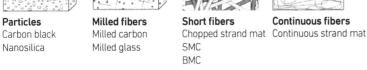

INCREASING LENGTH INCREASED ORIENTATION

Aligned discontinuous fibers
Stretch-broken carbon fiber
Directed fiber preforms etc.

Aligned continuous fibers
Unitape
Pultrusion

Woven biaxial fabric
Plain weave
Twill etc.

Woven triaxial fabric

Triaxial braid

Oriented fibers stacked laminate
Biaxial, triaxial, quadraxial noncrimp fabrics
Prepreg layup laminate
Dry fiber infused resin etc.

Filament winding

INCREASING LENGTH RANDOM ORIENTATION

Particles
Carbon black
Nanosilica

Milled fibers
Milled carbon
Milled glass

Short fibers
Chopped strand mat
SMC
BMC

Continuous fibers
Continuous strand mat

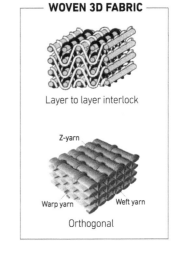

WOVEN 3D FABRIC

Layer to layer interlock

Z-yarn
Warp yarn   Weft yarn
Orthogonal

FIGURE 3-17. Spectrum of fiber length and orientation in composite reinforcements.

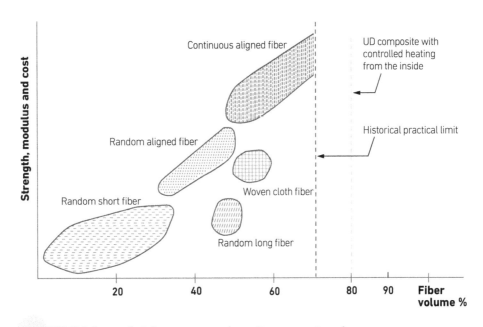

FIGURE 3-18. Influence of reinforcement type and quantity on composite performance.
(Source: "Structural Composite Materials," F.C. Campbell, 2010 ASM International, www.asminternational.org)

packed perfectly, touching edge-to-edge, is 0.907, or roughly 90%.[6] This has been difficult to reach because fibers are not perfectly round and perfect fiber packing is a challenge in actual parts production. However, by heating the composite through its volume from the inside and controlling this digitally, unidirectional carbon fiber composites with a volume fraction of 80% have been achieved. The performance of such composites, however, depends upon even distribution of resin around the fibers.

With increased length, the cost of reinforcement increases, as does the resulting composite with particles being the cheapest and long, continuous fibers the most expensive.

Composite properties (and cost) also increase with increased reinforcement *orientation*. Particles and discontinuous fibers tend to be randomly oriented, which does produce more **isotropic** materials—same properties in all directions—but limits strength and stiffness. Much higher strength and modulus (stiffness) are possible by aligning reinforcement in one direction. This results in **anisotropic** properties; that is, greater strength and modulus in the primary fiber direction of alignment versus at 45° or 90° degrees.

Continuous fiber composites are typically made into laminates by stacking single layers (plies) of fibers and matrix in different orientations (e.g., 0°, 90°, -45°, +45°, 30°, 60°, etc.). Thus, composites are infinitely tailorable by choosing the type of reinforcement, length, fiber volume, orientation and stacking sequence, and that is without even considering options in the matrix material.

❱ DISCONTINUOUS FIBER

As stated in this chapter's introduction, short fibers are typically used in non-structural composite applications. Though definitions vary by industry, as a general rule, milled fibers are typically ¹⁄₃₂ to ¼ inch (0.8 to 6 mm) and chopped fibers are typically ¼ to 1 inch in length. Fibers used in long fiber injection (LFI) molding range from ½ inch to 4 inches (12.5 mm to 100 mm). Lengths of ½ inch to 6 inches (12.5 mm to 152 mm) have been noted for stretch broken carbon fiber prepreg.

The strength of discontinuous fiber composites can approach that of continuous fiber composites if the aspect ratio of the fibers is great enough and if the fibers are aligned. However, it has been difficult to maintain good alignment in processes to date. Composite processes that typically use discontinuous fibers include injection-molding and overmolding, compression molding of sheet molding compound (SMC) and bulk molding compound (BMC) materials, hand layup using chopped and/or continuous strand mat and veils, and 3D printing using filament made from particle and/or short fiber reinforced thermoplastic polymers. Novel processes are being developed and software is continuously being improved to enable better prediction and control of fiber alignment in injection-molding processes.

Most often, discontinuous fibers are randomly oriented and are much less costly than continuous fibers. Thus, discontinuous fibers are often used where cost is the driver and continuous fibers are used where higher strength and stiffness are required. (Figure 3-19)

❱ CONTINUOUS FIBER

The highest performing composite reinforcement is continuous fiber with a unidirectional orientation. Historically, in aerospace applications, like primary structures in aircraft, missiles and satellite structures, the material of choice has been unidirectional prepreg, often slit into narrow tapes called unitape for application by an automated machine or robot, referred to as automated tape laying (ATL).

6 *Principles of Composite Material Mechanics*, 4th Edition, Ronald F. Gibson; CRC Press 2016.

Before unidirectional materials became popular, the most common form of continuous fiber reinforcement was woven fabric. Knitting and braiding are two other textile processes used to prepare continuous fibers for building a stacked composite laminate. These will be discussed in "Textile Technology," starting on page 98.

In processes like pultrusion and filament winding, which date back to the 1950s and 70s respectively, spools of fiber are drawn directly into the fabricating machinery. This approach was used to modify ATL so that spools of fiber are fed into the equipment instead of rolls of tape. This modified process is called automated fiber placement (AFP).

FIGURE 3-19. Examples of short fiber-reinforced composites.

[a] Owens Corning's PERFORMAX 249 30% chopped glass fiber-reinforced PP composite improved injection molding and composite strength and stiffness to achieve a thinner, stronger tub with 30% higher washing machine capacity. *(Courtesy of CompositesWorld)*

[b] This MAI Skelett project demonstrator of a BMW *i3* windshield frame/roof structure uses carbon fiber-reinforced PA6 pultrusions overmolded with short glass fiber/PA6 compound. *(Photo courtesy of SGL Carbon)*

[c] This air intake manifold with integrated liquid charge air cooler for Volkswagen's TSI engine was injection-molded using DSM's Akulon polyamide 6 resin with 30% short glass fiber, reducing number of parts, weight and cost by 20% compared to similar turbo systems. *(Courtesy of DSM)*

[d] High-performance loaded bracket is made with a lower melt temp continuous fiber-reinforced VICTREX PAEK unitape substrate, overmolded with VICTREX PEEK 150CA30 30% short carbon fiber-reinforced compound, to cut weight 60% vs. comparable metal components. *(Courtesy of Tri-Mack Plastics and TxV Aero Composites)*

Fabrics have also been updated. Multiaxial fabrics—also referred to as noncrimp fabrics (NCF) or stitchbonded fabrics—are produced as stacks of discrete unidirectional 0°, 90° and/or ±45° layers held together by stitching or knitting (*see* pages 103–113). Advantages include straighter fibers (versus crimp in the fibers from weaving) and tailored fiber architectures for higher performance laminates as well as labor savings because the laminate stack is prefabricated and applied as one layer in a single step.

3D textiles are the most complex continuous fiber reinforcements, offering reinforcement through the thickness, i.e., in the z-direction, for increased damage tolerance and high performance in applications like jet engine parts (e.g. fan blades, guide vanes), propeller blades and complex-shaped aerospace structures. (Figure 3-20)

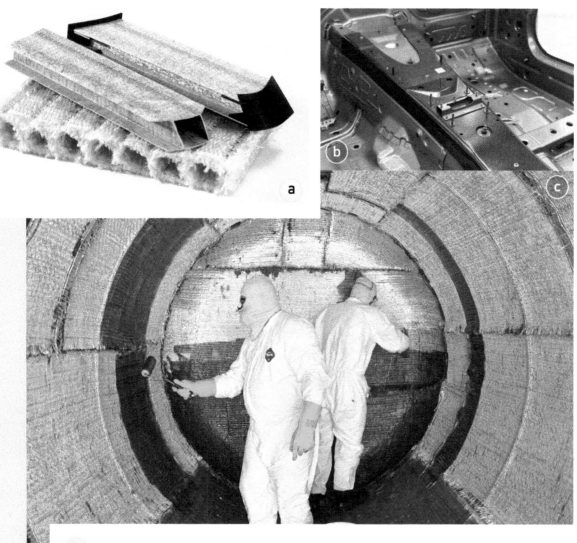

FIGURE 3-20. Examples of continuous fiber-reinforced composites.

[a] The 3D-LightTrans project used commingled glass/PET fiber in a 3D woven textile to thermoform low-cost, thermoplastic composite demonstrator automotive components. *(Photo courtesy of 3D-LightTrans Project)*

[b] BMW uses wet pressing to mold non-crimp fabrics into the CFRP tunnel which forms the backbone of the BMW 7 Series Carbon Core body-in-white (BIW). *(Courtesy of BMW Group)*

[c] Workers wet-out a 3D woven glass fabric with resin as part of the GENESIS retrofit system for underground storage tanks, which enables compliance with government regulations for double-walled containment of fuels and other chemicals. *(Courtesy of Poly Lining Systems, Inc.)*

) MODIFIED FORMS OF FIBER

Fibers can be modified to aid in processing as well as provide specific properties or property increases. These modifications can include applying different coatings to the outside of the fiber, mixing of materials that make up the fiber, and changing the shape of the fiber.

Sizing

Key to performance in all composites is the behavior at the resin-to-fiber interface. *Size* applied to reinforcing fiber affects a composite's mechanical properties (e.g., tensile strength, fatigue resistance, and impact resistance) and also its chemical and material properties such as resistance to water, corrosion and heat. Size formulations can consist of one or more films to apply, or come as a lubricant, coupling agent, or as a range of additives such as rheology modifiers, antifoams, adhesion promoters, plasticizers, anti-static agents and surfactants.

Back on page 68 of this chapter, the effect of size versus finish for glass fibers was compared, and carbon fiber sizing was discussed on page 75. Look back to page 86 in Table 3.15 to review several types of natural fibers shown to have improved properties with the application of silane coupling agents.

Specific size formulations have been developed for glass, carbon and other fibers to be used with specific thermoset resins such as epoxy and phenolic, and with specific thermoplastics including polyamide, polypropylene and polyolefins.

Commingled

Commingled tows or yarns are manufactured by intimately blending thermoplastic fibers with reinforcing fibers. This is done by commingling the filaments, where you end up with one single end tow or yarn of commingled fibers. Examples include reinforcing fibers such as carbon, glass or aramid combined with thermoplastic fibers made from PEEK, PPS, PE or PP. (Figure 3-21)

Commingled tow or yarn is often used for pultrusion and filament winding. It can also be woven into a fabric for use in layup and vacuum-bagged autoclave molding, as well as compression molding and other thermoforming processes.

More pliable than prepreg, commingled forms generally have a lower cost and are suited for high volume industrial processes. Unlike prepreg tapes, the fiber in this product form is not impregnated, however the proximity of thermoplastic fibers to reinforcing fibers allows for quick and easy wet out of the reinforcement fibers during thermal processing.

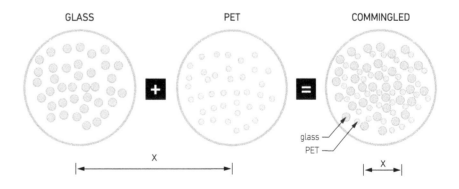

FIGURE 3-21. Commingled fiber.

Commingled fibers intimately blend thermoplastic fibers with reinforcing fibers, reducing the distance **(x)** between the two for quick, easy wet out and excellent fiber-to-matrix distribution.

TABLE 3.18 Commingled fiber products

Company	Polymers	Fibers
Coats/SHAPE Product:	PA6, PA66, PA12, PET, PEEK, PP, PPS, UHMWPE Synergex	Glass, carbon (3K-50K), aramid
Comfil	LPET, PA6, PBT, PC, PEEK, PEI, PET, PP, PPS	Glass, carbon, aramid, LCP, steel
Concordia	PA6, PA12, PEEK, PEI, PET, PP, PPS, PVDF	Carbon (3K-24K)

❱ SPREAD TOW

"Spread tow" refers to the practice of spreading a fiber into thinner, flatter reinforcements. For instance, a common method is 5-mm-wide 12K high-strength (HS) carbon fiber tow spread to a 25-mm width tape. (*See* Figure 3-2, view *[h]*, on page 64 of this chapter under "Reinforcement Terminology.") This unidirectional tape can then be used in automated tape laying (ATL) and automated fiber placement (AFP) processes. It can also be used to produce thin woven fabrics or multilayered noncrimp fabrics.

Though glass, aramid and polymer fibers may be spread, carbon fiber is the most common. Spread-tow reinforcements can weigh as little as 15 g/m^2 with a thickness of only 0.02 mm. Advantages of spread tow materials include:

- Same fiber volume in a thinner cross-section.
- More efficient load-carrying capability.
- Tailoring of *fiber areal weight* (the weight of fiber per unit area of fabric or tape).
- Straighter fibers and less crimp compared to conventional woven fabrics.
- Improved impact resistance versus noncrimp fabrics.
- Higher resistance to crack propagation via increased number of thinner layers.
- Increased fabric closure reduces resin pooling at fiber interstices.
- Improved surface finish.
- Unique aesthetics.

The fiber areal weight (FAW) of fabrics made with smaller diameter 1K and 3K conventional tow can be matched while using larger, less expensive tow sizes, as illustrated in Figure 3-22. Spread tow also produces fabrics with increased closure, meaning less gaps at fiber interstices for trapping resin. This, in turn, helps to improve the surface finish of molded parts.

Spread tow's ability to provide the same fiber volume in a thinner cross section is listed as a benefit above, but alternatively, it can be used to increase fiber volume in the same cross section. Testing has also confirmed that these spread tow, thin-ply laminates can withstand significantly greater stress compared to conventional thick-ply laminates before first-ply and last-ply failure, as well as provide improved damage tolerance after impact. (Figure 3-23)

Applications for spread tow include: canoes/kayaks, boat floors and bulkheads, boat hulls, sail masts and other tubes, golf club shafts and heads, hockey sticks, tennis rackets, bicycle frames and wheels, skis, snowboards, kiteboards, car parts, aircraft seats and structural components, helmets, aesthetic consumer goods and more. (Figure 3-24)

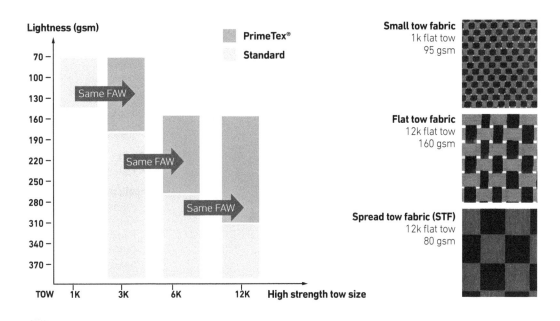

FIGURE 3-22. Spread tow effect on fabric areal weight and closure. *(Diagram source Hexcel PrimeText)*

FIGURE 3-23. Thin-ply laminates. *(Source: North Thin Ply Technology)*

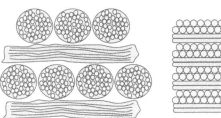

Thick ply laminate (300 g)

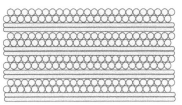

Thin ply laminate (300 g)

FIGURE 3-24. Spread tow applications. *(Photos courtesy of [a] TeXtreme, [b] Hexcel, and [c] North Thin Ply Technology)*

TABLE 3.19 Spread Tow Manufacturers, Materials, Forms and Benefits

COMPANY		
TeXtreme	**Chomarat**	**Sigmatex**
Products/Forms - Unidirectional (UD) tape - Spread tow fabrics	**Products/Forms** - C-Ply - C-Tape	**Products/Forms** - sigma*ST* - Unidirectional - Spread tow fabrics
Fibers - High Strength (HS) carbon fiber - Intermediate Modulus (IM) carbon fiber - High Modulus (HM) carbon fiber - Hybrids available	**Fibers** - HS carbon fiber - IM carbon fiber **Applications** - Sporting goods/leisure - Military - Transportation - Aerospace - Wind energy - Prepreg - Infusion - HLU - RTM	**Fibers** - HS-HM carbon fiber **Applications** - Aerospace - Automotive - Construction - Defense - Industrial - Marine - Space - Sporting goods
Applications - Sporting goods - Automotive - Transportation - Aerospace - Wind energy		
Benefits - Thin fiber spread tow layers enhance laminate mechanical properties - Lighter – weight savings - Increased stiffness per ply layer/ thickness - Increased strength per ply layer/ thickness	**Benefits** - Multiaxial fabrics down to 70 gsm per ply. - Gap-free technology, no resin-rich corners. - Noncrimp fabric, better mechanical properties. - Optimization of layer construction, cost savings.	**Benefits** - Lightweight materials - Very thin materials - Very low crimp materials - Additive-free - Volume production solution - Aerospace capability

❱ TEXTILE TECHNOLOGY

Continuous fibers used in composites are often processed into some type of textile. The principal methods of manipulating fibers or yarns into textile fabrics include weaving, braiding (or twisting) and knitting. Nonwovens may also be used.

Interweaving - two sets of straight threads intersecting at right angles *(weaving)*.

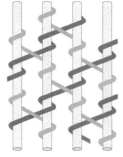

Intertwining - threads intertwined with each other at any set angle *(braiding and twisting)*.

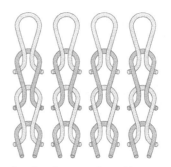

Interlooping - yarn formed into loops and loops intermeshed into a structure *(knitting)*.

FIGURE 3-25. Textile forms.

From *Knitting Technology* by David J. Spencer; 1983 Pergamon Press (available at www.elsevier.com/books)

TABLE 3.19 Spread Tow Manufacturers, Materials, Forms and Benefits *(continued)*

COMPANY			
North Thin Ply Technology	**Hexcel**	**Vectorply**	**Gernitex**

Products/Forms
- Thin ply prepreg
- Custom prepreg preforms
- Software
- Composite tubes
- Composite blocks
 	
Fibers
- HS carbon fiber
- IM carbon fiber
- HM carbon fiber
- Glass fiber
- Quartz fiber
- Aramid fiber
- PBO & other synthetics
 	
Applications
- Motor sports
- Sporting goods
- Luxury watches/jewelry
- High performance tubes
- Aerospace
- Space/satellites
- High-end marine
- Automated tape layup
 	
Benefits
Fabric weights reduced to:
- 30 gsm for PAN-based fibers
- 15 gsm for fibers such as MR70 12K
- 40-90 gsm with pitch fiber (YSH70 & K63712)
- 48 gsm with S-glass 9 microns
- 30 gsm with 14-micron quartz fiber
- 20 gsm with aramid
	Products/Forms
	- Unidirectional
	- Biaxial
	- Triaxial
	- Quadraxial
	- PrimeTex woven
	
	Fibers
	- HS-HM carbon fiber
	- Glass fiber
	- Quartz fiber
	- Aramid fiber
	
	Applications
	- Wind energy
	- Sporting goods
	- Automotive
	- Marine
	- Industrial
	- Infusion
	
	Benefits
	- No resin-rich areas facilitates a higher fiber volume.
	- Non-crimped fibers produce higher tensile and flexural properties in the finished laminate.
	- Reduced print-through, critical for boat hulls and automotive applications.
	- Spread tow plies are stitched together, thus easier to handle.
	- Heavier combinations are possible, meaning higher deposition rates.
		Products/Forms
		- Unidirectional
		- Noncrimp fabrics
		Fibers
		- C-BX 0300 carbon fabric
			Products/Forms
			- Unidirectional tape
			- Spread tow fabrics
			  • Bidirectional
			  • Triaxial
			
			Fibers
			- HS T-700 carbon fiber
			
			Applications
			- Aerospace
			- Automotive
			- Wind
			- Space
			- Sporting goods
			
			Benefits
			- Resistant to inplane stresses
			- Quasi Isotropic 60/0/-60
			- High fiber to matrix
			- Low crimp fabrics

❱ NONWOVEN MATERIALS

Nonwoven reinforcements used in composites include materials referred to as "tissue", "mat", or "veil." They may use long or short discontinuous fibers (e.g., chopped strand mat) or continuous fibers (e.g., continuous strand mat). They are typically used as surface ply materials in structural laminates, as reinforcements in semi-structural applications, or to promote permeation in liquid molding (infusion) processes. Nonwoven materials can use a variety of fibers including carbon, recycled carbon, glass and aramid fibers. They can also be attached to unidirectional, bidirectional, or multi-axial fabrics and tapes if desired.

FIGURE 3-26.
Nonwoven material.

Nonwoven veils made from nylon fiber are also used in composites, but not as reinforcements. These products (e.g., Spunfab) are used to aid resin flow during infusion of tightly packed carbon fibers and also provide a significant increase in laminate toughness.

Nonwoven roving is a reinforcement composed of continuous fiber strands loosely gathered together. This is basically another name for continuous strand mat.

❱ WOVEN FABRICS

Figure 3-27 is an illustrated glossary of common terms used for woven fabric materials.

[a] **Warp yarn**—Run the length of the roll in all conventional bidirectional woven fabrics. The warp yarn direction of the fabric is used for orientation purposes in composite panel manufacturing.

Fill or weft yarn—Run across the roll, perpendicular to the warp yarns. They are bound at the edges of the fabric with a selvage band or are woven in a loop-end fashion.

[b] **Selvage**—The selvage is the stitched or looped edge that runs along both sides and along the length of the roll. It prevents the fabric from raveling and is usually woven with a contrasting color fiber for best visibility. It is typically not used in the construction of a part. (May also be spelled "selvedge"; either spelling is correct.)

[c] **Warp face**—Fabric that has one surface or "face" in which the most-visible yarns are running in the warp direction. The opposite face has the most-visible yarns running in the fill direction. The side that is warp-dominant is called the "warp face" of the fabric.

　　The warp face of a harness satin fabric is always identified and is positioned either "warp up" or "warp down" in a layup, according to the laminate design. Warp face recognition is critical to proper **nesting** and **symmetry** in a layup using harness-satin weave fabrics, and to panel stiffness properties in the laminate design.

[d] **Tracers**—Contrasting-colored threads woven into fabric to identify the warp and fill direction of a fabric. They are also used to show the fiber-dominant surfaces of harness satin weave fabrics. On the warp surface they are typically spaced at 2-inch intervals and run in the warp direction. On the fill surface they are typically spaced at 6-inch intervals and run in the fill direction. Tracers are not standard in all weaves and usually require a special order from the fabric vendor. In some cases, tracers may be undesirable in a laminate as they are usually woven with a different type of yarn or thread.

Yarn count—The number of yarns per warp or fill direction, typically listed in yarns per inch.

Symmetric fabric—Has an equal number of identical yarns in both the warp and fill directions. It will be, for all practical purposes, equally strong in both directions.

Asymmetric fabric—Has a different number of yarns per inch in the warp and fill directions. Fabrics having a different yarn size or yarn type in each of the two directions or combinations of the above, may also be called an asymmetric-weave fabric, and will not necessarily have equal strength in both directions. Typically, asymmetric-weave fabrics have a greater number of warp yarns than fill yarns per inch and thus require specific warp fiber orientation when placing in a laminate.

fill face

warp face

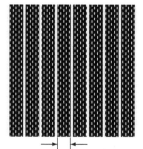

Tracers on 2 inch centers
in warp direction

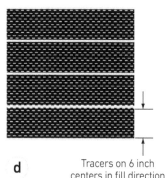

d

Tracers on 6 inch
centers in fill direction

FIGURE 3-27. Woven fabric terminology.

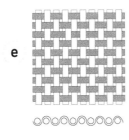

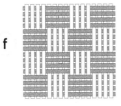

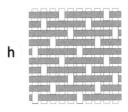

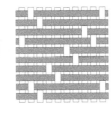

[e] Plain weave—Consists of yarns interlaced in an alternating fashion, one over and one under every other yarn. The plain weave fabric provides stability, but is generally the least pliable and least strong, due to a lower yarn count and a higher number of crimps than other weave styles. Plain weave fabrics tend to have significant open areas at the numerous yarn intersections. These intersections will be either resin rich pockets or voids in a laminate made with this fabric style. There is no warp or fill-dominant face on a plain weave fabric.

[f] Basket weave or woven roving—A heavy plain weave fabric that uses, instead of one warp yarn and one fill yarn, two or more warp yarns and two or more fill yarns (or roving) alternately interlaced over and under each other.

[g] 4-Harness satin (4HS) weave—One fill yarn floats over three warp yarns and under every 4th yarn in the fabric. 4-Harness satin is more pliable than plain weave and therefore is easier to form around compound curves. As with all harness-satin weaves, there is a dominant warp and fill face to consider in layup. (Also known as crowfoot weave.)

[h] 5-Harness satin (5HS) weave—One fill yarn floats over four warp yarns and under every 5th yarn in the fabric. The 5HS is slightly more pliable and will have a higher yarn count than the 4HS or plain weave fabrics.

[i] 8-Harness satin (8HS) weave—One fill yarn floats over seven warp yarns and under every 8th yarn in the fabric. This is a very pliable weave style with good drape, especially adaptable to compound curved surfaces. It is typically a more expensive weave, as a higher yarn count is attainable. (8HS carbon fabrics often have twice the yarn count of a carbon plain weave fabric of the same yarn size.)

[j] Twill weave—Characterized by a visual diagonal rib caused by one fill yarn floating over at least two warp yarns, then under two (e.g., 2 x 2 twill, left). Like a plain weave, there is no warp or fill-dominant face to be concerned with in layup. However, twills are more pliable than plain weave fabrics, allowing good conformance over compound curves. They have also demonstrated good impact resistance and damage tolerance.

[k] Hybrid fabrics—Those made using two or more different types of fibers. Glass/aramid, carbon/aramid and carbon/glass are all common examples. There are also hybrids with carbon and Dyneema® fibers and other more exotic combinations. Hybrid fabrics may contain the different fibers as discreet layers in a stitched multiaxial or may weave two or more fiber types together in a single woven fabric.

k

Woven carbon/aramid hybrid twill

FIGURE 3-27. Woven fabric terminology.

TABLE 3.20 Common Hybrid Fabrics

Manufacturer	Product	Warp Fiber	Fill Fiber	Weave
BGF Industries	94911	3K Carbon	ECG 75 1/0	Plain
	94990-94995	3K Carbon	1500 denier Kevlar (yellow, orange, red, blue or green)	2x2 Twill
Carr Reinforcements	Style 38172	3(Carbon-3K) 1(Kevlar 158)	3(Carbon-3K) 1(Kevlar 158)	Plain
	Style 38331	1(Carbon-3K) 1(Kevlar 158)	1(Carbon-3K) 1(Kevlar 158)	3x1 Twill
	Style 38175	1(Carbon-3K) 1(Kevlar 158)	1(Carbon-3K) 1(Kevlar 158)	2x2 Twill
	Style KG390T	1(Kevlar 240) 1(Glass 600)	1(Kevlar 240) 1(Glass 600)	3x1 Twill
	Style 38364	2(Carbon-3K) 2(Dyneema 177)	2(Carbon-3K) 2(Dyneema 177)	2x2 Twill
Fabric Development	1151	3K Carbon	1420 denier Kevlar	Plain
	4566	12K Carbon	330 yield Glass	4x4 Twill

Manufacturer	Product	Multiaxial Description		
Vectorply	KE-BX 1200	Kevlar/E-glass Double Bias		
Gurit	QEA 1201	Kevlar/E-glass Quadraxial		

Symmetric Fabric Considerations in Layups

Symmetric plain or twill weave fabrics can often be oriented at either 0° or 90° (or ±45°) and have the same properties in each direction. However, symmetric **harness-satin weave** fabrics inherently have a warp face and a fill face that are essentially warp fiber dominant on one side, and 90° opposite on the fill side of each ply. Although they can be oriented at either 0° or 90° in a layup (or at other angles) and each ply will have the same strength in each direction, the axial direction (orientation) of the fibers on the dominant face of the outer-most surfaces of the laminate will govern the flexural stiffness of the laminate.

Thus, unlike symmetric plain weave and twill fabrics, which have equal flexural stiffness in both directions, symmetric harness-satin weave fabrics have dominant fiber directions at each face and cannot be oriented at either 0° or 90° (or ±45°) with equal flexural stiffness properties in each direction.

Note: Symmetry in a fabric is different than symmetry of a laminate. For more information on laminate symmetry, *see* "Common Layup Terms and Conditions" ahead in Chapter 6 (page 160).

Non-Crimp Fabrics

Stitched multiaxial fabrics consist of one or more layers of unidirectional reinforcement, mat, or cloth, which are stitched together. They are sometimes referred to as stitch-bonded, stitched, knit, or non-crimp fabrics (NCF). Stitched fabrics are not woven and therefore the yarns are not kinked or crimped like traditional woven fabrics. This provides higher load-carrying properties and enables multiple fiber orientations to be achieved in a single ply. It also enables very heavy weight fabrics that are not possible with standard weaving processes. Fabric production is faster and laminate production time can also be reduced, as multiple steps to lay 0°, 90°, +45° and -45° plies can be replaced by a single placement step of a quadraxial (0°, 90°, ±45°), for example. Note, however, care must be taken in both design and fabrication to ensure laminate symmetry in the final product. (Figure 3-28)

Stitching

The stitching thread is usually polyester due to its combination of low cost and ability to bind the layers together. However, some resins do not bond to it readily, so good resin encapsulation can be an issue.

Courses is the term for the number of stitches of yarn per inch in a stitched fabric measured in the longitudinal direction. Varying from 4 to 30, it affects the fabric's drape and wettability. **Gauge** is the number of stitches of yarn per inch measured in the transverse direction and is usually a set value determined by the manufacturer.

Binding the layers is typically achieved using a tricot stitch, which is a special type of warp knitting that zigzags back and forth across the top of the fabric. It is often used for fabric using 0° reinforcements. A combination of stitches gives the best overall handling properties, with tricot on the top of the fabric and a chain stitch on the bottom of the fabric, running in a straight line down the warp direction, which enables drapeability.

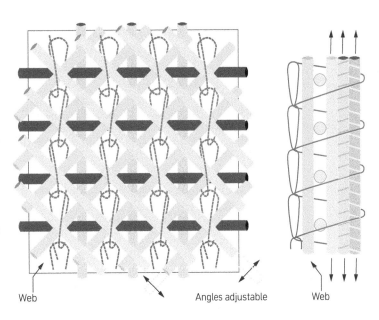

FIGURE 3-28. Stitched multiaxial fabric.

The oriented unidirectional plies of multiaxial fabrics are actually knitted together, hence their original name: knitted fabrics. *(Diagram source info: NPTEL nptel.ac.in)*

Web   Angles adjustable   Web

bottom of fabric   top of fabric

Multiaxial Fabric Types

Multiaxial fabrics were first produced using the common angles that comprise a quasi-isotropic stacking sequence, namely 0°, 90°, +45°, -45° or some combination of these. These are still the most common NCF configurations, but the angles possible have since changed. *See* "Thin-Ply Bi-Angle Fabrics" on page 106.

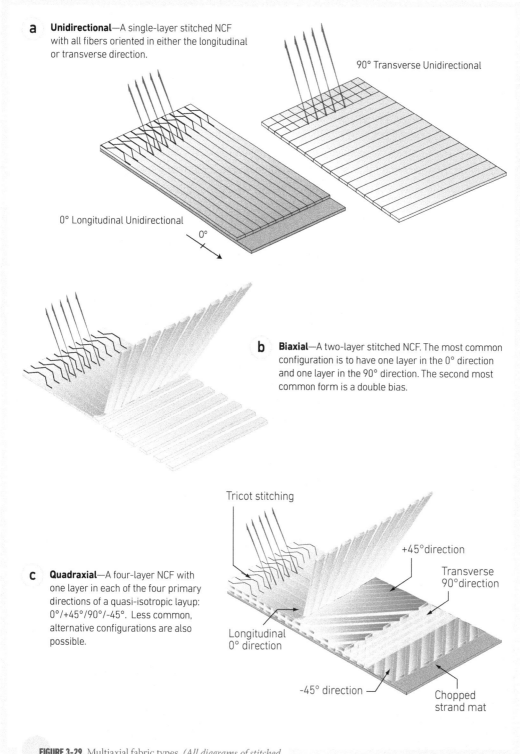

a **Unidirectional**—A single-layer stitched NCF with all fibers oriented in either the longitudinal or transverse direction.

90° Transverse Unidirectional

0° Longitudinal Unidirectional

0°

b **Biaxial**—A two-layer stitched NCF. The most common configuration is to have one layer in the 0° direction and one layer in the 90° direction. The second most common form is a double bias.

Tricot stitching

c **Quadraxial**—A four-layer NCF with one layer in each of the four primary directions of a quasi-isotropic layup: 0°/+45°/90°/-45°. Less common, alternative configurations are also possible.

+45° direction

Transverse 90° direction

Longitudinal 0° direction

-45° direction

Chopped strand mat

FIGURE 3-29. Multiaxial fabric types. *(All diagrams of stitched multiaxial fabrics are courtesy of Vectorply Corporation)*

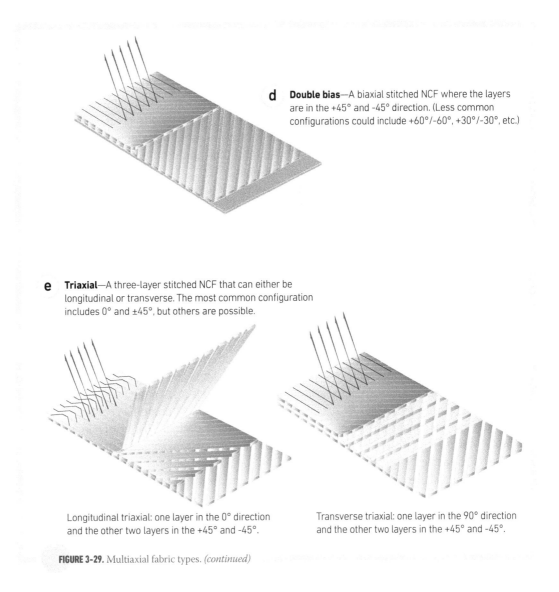

d **Double bias**—A biaxial stitched NCF where the layers are in the +45° and -45° direction. (Less common configurations could include +60°/-60°, +30°/-30°, etc.)

e **Triaxial**—A three-layer stitched NCF that can either be longitudinal or transverse. The most common configuration includes 0° and ±45°, but others are possible.

Longitudinal triaxial: one layer in the 0° direction and the other two layers in the +45° and -45°.

Transverse triaxial: one layer in the 90° direction and the other two layers in the +45° and -45°.

FIGURE 3-29. Multiaxial fabric types. *(continued)*

Numbering System

The numbering system commonly used with these fabrics was developed by Knytex when it introduced what it called "knit multiaxials" in 1975. Even if a fabric is named with an unrelated product number by the manufacturer, end-users may commonly still refer to it as an 1808 or 2408, for example.

First 2 Numbers—the weight of the fabric in oz/yd^2.
Second 2 Numbers—the weight of any chopped strand mat (CSM) in oz/ft^2 x 10.

TABLE 3.21 Common Stitched Multiaxial Glass Fabrics

Product	1808	2400	2408	2415	3208	3610	5600
Fabric Weight (oz/yd^2)	18	24	24	24	32	36	56
Mat Weight (oz/ft^2)	.8	-	.8	1.5	.8	1.0	-

Note: Vectorply has a 10800 product with a fabric weighing 108 oz/yd^2 and no mat.

Thin Ply Bi-Angle Fabrics

As stated above, the fiber angles possible in NCF changed with the introduction of C-PLY, a thin-ply (spread tow), low-angle (e.g., 0°/20° or 0°/30°) asymmetric NCF developed by Dr. Stephen Tsai and reinforcements supplier Chomarat. Launched in 2010, Dr. Tsai's goal was to enable the design of laminates that withstand significantly greater stress before first-ply failure and last-ply failure.[7] Asymmetric, thin-ply NCF provide a great deal of flexibility in laminate design for optimizing a structure's stiffness, strength, and damage tolerance. However, it is important to note that these forms can also contribute to asymmetry in a laminate that can cause twisting in the final structure.

The development of C-PLY has since led to weavers offering a wide range of angles, so that NCF are no longer limited to 0°, 90°, +45° and -45°, but may use 10°, 35°, 60°, 75° or other angles, depending upon each manufacturer's weaving equipment and capabilities.

Braided Fabrics

Braiding intertwines three or more yarns so that no two yarns are twisted around one another. In practical terms, braid refers to fabrics continuously woven on the bias. The braiding process incorporates axial yarns between woven bias yarns, but without crimping the axial yarns by the weaving process. Thus, braided materials combine the properties of both filament winding and weaving, and are used in a variety of applications, especially in tube construction.

Tubular braided structures feature seamless fibers from end to end and tend to perform well in torsion, which makes them well-suited for drive shafts and similar applications. In addition, braided fibers are mechanically interwoven with one another and provide exceptional impact properties as a result. Braids also provide good "channel" properties between the fiber bundles, which enhances resin distribution throughout a preform used in a vacuum infusion or resin transfer molding process. They are available in a wide variety of fibers, sizes and designated fiber angles.

There are also different types of braid architecture. Biaxial is the most common, with two yarns crossing under and over each other typically in the ±45° orientation, but also in lower angles. Triaxial braids add a third set of yarns in the axial direction, offering unidirectional and off-axis reinforcement in a single layer. (Figure 3-30)

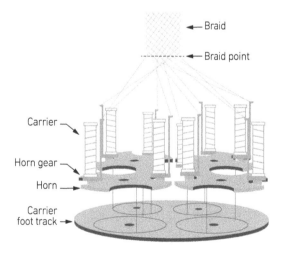
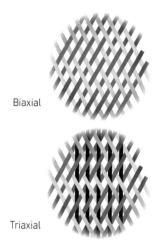

FIGURE 3-30. Biaxial vs. triaxial braids. *(Source: A&P Technology)*

---

7   Source: Tsai-Wu failure criterion, published in "A General Theory of Strength for Anisotropic Materials," *Journal of Composite Materials* 1971, Vol. 5, pp. 58-80.

There are several main types of braided forms:

- Sleevings
- Flat braids
- Wide braided fabrics
- Overbraiding
- Radial braiding
- Braided fillets or noodles

Sleevings

These braid forms feature two sets of continuous yarns, one clockwise and the other counterclockwise, where each fiber set Is braided in a continuous spiral pattern. A&P Technology's Sharx™ braided sleevings easily and repeatedly conform to the shape of products with changing geometries like prosthetics and hockey sticks, improving overall performance, minimizing weight, and maximizing strength.

A sleeving takes on the exact shape and dimensions of the part over which it is shaped. When the sleeving is pulled over a mandrel or part with changing cross-sections the fiber orientation, the thickness, and the yield of the braid vary at each point along the length of the part. These variations are predictable and repeatable and lend themselves to easy and precise manufacture of composite parts.

Flat Braids

Flat braids use only one set of yarns, and each yarn in the set is interwoven with every other yarn in the set in a zig-zag pattern from edge to edge.

Wide Braided Fabrics

Also called braided broadgoods, A&P Technology produces sleevings in excess of six inches that can be slit open to create a line of wide braided fabrics. (Figure 3-31) These fabrics can be produced as double bias biaxial fabrics, triaxial fabrics such as 0°, +45°, -45° or 0°, +60°, -60° fiber axes within one layer. The QISO fabrics noted below fall into this category (*see* next page).

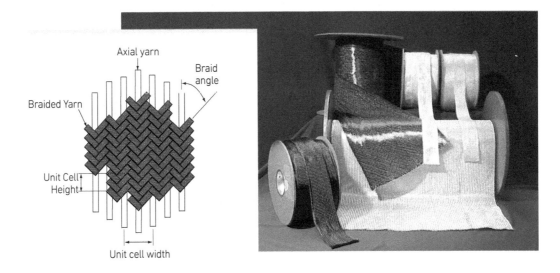

FIGURE 3-31. Braided fabrics.

In addition to sleevings, braided fabrics are also available as flat braids and wide fabrics/braided broadgoods. *(Photos courtesy of A&P Technology)*

QISO is a trademark brand of **quasi-isotropic (QI)** flat braid fabric from A&P Technology, Inc. It has three axes of orientation; 0°, 60°, -60° in a uniformly symmetric braided weave that can produce a quasi-isotropic, balanced, and symmetric laminate if the plies are all cut from the roll in the same direction. This form eliminates the need for cutting off-angle plies from the roll-stock, potentially saving a great deal of scrap material and labor in manufacturing. (Figure 3-32)

Overbraiding and Radial Braiding

Overbraiding is a technique where fibers are braided directly onto cores or mandrels that will be placed within a molding process. It is often used when a **preform** with very high bias angles or a contoured triaxial design is required, or when it is desirable to include circumferential windings in a preform's architecture.

Radial braiding is a machine configuration used in overbraiding, where the carrier and fiber spools are oriented at a 90° angle to the produced braid (*see* Figure 3-33) as opposed to being aligned in parallel for most braiding machines. This radial configuration reduces fiber damage and provides more space for overbraiding curved mandrels.

2D, 2½D and 3D Braids

Braiding machine supplier Herzog has established the following terminology to better distinguish between different types of braided preforms (*see* Figure 3-34).

Applications

Braided reinforcements are used in a variety of applications, including:

Aircraft—Helicopter driveshafts, propellers, jet engine fan cases/containment cases, stator vanes and engine dressing (the myriad pipes and tubing on the outside), fuselage frames, wing flaps.

Sporting goods—Bicycle frames/handlebars, hockey and lacrosse sticks, prosthetic devices, canoe hulls and gunwales.

Tanks and vessels—Pressurized compressed natural gas (CNG) and hydrogen tanks and composite overwrapped pressure vessels (COPVs).

Construction—Arch support structures for composite shelters and bridges.

Automotive—Driveshafts, wheels, roof bows, roof rails, bumper beams, crash cans/pillars, suspension components, side impact structures/side rails/side sills.

FIGURE 3-32. QISO wide braided fabric. *(Photo courtesy of A&P Technology)*

FIGURE 3-33. Large scale overbraiding with circumferential windings to aid in positioning and debulking. *(Photo courtesy of A&P Technology)*

Another type of application is the use of braided fillets or "noodles" at the intersection between the flange and web of structural stiffeners such as L- or T-section stringers and I-beams. The "noodle" provides fiber reinforcement for load transfer, prevents voids and achieves an accurate radius for the desired application. (Figure 3-36)

3D BRAIDING

- Real 3D shaping
- Interlocking of layers
- Specialized braiding techniques

2.5D BRAIDING

- Direct overbraiding
- Complex forms possible
- Braid thickness can be built by overbraiding with several layers

2D BRAIDING

- Rope or sleeve
- Round and flat braid possible
- Horizontal or vertical machine configuration

FIGURE 3-34. Braided preform types. *(Photos courtesy of Herzog)*

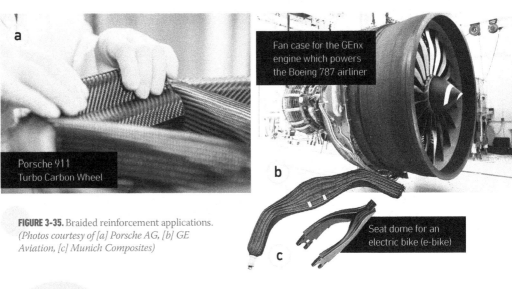

a

Porsche 911
Turbo Carbon Wheel

Fan case for the GEnx engine which powers the Boeing 787 airliner

b

Seat dome for an electric bike (e-bike)

c

FIGURE 3-35. Braided reinforcement applications. *(Photos courtesy of [a] Porsche AG, [b] GE Aviation, [c] Munich Composites)*

FIGURE 3-36. Braided fillet. *(Photo courtesy of A&P Technology)*

❱ 3D FABRICS[8]

3D woven fabrics have yarns interwoven in x, y and z directions. The major differentiator from 2D fabrics is yarn woven in the z direction, which ties the fabric plies together and improves resistance to delamination and damage tolerance.

These products have changed significantly over the decades. In the 1980s and 90s, 3D woven fabrics were sometimes referred to as technical textile preforms, 3D textile preforms or complex 3D fabrics. They were typically designed for a specific application that required pre-shaping to help meet demands in the manufacturing process and/or advanced structural application. These pre-shaped textiles were called preforms (e.g., shaped for a rocket nozzle, impeller, egg crate structure, etc.), and were typically expensive to manufacture. Preforms today are made using highly automated processes and a wide range of reinforcements (e.g., nonwoven mat, unidirectional prepreg tape, thermoplastic organosheet, 3D woven fabric). For more details, *see* the "Preforms" section in Chapter 8 (page 224).

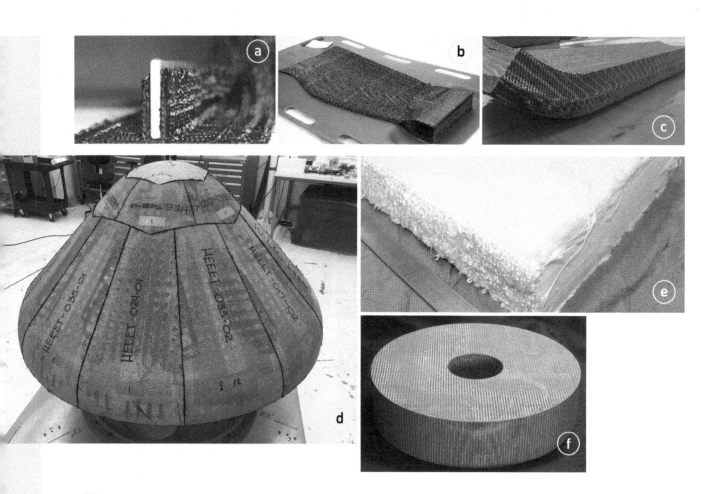

FIGURE 3-37. Examples of 3D fabrics.

[a] 3D woven Pi preforms provide an effective means for reinforcing joints in composite structures. [b] The LEAP aircraft engine features CFRP fan blades made with a 3D woven preform for extreme impact resistance at light weight. *(Photos courtesy of Albany Engineered Composites, Inc.)* ▪ [c, d] Dual-layer carbon/phenolic fiber orthogonal 3D woven fabric enables NASA's HEEET family of spacecraft heatshields. *(Photo courtesy of NASA Ames)* ▪ [e, f] 3D woven quartz fiber fabric is used for the compression pads in the Orion spacecraft. *(Photo courtesy of Bally Ribbon Mills and NASA Ames)*

8   This section draws information from "3D Woven Fabrics" by Pelin Gurkan Unal, Associate Professor at Namik Kemal University Department of Textile Engineering, Tekirdağ, Turkey. Published by *InTechOpen*.

3D woven fabrics are now much more automated, using digital looms and a variety of weaving techniques and thus are much cheaper than decades past, though still often more expensive to produce than 2D fabrics. However, these fabrics often eliminate multiple cutting and kitting steps, so some of this cost is offset during part manufacture. The 3D woven carbon fiber preform used for the LEAP engine fan blades in Figure 3-37[b] enables a very thick root contiguous with the shaped blade, as well as build-up of fibers in areas requiring higher strength and stiffness.

The three most common types of 3D woven fabrics are described in Table 3.22 below.

TABLE 3.22 Three Main Types of Woven 3D Fabrics

Multilayer	Multiple layers, each with its own warp and weft yarns, connected by existing or external yarns	
Weaving equipment:	Can be produced with conventional 2D weaving looms	
Angle/Warp Interlock	At least warp and weft yarns and may also include stuffer yarns. Two main types: • Through-thickness—warp yarns travel from top to bottom surface holding all fabric layers together	
	• Layer-to-layer—warp yarns travel from one layer to adjacent layer and back	
Weaving equipment:	Can be produced with conventional 2D weaving looms	
Orthogonal	Consists of 3 sets of yarns (X, Y and Z) which are perpendicular to each other and Z yarns interconnect all warp and fill yarns to solidify fabric	
Weaving equipment:	Requires specially designed 3D weaving loom	

Within the three main 3D woven fabric types, different weave patterns are possible, as is the use of different fibers. For example, in the orthogonal dual layer 3D woven fabric for the HEEET thermal protection system developed by NASA and weaver Bally Ribbon Mills, the outermost layer is 100% carbon fiber and dense while the inner layer is a hybrid of carbon and phenolic fiber and less dense. These fabrics may also feature tailored drape, impact resistance, structural properties, and permeability to improve resin flow during infusion and RTM processes. Figure 3-38 shows a select few of the nearly limitless variations possible with 3D woven fabrics.

Applications for modern 3D woven textiles can be seen above in Figure 3-37 and include:

- Fan blades and containment cases for turbofan jet aircraft engines
- Thermal protection system (TPS) panels for spacecraft heatshields
- T-stringer, I-beam, X-beam and other shaped stiffeners
- Wind blade spars
- Ballistic protection
- Rocket engine exit and tail cones, nozzles
- Hydraulic fracturing plugs
- Fins, impellers, rotors.

Knitted Fabrics

In contrast to woven fabrics, where warp and weft threads are interlaced at a 90° angle, knitted fabrics comprise consecutive rows of interlocking loops (Figure 3-39, view [d]). All types of fibers can be knit, including glass, carbon and Kevlar® fibers and metal wire. Knits are highly elastic especially along the vertical axis, though this can be constrained if necessary. They can also be bent or curved around a surface without being distorted, which makes them amenable to 3D preforms. However, because the fiber is looped, its load-carrying capacity is much lower than a straight, unidirectional fiber. Some knitting machines are able to insert unidirectional fibers to increase structural properties.

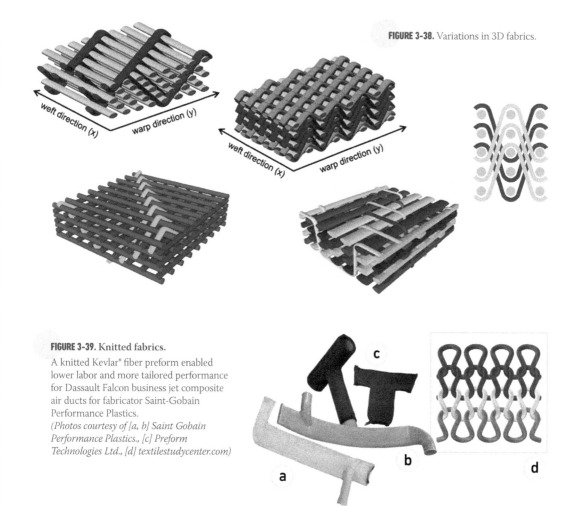

FIGURE 3-38. Variations in 3D fabrics.

FIGURE 3-39. Knitted fabrics.
A knitted Kevlar® fiber preform enabled lower labor and more tailored performance for Dassault Falcon business jet composite air ducts for fabricator Saint-Gobain Performance Plastics.
(Photos courtesy of [a, b] Saint Gobain Performance Plastics., [c] Preform Technologies Ltd., [d] textilestudycenter.com)

Knitting is faster than braiding but slower than weaving or twisting. However, it does not require respooling fiber onto special packages as do weaving, braiding and twisting, which reduces total production time. There are two kinds of knitting: warp knitting and weft (flat) knitting.

Recycled Fiber

❭ CARBON FIBER

Recycled carbon fiber (RCF or rCF) has become a commercial product to reduce production waste and to close the life-cycle loop for end of life (EOL) composite products. It typically offers 30–50% lower cost and very similar properties compared to virgin carbon, but with fiber length shortened depending upon the recycling method used. RCF products may be made from waste during production of carbon fiber, textiles and parts (e.g., scrap PAN precursor, fabric selvage, preform trimmings, and prepreg skeletons from automated cutter operations) or from cured parts such as bicycle frames.

The most common methods for recycling carbon fiber include **pyrolysis**—decomposition via high temperature—and **solvolysis**, also called chemical recycling, which is achieved by immersion in a solvent. (Figure 3-40)

Pyrolysis

In pyrolysis, dry fiber, prepreg scrap and cured parts are processed in an oven at 350°C–500°C or higher. The high temperature breaks down the resin molecules (as well as sizings and binders on the fibers) and draws them off via ventilation. Fibers emerge clean and undamaged, though many recycling operations shred incoming waste to maximize bulk for the pyrolysis furnace. Thus, processed fibers will be shorter in length.

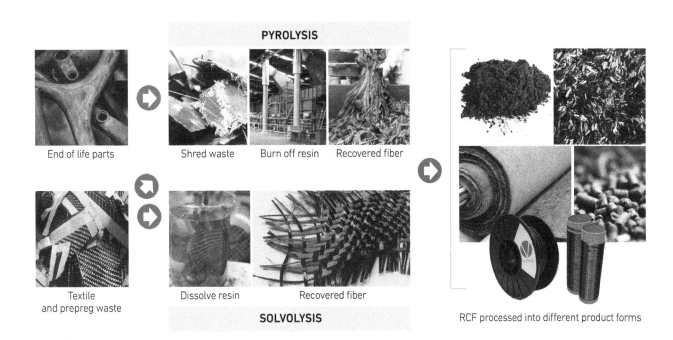

PYROLYSIS

End of life parts   Shred waste   Burn off resin   Recovered fiber

Textile and prepreg waste   Dissolve resin   Recovered fiber

SOLVOLYSIS

RCF processed into different product forms

FIGURE 3-40. Carbon fiber recycling process chain.

(Photos courtesy of Carbon Conversions, Connora Technologies, ELG Carbon Fibre, and Vartega.)

After pyrolysis, fibers may receive custom conditioning such as fiber sizing tailored for certain polymers and/or binders applied for a specific reuse. They may be further processed into milled or chopped carbon fiber and also compounded into reinforced thermoplastic pellets for injection molding or 3D printing filament. Nonwoven mats are also a standard rCF product, including variants hybridized with thermoplastic fibers for compression molding. (Figure 3-41)

Solvolysis

Solvolysis is typically a lower-temperature process than pyrolysis (*see* Table 3.23), and uses solvents like acetone, alcohol and acetic acid (vinegar) to remove resin from prepreg and cured composites. It can produce clean, undamaged fibers with properties suitable for reuse, but offers the possibility of maintaining the fiber's original length and textile architecture.

TABLE 3.23 Comparison of Recycling Methods for CF/Epoxy Composites

	Pyrolysis	**Hitachi**	**CAS***	**Vartega**	**Connora Technologies**
Method	Thermal	K_3PO_4, BZA	H_2O_2, Acetone	Proprietary	Acetic Acid, Ethanol (alcohol), H_2O
Temperature	500-700°C	200°C	120°C	20°C	70°C
Pretreatment	Crushing/ Shredding	-	Acetic Acid	-	-
Status	Commercial	Pilot	Lab	Commercial	Commercial
Application	Conventional CFRP	Conventional CFRP	Conventional CFRP	Uncured prepreg	CFRP using Recyclamine® hardener
Recycled Products	Chopped Carbon Fiber	Long Carbon Fiber	Long Carbon Fiber	Chopped and Long Fiber, Milled, Fabric	Carbon Fiber Fabric TP Epoxy Resin
	1	2	3	4	

*Chinese Academy of Sciences (Beijing)
SOURCE: Connora Technologies report: "Third Party Evaluation of Recycling CF/Epoxy Composites" with Vartega information added from Vartega.

Milled fiber (left), chopped fiber (center) and fiber compounded with thermoplastic polymer into pellets for injection molding (right). *(Photos courtesy of ELG Carbon Fibre)*

Nonwoven mats and veils made with fiber alone or combined with thermoplastic fibers. *(Photos courtesy of (left) CFK Valley Recycling, (center) Technical Fibre Products Optiveil®, (right) Carbon Conversions)*

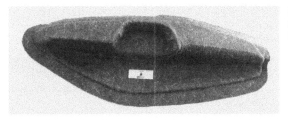

Preform: preshaped nonwoven mat for molding the door bolster support for the Ford Escape. *(Photo courtesy of Carbon Conversions)*

FIGURE 3-41. Recycled carbon fiber products.

TABLE 3.24 Suppliers of RCF Products *(not exhaustive)*

Process	Capacity (metric tonnes/yr)	Products
COMMERCIAL rCF		
Carbon Conversions (Lake City, SC)		
Pyrolysis + 3DEP preforming	1,400-2,300 mt/yr 3-5 million lb/yr	3D preforms, mat
Carbon Fiber Remanufacturing (Whitewater, KS)		
Process reclaimed virgin CF into nonwoven materials	1,800 mt/yr 4 million lb/yr	3D preforms, precision cut & chopped fiber
CFK Valley Stade Recycling (Wischhafen, Germany)		*Tradenames: CarboNXT*
Pyrolysis + oxygen	1,000 mt/yr rCF 2.2 million lb/yr	Chopped fiber, milled fiber, mat

(continued)

TABLE 3.24 Suppliers of RCF Products *(not exhaustive) (continued)*

Process	Capacity (metric tonnes/yr)	Products
COMMERCIAL rCF		
ELG Carbon Fibre Ltd. (Coseley, UK)		*Tradenames: Carbiso*
Pyrolysis	2,000 mt/yr scrap 1,000 mt/yr fiber	Chopped, DLFI pellets, mat, milled fiber
Hadeg Recycling (Stade, Germany)		*Tradenames: Recyclat A*
Pyrolysis	36–48 mt/yr rCFF 79,000–106,000 lb/yr	Milled fiber
Karborek (Martignano, Italy)		
Pyrolysis + oxygen	1,500 mt/yr scrap 3.3 million lb/yr	Chopped fiber, felt, milled fiber
Procotex (Dottignies, Belgium)		
Pyrolysis + oxygen	1,000 mt/yr rCF 2.2 million lb/yr	Chopped fiber, granulate, milled fiber
R&M International		
Solvolysis	Scaling to 454 mt/yr rCFF 1 million lb/yr	Bulk molding compound, mat/ nonwovens, pellets
Toray/Toyota Chemical Engineering Co., Ltd. (Handa, Japan)		
Pyrolysis	1,000 mt/yr rCF 2.2 million lb/yr	Chopped fiber, mat
PRODUCTS REUSING RCF and IN-HOUSE PRODUCTION WASTE		
BMW (Munich, Germany) **SGL Automotive Carbon Fibers** (Wackersdorf, Germany)		*REUSE IN-HOUSE PRODUCTION WASTE*
Reprocess textile mfg. and kitting scrap into stitched NCF	10% of CFRP in *i3* and *i8* models uses rCF	CF/epoxy roof CF/PUR seat
Composite Recycling Technology Center (Port Angeles, WA)		
Developing methods and products for uncured waste prepreg		Pickleball paddles Park benches
Polynt (Lincolnshire, UK)		*Tradenames: RECarbon®*
Purchased RCF veils, mats into SMC		Epoxy and vinylester molding compounds
Sigmatex (Cheshire, UK)		*REUSE IN-HOUSE PRODUCTION WASTE*
Reprocess textile mfg. and kitting scrap into commingled PET/CF		Tape, Noncrimp fabric
Technical Fibre Products (Kendal, UK)		
		Veil products
Vartega (Golden, CO)		
RCF from solvolysis		3D printing filament

4

Nanocomposites

CONTENTS

Nanomaterials Overview

Nanocomposites are a mixture of two or more materials that retain their distinct characteristics when combined at the nanometer (nm, 10^{-9} or 1 billionth of a meter) scale. Nanoparticles can be organic (carbon-based) or inorganic (e.g., minerals), naturally occurring or man-made. It is possible to synthesize nanocomposites using polymer, ceramic or metal **matrix materials.**

Because nanomaterials are sized between 1–100 nm, they have very large surface areas compared to their volume and a very high aspect ratio as a matrix reinforcement. The interface area between the matrix and nanoparticles/nanofibers is an order of magnitude (i.e., 10X) greater than for conventional composites. Thus, a relatively small amount (e.g., less than 5% by weight) of nanomaterial can have a significant effect on macroscale composite properties—increasing them as much as 20% to 50%. (Figures 4-1, 4-2)

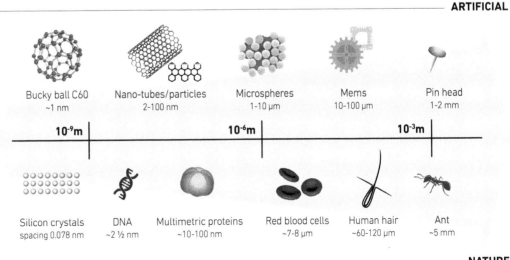

ARTIFICIAL

Bucky ball C60
~1 nm

Nano-tubes/particles
2-100 nm

Microspheres
1-10 μm

Mems
10-100 μm

Pin head
1-2 mm

10^{-9}m 10^{-6}m 10^{-3}m

Silicon crystals
spacing 0.078 nm

DNA
~2 ½ nm

Multimetric proteins
~10-100 nm

Red blood cells
~7-8 μm

Human hair
~60-120 μm

Ant
~5 mm

NATURE

FIGURE 4-1. The scale of things—nanometers and more.

Small cells are more efficient
Surface area to volume ratio must remain high

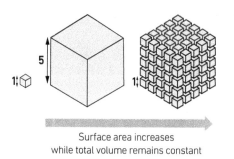

Total surface area (height x width x number of sides x number of boxes)	6	150	750
Total volume (height x width x length x number of boxes)	1	125	125
Surface-to-volume ratio (area ÷ volume)	6	1.2	6

Surface area increases
while total volume remains constant

FIGURE 4-2. Surface-to-volume.

Mechanical properties such as strength, stiffness, and toughness can also be enhanced, as can barrier properties (i.e., resistance to water, chemicals, UV radiation, etc.), magnetic properties, thermal stability, flame resistance, electrical and thermal conductivity, optical clarity, light reflection/absorption and surface appearance.

Popular and promising materials being used in nanocomposites include carbon nanotubes (CNTs), graphene, nanoclays, nanosilica, nanocalcite, metal oxides and various types of nanofibers. Nanocomposite applications currently being developed include:

- Solar cells
- Fuel cells and energy storage devices
- Higher strength-to-weight structures
- More damage-tolerant structures
- Reinforced plastics that are easier to process and paint
- Fire and heat-resistant components
- Multifunctional structures (e.g., morphing, sensing, energy storage) and lightweight sensors
- Spacecraft structures that can fold, unfold and self-assemble
- Bioimplants and faster healing
- Improved filtration
- Food packaging
- Armor and protective clothing
- Electrostatic charge dissipation
- Electromagnetic radiation and absorption
- Self-healing composites
- Ever-smaller information storage and electronics devices

Nanocarbon

CNTs, graphene and fullerenes are nanomaterials that are **allotropes** of carbon; they are the same chemical element, but with a different arrangement of atoms. These different atomic arrangements result in different physical forms and properties. For example, diamond is a carbon allotrope that does not conduct electricity, while graphene and carbon nanotubes typically feature high electrical and thermal conductivity. Table 4.1 shows various carbon allotropes.

TABLE 4.1 Allotropes of Carbon

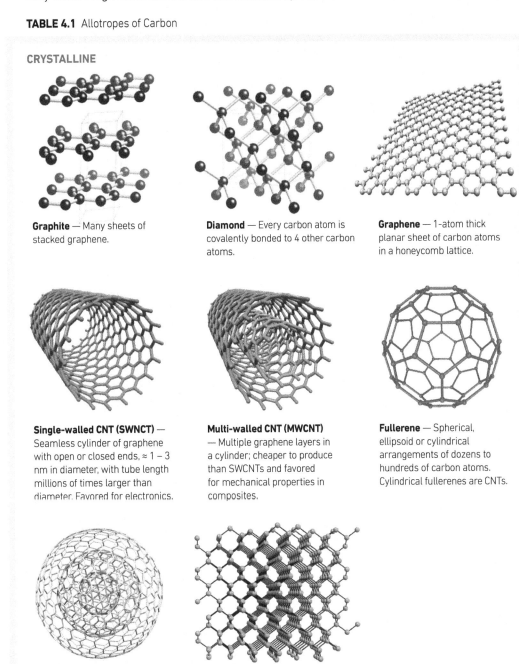

CRYSTALLINE

Graphite — Many sheets of stacked graphene.

Diamond — Every carbon atom is covalently bonded to 4 other carbon atoms.

Graphene — 1-atom thick planar sheet of carbon atoms in a honeycomb lattice.

Single-walled CNT (SWNCT) — Seamless cylinder of graphene with open or closed ends, ≈ 1 – 3 nm in diameter, with tube length millions of times larger than diameter. Favored for electronics.

Multi-walled CNT (MWCNT) — Multiple graphene layers in a cylinder; cheaper to produce than SWCNTs and favored for mechanical properties in composites.

Fullerene — Spherical, ellipsoid or cylindrical arrangements of dozens to hundreds of carbon atoms. Cylindrical fullerenes are CNTs.

Nano-onion — Concentric spheres of graphene or carbon atoms with potential for energy storage devices.

Nanodiamond — Detonation or ultrasound is used to convert mm-size graphite flakes into diamond lattices ≈ 5 nm in diameter.

(continued)

TABLE 4.1 Allotropes of Carbon *(continued)*

AMORPHOUS

Carbon black — Produced by the incomplete combustion of heavy petroleum products.

Charcoal — The hydrocarbon residue produced by heating wood or other substances in the absence of oxygen, known as pyrolysis.

❱ GRAPHENE

Graphene is a single layer of carbon atoms arranged in a hexagonal lattice with extraordinary properties:

- Lightest material known at 0.77 mg/m^2
- Strength 100–300 times that of steel at 1,100 GPa
- High stiffness with Young's modulus of 0.5 TPa
- Best known conductor of heat at room temperature with thermal conductivity of 5,000 W/mK
- Best known conductor of electricity (lowest resistivity)
- 97.3% transparent

Graphene is available as materials that are up to ten layers thick, with properties much different to those of a pure single layer of graphene. A sheet of graphene can vary in size from nm to cm. As discussed below with CNTs, graphene is often **functionalized**, which enables it to better disperse within a material, improving the composite's properties.

Graphene may be used as a powder additive, in solution as a coating and as a paper or film. It may also be grown using **chemical vapor deposition (CVD)** onto a substrate such as metal sheets or foam.

Graphene Oxide

Graphene oxide is not pure graphene, but it can be used to make graphene. Graphene oxide is an oxidized form of graphene laced with oxygen-containing groups. It was developed as a graphene derivative that is easier and lower-cost to make and also easier to process into other materials and parts because it is dispersible in water and other solvents. It is not a good conductor but there are processes to augment its properties.

Graphene oxide films can be made conductive and are being explored for use in flexible electronics, solar cells, chemical sensors, batteries and touch screens. Graphene oxide powder is readily mixed with polymers, ceramics and metals, improving the tensile strength, elasticity, conductivity, heat resistance and other properties of that composite versus the unreinforced materials.

Graphene Nanoplatelets

Graphene nanoplatelets are nanoparticles consisting of short stacks of graphene sheets having a platelet shape. They range in length from 1–20 micrometers (1,000–20,000 nm) and have a thickness of 0.34 nm, that of a single graphene layer, up to 100 nm. Most commonly, they are 1–15 nm thick. Graphene nanoplatelets may be produced by intercalation (the insertion of a molecule or ion into compounds with layered structures) and exfoliation (separation of a material's layers), typically from graphite. Oxygen is often introduced during this process. (Figure 4-3)

Graphene in Composites

Because of this, graphene is being used to develop composites for wide-ranging applications:

- Automotive body components
- Aircraft structures
- Multifunctional composites (conducting, sensing, electrical shielding, lightning strike protection, deicing, energy storage, etc.)
- Fuel cells and batteries
- Explosion-resistant tanks and pipes, fuel tanks

- Impact-resistant helmets
- Ballistic armor
- Antennas
- Biomedical applications
- Sporting goods (e.g., fishing rods, tennis rackets, etc.)
- 3D printing filaments and printed parts ...and much more.

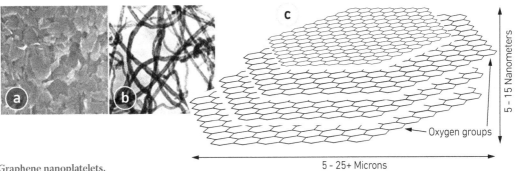

FIGURE 4-3. Graphene nanoplatelets.
Micrographs of [a] graphene nanoplatelets and [b] carbon nanotubes (CNTs) show their different structures. [c] Being a form of graphene, nanoplatelets—also referred to as nano graphene platelets (NGPs)—offer higher performance than CNTs and other nanomaterials. *(Source: "Nano Graphene Platelets (NGPs)" article at azonano.com)*

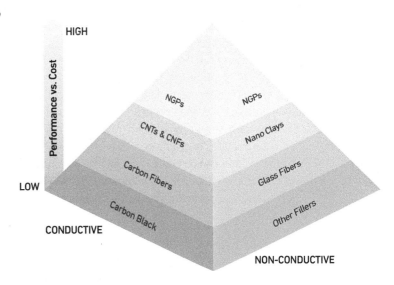

Examples of graphene development in composites include Haydale's partnerships with Huntsman to develop graphene-enhanced Araldite® epoxy resins, and with Flowtite Technology AS to use graphene in glass-fiber reinforced plastic (GRP) pipes. Applied Graphene Materials is working with multiple companies as well:

- SHD Composites Ltd., to develop graphene-enhanced epoxy prepregs;
- Airbus Defense and Space, to develop a graphene composite application for satellites;
- Spirit AeroSystems, to explore the benefits of graphene nanoplatelets in resin-infused aerospace structures;
- Century Fishing Rods, to develop new products using graphene nanoplatelets.

TABLE 4.2 Selected Graphene Suppliers/Producers Active in Composites

Company Location	Products/Technology Development
Angstron Materials Dayton, OH	Graphene and graphene oxide materials, nanocomposites
Applied Graphene Materials Cleveland, UK	Graphene nanoplatelets and dispersions
Graphene Technologies Novato, CA	Patented method for synthesizing graphene CO_2. Developing graphene foam-based composites
Haydale Ammanford, UK	HDPLAS® functionalization process for graphene and other nanomaterials
Imagine Intelligent Materials Pty Ltd North Geelong, Australia	Imgne® graphene materials, multifunctional composites
Imagine Intelligent Materials Oy Espoo, Finland	
Vorbeck Jessup, MD	Vor-x® graphene Vor-flex™ thermoplastic elastomers Vor-x® reinforced thermoplastics

❯ CARBON NANOTUBES

CNTs are basically graphene rolled into tubes and they are another carbon allotrope that, per unit weight, are 10 to 100 times stronger than steel. They derive their strength from the nature of the **covalent bonding** of the carbon atoms, which is much stronger than the covalent bonding in graphite. The modulus of multi-walled carbon nanotubes (MWCNTs) is typically about 1.0 TPa, with tensile strength of 10–60 GPa (compared to stainless steel at 3 GPa). This structure also makes CNTs highly electrically conductive, with an electrical current density of 4 x 109 A/cm^2, more than 1,000 times greater than copper.

CNTs are not the same as carbon nanofibers (CNFs). In general, CNFs are arranged as stacked cones, cups or plates of graphene, while CNTs are CNFs in which graphene has been rolled into a cylinder. CNFs can be hollow as well. The dimensions of CNFs are similar to multi-wall CNTs, with diameters of 50–200 nm compared to MWCNTs at 30–100+ nm. Single-wall CNT diameters are much smaller, mostly within single digits (*see* Table 4.1).

Production

CNTs are produced using three main methods:

Arc discharge—CNTs are created through arc-vaporization of two carbon rods placed end to end in an inert gas at low pressure. The discharge vaporizes the surface of one carbon electrode and forms a deposit on the other electrode. Purification is required to separate the CNTs from the soot and residual catalytic metals in the deposit.

Laser ablation of graphite—A laser is used to vaporize graphite rods in a catalyst mixture (e.g., cobalt, nickel) and inert gas (e.g., argon), followed by heat treatment in a vacuum at 1,000°C.

Chemical vapor deposition (CVD)—A metal catalyst (e.g., iron) is reacted with carbon-containing gases (e.g., hydrogen or carbon monoxide) inside a high-temperature furnace to form CNTs on the catalyst.

Though all three methods require the use of metals (e.g., iron, cobalt, nickel) as catalysts, the CVD process is currently favored because it produces larger quantities of CNTs, is more easily controlled and thus offers lower cost.

Agglomeration and Functionalization

CNTs, graphene, nanodiamonds and other nanoparticles have typically been added to polymer, ceramic or metal matrix materials to form nanocomposites. The success of this approach has been limited by the tendency of nanomaterials to **agglomerate** or group together in solution, raising viscosity. For example, the addition of CNTs at only 1% by weight can potentially convert a typical epoxy resin system (832 cP) into a thick, viscous peanut butter-like paste (272,000 cP).[1]

Various technologies have been developed to mitigate this issue, including **functionalization** of the nanomaterial surface. Common functionalization techniques, which are often combined, include modifying the electric charge of the nanomaterial or solution, modifying the pH, and/or using a surfactant (substance that reduces the surface tension of a liquid). Note that functionalization may also increase the toxicity of CNTs and other nanomaterials precisely because they are now more easily dissolved into fluids (e.g., blood) and biological structures (e.g., tissues and cells). (*See* Chapter 14, page 383.)

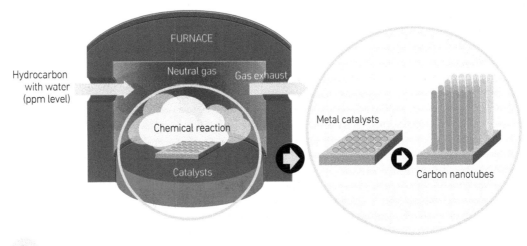

FIGURE 4-4. Chemical vapor deposition production of CNTs.

---

1   "The impact of different multi-walled carbon nanotubes on the X-band microwave absorption of their epoxy nanocomposites" in *Chemistry Central Journal*, (2015) Bien Dong Che, et al. (*See* Table 3) (https://www.ncbi.nlm.nih.gov/pmc/articles/PMC4353877/)

Once CNTs have been functionalized, then they may be mixed using sonication (applying sound energy to agitate particles) or high-shear mixing, though the latter may damage or alter the nanostructure.[2]

One example of functionalization is the SP1 protein developed by SP Nano Ltd. (Yavne, Israel), which binds tightly to CNTs forming a stable SP1/CNT complex that prevents CNT agglomeration and allows homogenous mixing in a resin matrix. Applied as a size-like coating to dry carbon fiber at loadings of 0.3–0.4% weight, SP1/CNT incorporated into carbon fiber/phenolic composites have exhibited a 47% increase in interlaminar shear strength and a 176% increase in through-thickness tensile strength compared to the same composites without SP1/CNT.

CNTs have also been successfully mixed into resins and adhesives by Zyvex Technologies (Columbus, Ohio). Its ZNT-fuse™ epoxy adhesive and Aroply™ epoxy film adhesive are reported to have higher properties than commonly used alternatives for composites. Zyvex also reports a 30–50% increase in stiffness and strength for its Arovex prepreg, compared to carbon prepreg without CNTs.

Vertically Aligned CNTs

A different approach is being pursued by N12 Technologies (Cambridge, Massachusetts), which grows vertically aligned CNTs (VACNTs) onto films or radially from the surface of carbon fibers using a CVD process. The VACNT film NanoStitch® products mechanically bridge ("stitch together") adjacent plies in a composite laminate and can be applied as a film during prepreg-

FIGURE 4-5. NanoStitch® vertically aligned CNTs.

Vertically aligned carbon nanotubes (VACNTs) are grown onto films and available for interleafing between plies in a composite laminate to prevent delamination and improve properties. *(Photos courtesy of (bottom) N12 Technologies; (top two) Dr. Enrique Garcia for MIT's "necstlab")*

2   "Processing of CNT/Polymer Composites" *CNT Composites*. (https://sites.google.com/site/cntcomposites/processing-of-cnt-polymer-composites).

ging or added between plies during layup. NanoStitch® is being used to prevent delamination in composite structures and/or to make them thinner and lighter.[3] (Figure 4-5)

Designed to be handled, cut and used in automated layup machines just like prepreg, NanoStitch® has increased interlaminar shear strength by more than 30% in a variety of composite laminates at less than 2 g/m^2 in additional mass per ply, or less than 1% volume fraction. N12 Technologies has also demonstrated a 100% increase in fatigue life and a 100X improvement in shear fatigue life for NanoStitch®-containing carbon fiber composite laminates. Other benefits from incorporating VACNT film into unidirectional carbon fiber composite laminates have been confirmed by MIT, including: 30% increase in tension-bearing (bolt pull out) critical strength, 14% increase in open-hole compression ultimate strength, and >25% increase in L-section bending energy and deflection.

CNT Fibers

CNTs are also being spun into continuous yarn-like fibers with lengths ranging from millimeters to a kilometer. Hydrogen and other industrial gasses are heated to > 1,000°C in a furnace. They react to form an elastic smoke (also called an aerogel or "CNT sock") containing billions of CNTs. This CNT aerogel is drawn out of the reactor and either collapsed and densified as a flat sheet, or pulled into a spinneret to form a yarn. For the latter, CNTs measuring 1–5 nm in diameter are spun into hair-sized diameter fibers via a continuous process at rates up to 100 m/min.

With a surface area thousands of times greater than a typical carbon fiber, these CNT fibers exhibit tensile strength similar to AS4 carbon fiber but with orders of magnitude higher electrical and thermal conductivity at roughly half the density. They also maintain 60%–70% of knot efficiency (the strength of a fiber with a single knot in it) similar to textile fibers like cotton and wool, as opposed to carbon fiber which loses 99% of its strength when knotted. (Figure 4-6)

Miralon® yarn is the family of CNT fiber products produced by Nanocomp Technologies. It has been used in communication cables, radiation shields and engine covers for satellites, and reduced the weight and thickness of ballistic soft armor by 30% and 20%, respectively, without compromising performance. The company claims that Miralon® yarn could slash the weight of data cable in a 787 aircraft from 8,000 to 2,400 lbs. Nanocomp Technologies has developed Miralon® honeycomb core for composite structures and is working with NASA to introduce a second-generation high-strength, high-modulus CNT fiber with IM7 type properties yet flexible and electrically conductive. (Figure 4-7, on the next page)

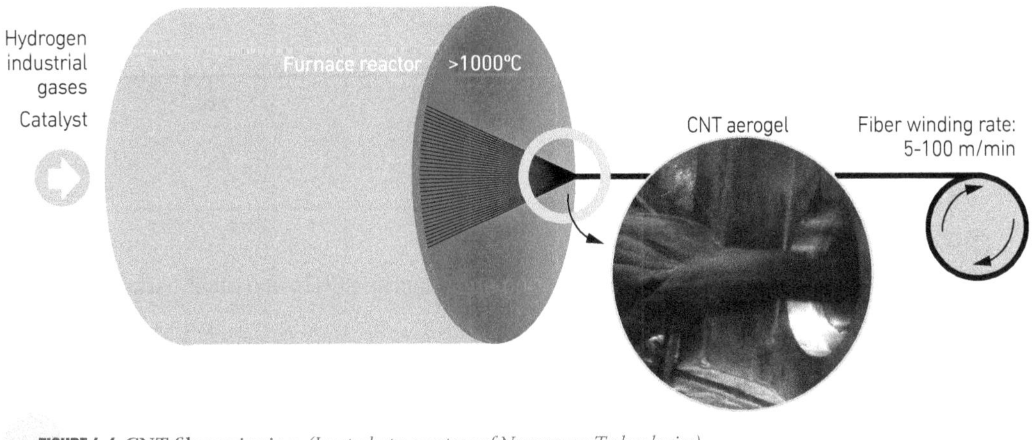

FIGURE 4-6. CNT fiber spinning. *(Inset photo courtesy of Nanocomp Technologies)*

3 Source: Dr. Brian Wardle, Department of Aeronautics and Astronautics, MIT (Cambridge, MA). (https://www.sciencedirect.com/science/article/pii/S026635381630687X)

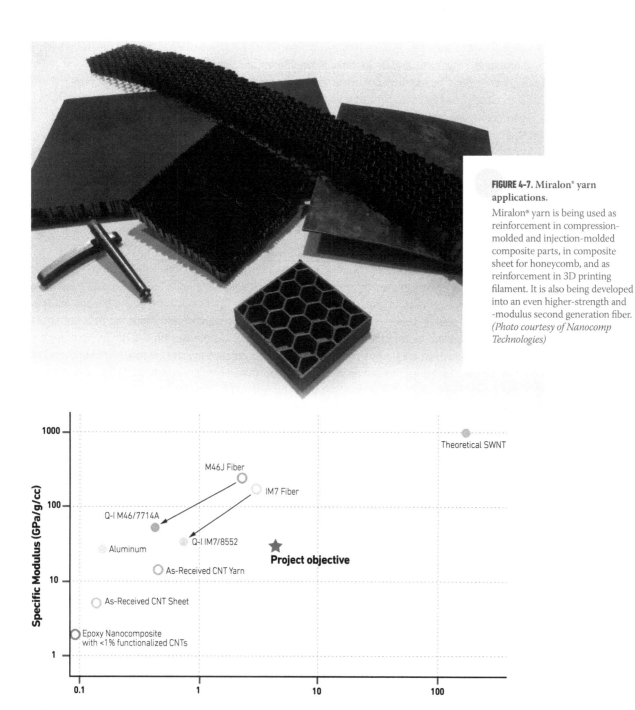

FIGURE 4-7. Miralon® yarn applications.

Miralon® yarn is being used as reinforcement in compression-molded and injection-molded composite parts, in composite sheet for honeycomb, and as reinforcement in 3D printing filament. It is also being developed into an even higher-strength and -modulus second generation fiber. *(Photo courtesy of Nanocomp Technologies)*

CNTs in Composites

There are at least 70 companies manufacturing and supplying CNTs and other carbon-based nanomaterials globally. *See* Table 4.3.

Applications in composites include higher performance sporting goods, anti-icing systems for wind blades and wing surfaces, improved toughness in polymer and ceramic matrix composites (CMCs), lighter-weight unmanned marine and aerospace vehicles, electrostatic discharge (ESD) protection in fuel lines, tanks and piping, electronics and spacecraft structures, infrared heat signature suppression for stealth and thermal camouflage, multifunctional aircraft structures, and a wide array of electrical conductivity applications including lightning strike protection and filled/reinforced plastics that are more easily painted with electrostatic coating processes.

TABLE 4.3 Selected CNT Suppliers/Producers Active in Composites

Company Location	Products/Technology Development
Arkema Colombes, France	Graphistrength® MWCNTs and thermoplastic masterbatches
Bayer Material Science Leverkusen, Germany	Baytubes® MWCNTs
Chasm Advanced Materials Boston, MA; Norman, OK	Acquired Southwest NanoTechnologies (SWeNT) SWCNTs and MWCNTs made using patented CoMoCAT™ process
Fibrtec Atlanta, TX	Licensed Ros-1™ short MWCNT technology for thermoplastic composites
General Nano Cincinnati, OH	Veelo™ CNT sheet, film, tape and coating products for lightning strike protection, heat blankets (deicing, composite curing), EMI shielding and emissive coating
Huntsman The Woodlands, TX	Acquired Nanocomp Technologies, maker of Miralon® CNT yarn
Hyperion Catalysis Cambridge, MA	FIBRIL™ MWCNTs
LANXESS Cologne, Germany	Rhenofit® CNT aqueous dispersions
N12 Technologies Cambridge, MA	NanoStitch® vertically aligned CNT layer products
OcSiAl Luxembourg; Columbus, OH	TUBALL™ SWCNTs
SP Nano Yavne, Israel	SP1/protein functionalizes CNTs SP1/CNT coating, SP1/CNT coated polyester, aramid and carbon fabrics
Zyvex Technologies Columbus, OH	Aroply™ CNT-modified epoxy film adhesive Arovex® CNT-reinforced epoxy prepregs
Nanocyl Sambreville, Belgium	NC7000™ MWCNTs PLASTICYL™ CNT thermoplastic masterbatch ELASTOCYL™ CNT elastomeric masterbatch EPOCYL™ CNT epoxy resin concentrates AQUACYL™ CNT waterborne dispersions Purified CNT products for research

Nanofibers

CNT fibers described in the Nanocarbon section above are not the only nanofibers being used in composites. Nanofibers are also being produced from a variety of metals, ceramics, polymers and natural materials such as chitin (the protein that makes up crustacean shells) and cellulose.

Nanofibers have much greater surface area and much higher aspect ratios, offering the potential for increased mechanical properties. They also enable multifunctional composites, providing additional functions such as electrical conductivity, ability to change optical properties or to convert light or stress into electrical charge, etc. Nanofibers are being investigated for use in higher-performance composites including aircraft and space structures, fuel cells, conductivity and electromagnetic shielding, smart/sensing structures, structural health monitoring (SHM) of composite structures and components that can change shape (morphing) and/or appearance (translucence, transparency, color and cloaking). (Figure 4-8)[4]

Polymer nanofibers have also been made into nonwoven veils which, when interleaved between plies in glass fiber/epoxy laminates, show up to a 50–60% reduction in impact damage, especially at higher energies. (Figure 4-9)[5]

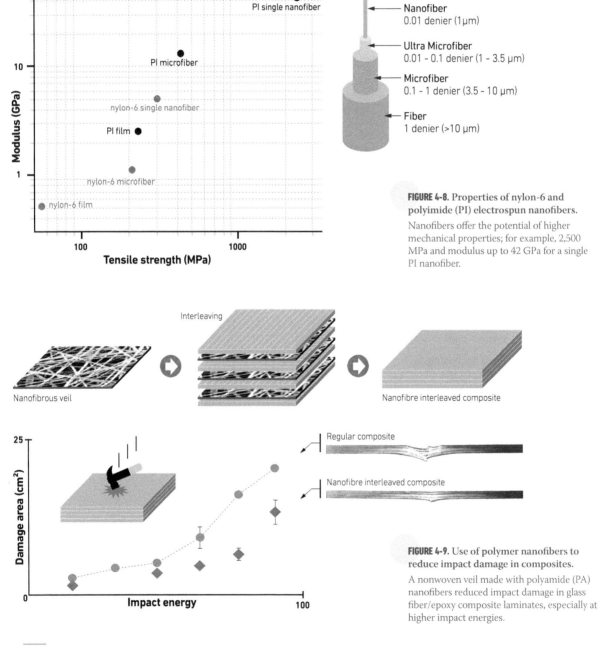

FIGURE 4-8. Properties of nylon-6 and polyimide (PI) electrospun nanofibers.

Nanofibers offer the potential of higher mechanical properties; for example, 2,500 MPa and modulus up to 42 GPa for a single PI nanofiber.

FIGURE 4-9. Use of polymer nanofibers to reduce impact damage in composites.

A nonwoven veil made with polyamide (PA) nanofibers reduced impact damage in glass fiber/epoxy composite laminates, especially at higher impact energies.

4   "Electrospun Nanofiber Reinforced Composites: Fabrication and Properties," Shaohua Jiang, *University of Bayreuth*, 2014 (pp. 54-55).
5   Source: Dr. Ir. Lode Daelemans, researcher, Centre for Textile Science and Engineering, Dept. of Materials, Textiles and Chemical Engineering (MATCH), Ghent University, Belgium.

Purification - Bleaching

Chemical pretreatment
Mechanical disintegration

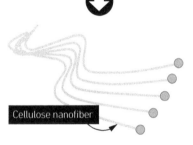

FIGURE 4-10. Cellulose nanofiber.
Cellulose nanofiber can be made from wood waste and used to produce transparent films and papers. *(Bottom photo)* Transparent nanofiber paper, made from optically transparent composites reinforced with CNF. *(Source: M. Nogi, Osaka Univ.)*

NANOFIBER PRODUCTION

Polymer, ceramic, metal and even CNT fibers can be produced directly by **electrospinning**. In this process, a solution is passed through a spinneret. As the solution emerges from the spinneret as a drop, an electrical charge is applied which causes the drop to change to a cone and then a jet of liquid that whips into thin fibers.

Though nanofibers offer great potential, much work remains to exploit this in macroscale, commercially produced composite components. Issues include commercially viable processes for producing high-quality nanofibers, mixing the nanofibers into composite matrix materials, achieving good nanofiber-matrix adhesion, and cost.

CELLULOSE NANOFIBERS

Cellulose is the most plentiful organic polymer on earth, found in plant walls and trees. Cellulose nanocrystals (CNCs) and cellulose nanofibrils (CNFs) are the main types of nanocellulose being commercialized. Typically, wood pulp, waste or byproduct is hydrolyzed (reacted with water) to produce rod-like CNCs which are 3–20 nm wide and 50–500 nm in length. CNFs are made using mechanical processes without chemical treatments, producing fibrils that are 4–50 nm wide and >500 nm in length.

Cellulose nanomaterials are lightweight, strong, and stiff. CNCs have piezoelectric properties, meaning they generate an electric charge in response to applied load or stress. They also have photonic properties, using light in the same way electronic devices use the electron. These photonic properties also enable engineered optics, including transparency and color manipulation. Various transparent films and papers have been made with cellulose nanofibers and nanocrystals, and they may offer a path toward transparent composites. (Figure 4-10)[6]

Somewhat like natural fibers, cellulose nanofibers offer the potential for low cost and improved sustainability and recyclability. They are being investigated for lightweight composites in ballistic armor, spacecraft, automobile body and interior components and aircraft interior and seat components. They have been incorporated into PE, PP and biopolymers (PLA and PHA), enhancing mechanical and abrasion resistance.

6   Sources: "Recent Advances in Nanocellulose Composites with Polymers: A Guide for Choosing Partners and How to Incorporate Them," Arindam Chakrabarty, Yoshikuni Teramoto, *Polymers*, 10(5) 2018, p. 517 (doi.org/10.3390/polym10050517); and "Transparent Conductive Nanofiber Paper for Foldable Solar Cells," Masaya Nogi, et al, *Scientific Reports* 2015, Vol.5, Article# 17254 (nature.com/articles/srep17254).

Nanoparticles

Nano-scale particles or spheres—a.k.a., *nanoparticles*—are usually defined as having a size of ≤ 100 nm, but larger size particles are used for select applications. Nanoparticles of various materials are used in polymers, metals, and ceramics to improve the physical properties of these materials.

❭ NANO-SILICA

In CFRP composite structures, crystalline silicon dioxide (silica) nanospheres are used to fortify the polymer matrix in prepreg materials and essentially fill in gaps between carbon fibers, while minimizing resin content. (Figure 4-11)

The resulting structures have a slightly higher density, increased modulus, higher tensile strength, better fracture toughness, increased hardness, lower coefficient of thermal expansion (CTE), lower cure exotherm, and a host of other desirable properties when compared to those made with conventional (control) resins, as illustrated in Table 4.4 below.

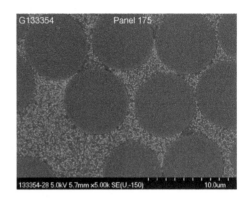

FIGURE 4-11. Field emission SEM micrograph of a polished carbon fiber composite cross-section that shows 7-micron carbon fibers with 3M Matrix Resin 3831 and 140 nm silica particles uniformly dispersed at ~35% by weight. The fiber volume is 60%. *(Photo courtesy of 3M)*

TABLE 4.4 Nano-silica Fortified Matrix Resin Properties

Property	Control Resin	Nanosilica Fortified Resin
Silica (wt%)	0	43
Density (g/cc)	1.23	1.51
Complex Viscosity @ 63 °C (Pa-s)	6.5	26.3
Minimum Complex Viscosity (Pa-s)	0.3	1.5
Cure Exotherm (J/g)	594	231
Tg (°C)	232	232
Linear Cure Shrinkage (%)	0.75	0.50
CTE (μm/m/°C)	46	36
Tensile Modulus (GPa)	3.1	6.3
Tensile Strength (MPa)	56.8	67.2
Tensile Strain (%)	2.6	1.0
Fracture Toughness (MPa-m½)	0.51	0.72
Barcol Hardness (H$_B$)	37	76
Nanoindentation Modulus (GPa)	4.9	7.3
Nanoindentation Hardness (GPa)	0.36	0.45

Source: 3M Aerospace

❭ CALCITE

Historically, various forms of calcium carbonate (calcite) have been used as fillers for polymers and low-cost fiber composites such as sheet molding compounds (SMC). Next generation surface-functionalized nano-scale calcite (nanocalcite) has demonstrated improved properties and exceptional processing performance in the manufacture of filament-wound torsion drive shafts and other structural composite applications. Nanocalcite enhanced matrix resins provide a lower cost alternative to nanosilica, with similar enhanced properties. Applications use a larger nanoparticle than those of nanosilica applications, with a nominal particle size of 400 nm.

❭ NOBLE METAL AND METAL OXIDE NANOPARTICLES

Noble metal and metal oxide nanoparticles are used for many different applications in polymer, metal, and ceramic matrix composites. For example, gold and silver nanoparticles may be used for thermal or electrical conductivity, iron oxide may be used for magnetic properties, while other metal oxides are catalysts and used to create or promote a chemical reaction within the polymer matrix.

❭ NANOCLAY

Interest in nanoclay composites was stimulated in 1992 by Toyota research group results which showed that adding 4–5% clay nanoparticles by weight to Nylon-6 thermoplastic polymer increased tensile modulus by 68%, tensile strength by 42%, heat distortion temperature by 82°C and water permeability resistance by 4%. These improvements are much higher than those achieved with greater quantities of micro-size fillers. However, improvements seem to plateau at this weight percentage.

Clay consists of layered silicates with the stacks held together by Van der Waals forces. Like CNTs, nanoclay particles tend to agglomerate, but this can be overcome by functionalizing the surface using a type of chemical reaction called ion exchange or cation exchange reactions. For this reason, cation-exchange *capacity* (CEC) is an important factor in a nanoclay's ability to be readily synthesized into polymer nanocomposites.

TABLE 4.5 Types of Clay vs. Use in Nanocomposites

	Composition	Structure	Use in nanocomposites
Type of clay Kaolin			
High alumina Low silica	One octahedral alumina layer bonded to one tetrahedral silica layer (O-T)	Low CEC Non-expanding	
Type of clay Montmorillonite-Smectite			
Low alumina High silica	Two tetrahedral silica layers sandwiching an octahedral alumina layer (T-O-T)	High CEC Expanding	Type most often used in nanocomposites
Type of clay Illite			
High K_2O (hydro-mica)	T-O-T	Low CEC Non-expanding	
Type of clay Chlorite			
Mg, Fe, Ni or Mn at the end of silicate lattice	T-O-T with Brucite in between each sandwich (T-O-T-BRUCITE-T-O-T)	Low CEC Swelling and surface area restricted	

The typical methods used to make a clay nanocomposite include **intercalation** (the insertion of a molecule or ion into compounds with layered structures) where polymer chains are inserted into the spaces between the silicate layers, and **exfoliation** (separation of a material's layers) where the silicate layers are completely separated and randomly dispersed into a polymer matrix. Exfoliation usually produces nanocomposites with the best property improvements because the clay becomes dispersible nanoparticles.

An individual clay platelet is only 1 nm thick, but its surface can span 100 to 600 nm, resulting in an unusually high aspect ratio that can dramatically alter the properties of a nanocomposite. Potential benefits include increased mechanical strength, improved dimensional stability, decreased gas permeability and improved barrier properties, decreased flammability, improved thermal stability, changes in electrical conductivity and enhanced transparency. Nanoclays have been investigated with numerous polymers including polystyrene, polyethylene, polypropylene, polycarbonate, polyurethane, polyamides, polyimides and epoxy resins. (Figure 4-12)

Nanoclay-reinforced polypropylene was used in the 2004 Acura TL seat back structures; nanoclay-reinforced polyolefin composites were used in the step-assists for the 2002 Chevrolet Astro van, body molding for the 2004 Chevrolet Impala, and cargo bed for the 2005 Hummer. Additional applications being researched include nanoclay-polymer composites for ballistic armor and electrical insulation components (e.g. transformer structures), as well as solar cells, fuel cells, and biosensors. (Figure 4-13)

In *CompositeWorld*'s 2011 "High-Performance Composites for Aircraft Interiors" conference (September 25–26, Seattle, Washington), Jaime Grunlan from Texas A&M University's Polymer NanoComposites Lab (College Station, Texas) presented a water-based nanocoating of montmorillonite (MMT) clay and chitosan, which when applied to polyurethane foam and cotton fabric, resisted 22 seconds of direct flame from a propane torch with only surface charring as a result. The coating reportedly provides a simultaneous heat shield and gas barrier and can eliminate melt-dripping in foams and ignition in cotton fabrics.

FIGURE 4-12. Nanoclay platelets.

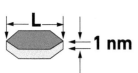

One clay platelet
L: 100 – 200 nm in case of MMT

Intercalated

Exfoliated

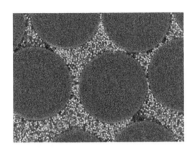

Nanosilica

Particle size 80–100 nm
$3–8/kg

Nanocalcite

Particle size 400 nm
$.04–10/kg

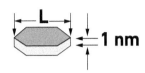

Nanoclay platelet

Thickness 1 nm
Length 100–600 nm
$0.3–0.7/g

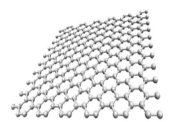

Graphene

Thickness 1–15 nm
Length 5–25 μm
Graphene oxide $5–$100/g
Monolayer film $400–$850 per 4" wafer

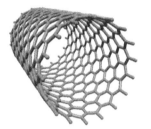

Single-wall CNT

Diameter 0.5–3.0 nm
Length 1–300 μm
$9 – $2,500/g

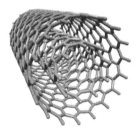

Multi-wall CNT

Diameter 30–100+ nm
Length 1–300 μm
$1 – $200/g

Carbon nanofiber (CNF)

Diameter 50–200 nm
Length 1–300 μm
$0.22 – $1.10/g

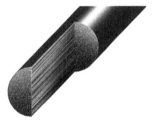

Carbon fiber

Diameter 10,000 nm
Length continuous
$10/lb

CNT fiber

Diameter 83 nm
Length continuous
—

FIGURE 4-13. Comparison of nanocomposite technologies and cost.

Sandwich Core Materials

CONTENTS

Why Use Sandwich Construction?

A sandwich structure consists of two relatively thin, stiff and strong faces separated by a relatively thick, lightweight core material. The face sheets are usually connected to the core material with an adhesive. (Figure 5-1)

In sandwich construction, the sandwich structure acts like an I-beam that can withstand great loads without bending (high stiffness) or failing (high strength). The faces of the sandwich panel act as flanges and the core acts like the I-beam's web, connecting the load-bearing skins. Depending on the type and properties of core used, it is possible to dramatically increase the bending stiffness and strength of a structure without increasing its weight. As shown in Figure 5-2, this increase is directly related to the thickness of the core.

Other reasons for using core may include additional benefits offered by specific core materials:

- Thermal insulation or thermal transfer.
- Dampening of vibration and noise.
- Filling of hollow spaces to limit water ingression.

FIGURE 5-1. Types of composite sandwich panels.
[a] Corrugated core panels, [b] foam-cored panels, and [c] honeycomb-cored panels, along with balsa-cored panels, are the most common types of composite sandwich construction.

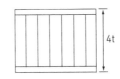

Relative stiffness	100%	700%	3700%
Relative strength	100%	350%	925%
Relative weight	100%	103%	106%

FIGURE 5-2. Sandwich structure stiffness increases with thickness. *(Courtesy of Hexcel)*

❯ SANDWICH VS. SOLID LAMINATE

Sandwich construction is just one type of structural composite design. Although cored construction has many benefits, it also has some disadvantages, which may dictate alternate construction for certain applications. Solid laminates offer better resistance to damage and better damage tolerance than do cored structures, but they also tend to be heavier. Solid laminates are very effective where the overwhelming need is for strength versus stiffness, in which case a thin solid laminate is more optimized to meet the high-strength loading demands using the least amount of material.

Skin-stringer, hat-stiffened or rib-stiffened structure can be described as a plate that is reinforced periodically by hollow, solid or cored beams. It is the most common alternative to sandwich construction in aircraft structures. It is also common in boat hulls.

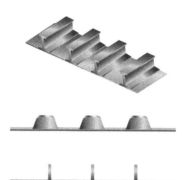

- Stringers are solid laminate reinforcing beams, usually in the form of I-beams or J-sections.
- Hat stiffeners are hollow or foam-cored beams having a rectangular or trapezoidal cross-section.
- Ribs are thin, solid plates or verticals used as reinforcing members.

By placing a few stiffeners or stringers where additional reinforcement is needed, skin-stringer construction achieves high stiffness with a minimal amount of material. It is similar to solid laminate construction both in damage performance and its ability to meet high strength requirements efficiently, and it is a very competitive alternative to sandwich structure.

Corrugated construction offers stiffness approaching that of sandwich construction combined with the damage tolerance of a solid laminate and perhaps the easiest and most cost-effective fabrication, especially in wide spans. It is well-suited to continuous processing. Common examples of corrugated construction include:

- Corrugated fiberglass sheets used in greenhouses, skylights, siding and roofs.
- "Sinusoidal stringers" or "sine-wave beams" used in aircraft spars and other stiffness-driven structures.
- Corrugated composite laminates used as a sandwich core in composite bridge decks and other heavily-loaded structures.

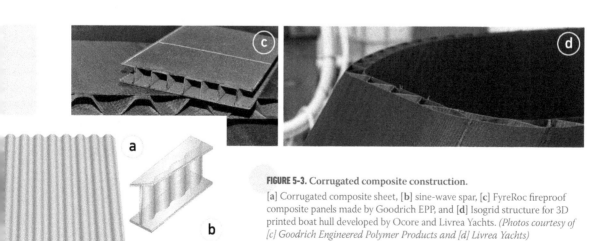

FIGURE 5-3. Corrugated composite construction.
[a] Corrugated composite sheet, [b] sine-wave spar, [c] FyreRoc fireproof composite panels made by Goodrich EPP, and [d] Isogrid structure for 3D printed boat hull developed by Ocore and Livrea Yachts. *(Photos courtesy of [c] Goodrich Engineered Polymer Products and [d] Livrea Yachts)*

Balsa Core

End-grain balsa is the most commonly used wood core in composite sandwich structures. Balsa is very light yet exceptionally strong. Although classified as a hardwood, its density ranges from only 4 to 20 pounds per cubic foot (pcf). Balsa gets its strength from its vascular system. Similar to honeycomb, this vascular system is capable of carrying tremendous compressive and shear stresses.

FIGURE 5-4. The vascular system of balsa wood, similar in structure to honeycomb, which enables it to withstand large compressive and shear stresses. *(Courtesy of I-Core Composites)*

End-grain balsa is produced by cutting harvested and kiln-dried balsa wood across the grain. These cut pieces are assembled into blocks and cut into sheets, or cut into smaller blocks which are then glued onto a scrim fabric to produce a drapeable sheet. The grain orientation of the final balsa core is in the z-direction, or perpendicular to the **face skins**, which gives the maximum strength and stiffness. (Figures 5-4, 5-5)

Balsa continues to be used in many cored composite applications due to its high stiffness, compressive properties and low cost. Although it was used in early applications of aircraft flooring, it is no longer common in aircraft due to increased stringency of smoke and flammability regulations as well as a general trend toward lighter core materials such as honeycombs and foam.

Balsa and foam core materials are typically used in boat construction, with foams traditionally being used more than balsa in Scandinavia and Europe. Balsa remains popular in cored boat hulls and decks, especially as resin infusion becomes an increasingly popular fabrication method. Open-celled materials such as honeycomb are not easily compatible with resin infusion, because the cells fill up with resin. Honeycomb and foam are also more expensive than balsa.

Foam Cores

Foam is made by adding foaming or blowing agents to polymer materials (e.g., resins or melted plastics), which causes gas pockets to form within the structure of the polymer as it cures or hardens. Foaming agents may be chemical or physical.

- Physical foaming agents include pressurized gases, such as nitrogen.
- Chemical foaming agents are materials that cause a chemical reaction which produces gas.

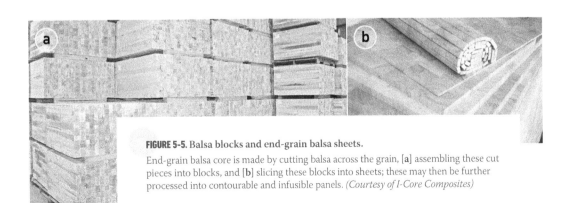

FIGURE 5-5. Balsa blocks and end-grain balsa sheets.
End-grain balsa core is made by cutting balsa across the grain, [a] assembling these cut pieces into blocks, and [b] slicing these blocks into sheets; these may then be further processed into contourable and infusible panels. *(Courtesy of I-Core Composites)*

Most cores used in composite sandwich structures are rigid and closed-cell, meaning that each gas cell is completely surrounded by cured resin and thus isolated from all other cells. A closed-cell structure prevents water migration through the core material, which assists in minimizing moisture ingress and absorption.

Foams are manufactured from a wide variety of polymers and can be supplied in densities from less than 2 pcf to 60 pcf.

TABLE 5.1 Foam Core Materials

Polymer	Trade Name	Manufacturer	General Applications
Polyvinyl Chloride (PVC)	Airex C70	Airex AG/3A Composites	Composite sandwich
	Divinycell	DIAB Group	
	Gurit PVC	Gurit	
	Stru-Cell	Polyumac	
Polyurethane (PUR)	Last-A-Foam	General Plastics	Composite sandwich, SIPs, buoyancy/flotation, cushioning, packaging, insulation, models
	Modipur	Hexcel	
	Air-Foam	Polyumac	
	Stepanfoam	Stepan Company	
PUR, fiber-reinforced	Airex PXc, Airex PXw	Airex AG/3A Composites	Composite sandwich
Polyester, crosslinked	Aircell	Polyumac	Composite sandwich
PET	Airex T10, T90, T92	Airex AG/3A Composites	Composite sandwich
	ArmaForm	Armacell	
	Divinycell P	DIAB Group	
Polyisocyanurate (ISO)	Elfoam	Elliott Company	Composite sandwich, insulation
Polystyrene (EPS or PS)	SecureTherm	ACH Foam	SIPs, refrigeration, insulation
	Styrofoam	Dow Chemical	
	EPS	Benchmark Foam	
Polyetherimide (PEI)	Airex R82	Airex AG/3A Composites	Structural cryogenic insulation and high fire resistance
Polymethacrylimide (PMI)	Rohacell	Evonik	Composite sandwich
Copolymer	**Trade Name**	**Manufacturer**	**General Applications**
SAN	Corecell	Gurit	Composite sandwich
Hybrid	Divinycell HD	DIAB Group	
	Ultracore	Futura Coatings	
Syntactic Foam	**Trade Name**	**Manufacturer**	**General Applications**
Polyester Epoxy Multiple resins Cyanate Ester	Spray-Core	ITW SprayCore	Composite sandwich, print barrier, sub. for thin honeycomb, sub. for faceskin plies
	Scotch-Core Syntactic Film	3M	
	Microply	YLA	
	BryteCor EX-1541	Bryte Tech.	
BMI	FM 475	Cytec	

❭ LINEAR VS. CROSSLINKED PVC

Linear foams are made from thermoplastic polymers and have no crosslinks in their molecular structure (*see* Figure 5-6). Thus, linear PVC (polyvinyl choride) foam is a purely linear foam, as opposed to crosslinked PVC which is a blend of thermoplastic PVC and thermoset polyurethane. The non-connected molecular structure of linear PVC foam results in higher **elongation** and **toughness**, but somewhat lower mechanical properties. It also enables significant deflection without failure, which means better impact resistance and energy absorption. Linear PVC foam is also easier to thermoform around curves; however, it is more expensive than crosslinked PVC foam, and has a lower resistance to elevated temperatures and styrene. Cell diameters range from .020 to .080 inches.

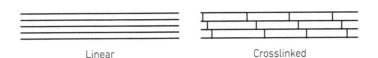

Linear   Crosslinked

FIGURE 5-6. Molecular structure of linear vs. crosslinked PVC.

Crosslinked foams incorporate thermoset polymers and have crosslinks between the molecular chains, producing higher strength and stiffness but less toughness. Elongation for crosslinked PVC foam is 10–20% versus 50–80% for linear PVC foam. Crosslinked PVC is more rigid—producing stiffer panels—but is also more brittle and prone to forming 45° cracks under impact. It is cheaper than linear PVC foam and less susceptible to softening or creeping at elevated temperatures. Cell diameters range from .0100 to .100 inches (compared to .0013 inches for balsa).

❭ POLYURETHANE VS. POLYISOCYANURATE FOAM

Polyurethane and polyisocyanurate foams are similar in that both are closed cell foams that have high **R-values** (typically between R-7 and R-8 per inch). These foams have higher R-values than most other foams because they trap small bubbles of gas during the foam manufacturing process. The gas is usually one of the HCFC or CFC gases, which have twice the R-value of air. Both kinds of foam can be sprayed, poured, formed into rigid boards and fabricated into laminated panels. Note that the high R-value can be detrimental to thick core structures during processing at elevated temperatures, where the outer surfaces of the core heat up at a faster rate than the inner core, causing internal stresses that result in through-thickness splitting at mid core.

Polyiso foams have several advantages over polyurethanes, however:

- Improved dimensional stability over a greater range of temperatures.
- More fire-resistant.
- Slightly higher R-value.

Because of its low mechanical properties, polyisocyanurate foam is not really used as a structural core material. Although it is used as a foam core in composite sandwich structures, it is primarily an insulation layer due to its low density and high insulative properties. The number one use of polyiso foam is insulation in construction, where it is used in more than half of new commercial roofing applications and nearly 40% of residential sheathing applications.

Polyiso foam was also used as insulation on the Space Shuttle *Orbiter*. One of the most easily recognizable parts of the Space Shuttle was the large, orange external fuel tank—the orange color is from a thin layer of polyisocyanurate foam covering the entire tank structure. The foam kept the liquid oxygen (LOX) fuel inside the tanks from boiling. LOX is a cryogenic fuel with a boiling point of around -200°F.

❱ POLYETHYLENE-TEREPHTHALATE (PET)[1]

PET is the most abundant polymer in the world and is the primary material for plastic bottles. Foam core made from PET is thermoplastic and often described as recyclable. Often it is made with recycled content, typically from waste bottles. As a structural foam, PET requires a slightly higher density to match the mechanical strength and stiffness of SAN and PVC thermoset foam cores, and a substantially higher density to match end-grain balsa.

❱ COPOLYMER AND PMI FOAMS

Corecell is a linear, styreneacrylonitrile (SAN) copolymer foam that was originally developed as a tougher alternative to PVC foam core for marine applications. Formerly manufactured by ATC Chemicals, Corecell is now produced by Gurit. Used as a structural core in composites, Corecell is available in densities from 3 to 9 pcf. It offers high shear elongation and impact strength, and is compatible with polyester, vinyl ester and epoxy resins. Applications include boat hulls, decks and other structures.

Rohacell® is a structural foam core material made from polymethacrylimide, or PMI. It is produced by mixing methacrylic acid and methacrylonitrile monomers, which react to form the PMI polymer. The monomer mixture, along with various additives are heated to a foaming temperature, which can be as high as 338°F depending on the desired density and grade of the foam. After foaming, the block is cooled to room temperature and the slick skins that form on the outer surfaces are removed. The block is then machined to produce foam sheets of various thicknesses.

The compacted, higher density Rohacell® foams have an oriented cell structure due to the manufacturing method used. Similar to end-grain balsa wood, higher density Rohacell® is cut into blocks with the cell orientation in the vertical direction and then bonded together. The resulting sandwich core has excellent properties in the thickness direction. Rohacell® is more water absorbent than other structural foam materials and it burns easily. However, it can withstand processing temperatures of up to 350°F. It is most commonly used in the aerospace industry in a variety of structural sandwich applications in commercial, military and general aviation aircraft, satellites, and helicopters.

❱ GROOVED, SCORED AND PERFORATED CORES

The use of foam as a structural core in sandwich construction has grown with the use of resin infusion processing. The closed-cell nature of foam means that it does not fill with resin, unlike the open cells of honeycomb. However, it has become common to use perforated and/ or grooved foam core (also called *knife-cut* or *scored core*) to aid resin flow during infusion. The holes in "perf-core" help to evacuate air as vacuum is applied, pulling resin from one side of the core to the other, which minimizes dry spots on the tool side of the laminate. Grooves can also be cut into the foam to act as resin flow channels and possibly eliminate the need for a separate flow medium. (Figure 5-7)

The downside is that these holes and cuts fill up with resin during infusion. SP Systems, now absorbed by Gurit, completed extensive testing on over 30 different types of core cuts for infusion processing. It documented significant weight gain with heavily cut foam, as well as a noticeable effect on mechanical properties and impact performance due to the additional resin cured in the numerous channels cut into the core.

According to the study, making too many cuts introduces too much resin at one time. The ideal process is referred to as "staged infusion"—just enough cuts to impel the resin front

1   From "Core for composites: Winds of Change," Jeff Sloan, *CompositesWorld* magazine, June 2010. www.compositesworld.com/ articles/core-for-composites-winds-of-change

forward, but also relies on multiple resin feed ports across the part. There are also a variety of surface treatments that foam and balsa core manufacturers have developed to reduce resin uptake in order to maintain light weight in resin-infused composite sandwich structures.

Both foam core and balsa core are also cut or scored in order to enable conformability for curved parts. Single-cut foam is typically used for shapes with single-direction curvature, while double-cut foam enables complex-curved parts. For both balsa and foam cores, maximum conformability is provided by cutting the core into blocks which are then bonded onto a fiber scrim.

❱ SYNTACTIC FOAMS

Syntactic foam core materials are made by mixing hollow microspheres of glass, epoxy and/or phenolic into fluid resin. Additives and curing agents are also combined to form a fluid which is moldable, and when cured forms a lightweight cellular solid.

Syntactic foam core can be produced from polyester, vinyl ester, phenolic and other resins, and is typically spray-applied in thicknesses up to 3/8-inch and in densities between 30 and 43 pcf. Syntactic cores and non-woven materials (or laminate bulkers) are often used in the same manner to build up laminate thickness and increase flexural properties at minimal cost. They are not intended as structural core materials.

Perforated

Allows resin to flow from one side of core to the other, aids infusion. 20mm between perforations is common.

Grooved - Single direction

Aids wet out and resin flow during infusion.

Grooved and perforated

One side is grooved and other is perforated. Developed for fast, reliable and robust infusion on flat or slightly curved surfaces, while minimizing imprints on the surface.

Grid-scored - Two directions

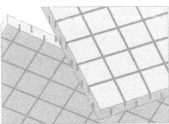

Cut in two directions on both sides of the core, with opposing sides offset by 50%. Where the grids intersect perforations are created. Used for parts with double curvature.

Grid-scored with scrim

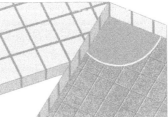

Cut in two directions for flexibility in forming curved surfaces. Scrim on one side for stability.

Grid-scored, grooved & perforated

Allows very fast, robust resin flow with curved geometries, ensuring wet-out on both sides of the core and one surface with good surface finish.

FIGURE 5-7. Perforated and cut foam core. (*Source: Diab Quick Guide to Finishing Options*)

Honeycomb Cores

Honeycomb, foam and balsa wood are all low density, cellular materials. Density is expressed in pounds per cubic foot (PCF). The cells in balsa and foam are really bubbles in the constituent material (wood or polymer resin, etc.). The cells in honeycomb are actually large spaces between formed sheets of paper, metal or other material. Thus, honeycomb is an "open-celled" material as the cells are open spaces, whereas most foams and balsa are described as "closed-cell" because the cells are sealed bubbles.

Hexagonal is the most common cell shape for honeycomb cores, but others are also used including tubular, triangular/sinusoidal, and corrugated. Within hexagonal cell honeycomb, there are several additional shape variations available. (Figure 5-8) These include over-expanded for conforming to simple or 2-D curvature, and specially-shaped "flex" cells for complex or 3-D curvature.

Over-expanded cell configuration is available from a variety of suppliers for Nomex® honeycomb—including M.C. Gill's OXHD honeycomb—and from Hexcel for its HRP impregnated fiberglass fabric honeycomb (sold as OX-Core®).

Flex-Core® is a patented technology developed by Hexcel. It enables easy forming of honeycomb into complex curvatures and shapes without loss of mechanical properties. It is offered for a number of Hexcel's honeycomb cores including those made from HRH-10 Nomex® paper, HRH-36 Kevlar® paper, HRP fiberglass cloth and for aluminum honeycombs made from 5052 or 5056 alloy foil. Double-Flex Core is only offered for Hexcel's aluminum honeycombs. UltraFlex is a flexible honeycomb cell shape offered by Ultracor Inc., which can be made with any nonmetallic honeycomb and cell size. It uses a different cell configuration than Hexcel's Flex-Core® product and is also not formed via an expansion process.

❱ L (RIBBON) VS. W (TRANSVERSE) PROPERTIES

Hexagonal honeycomb does not have the same properties in all directions. As a result of how hexagonal honeycomb is made, shear properties in the "L" or "ribbon" direction are much higher than those in the "W" or "transverse" direction (as shown in Figure 5-8). The ribbon

Standard honeycomb

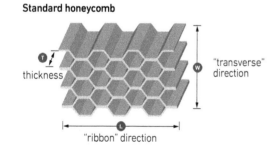

FIGURE 5-8. Variations in honeycomb cell configurations. *(Source: Hexcel)*

Flex-Core®

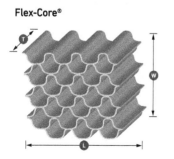

Ox-Core®

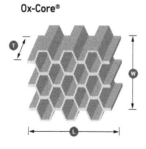

Double-Flex Core®

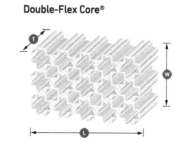

direction runs along the continuous sheets of web material that make up the honeycomb structure. The transverse direction runs perpendicular to the ribbon direction. It is the direction that runs across the sheets of web material as opposed to along the sheets. Another way of easily defining the W direction is that it runs from flat side to flat side across the hexagonal cells.

❱ HOW HONEYCOMB IS MANUFACTURED

Expansion (Figure 5-9) is a process in which web material is fed into machines that apply ribbons of adhesive. These ribbons of adhesive will form the *node-lines* for the expanded core material (nodes are where two sheets of web material are bonded together with adhesive). Sheets are then cut and stacked in layers, with every other sheet offset a specified amount, so that the resulting stack forms a honeycomb pattern when expanded.

This HOBE (HOneycomb Before Expansion) block is expanded and honeycomb blocks made from paper are then dipped in resin (typically phenolic). Multiple dippings are sometimes required to produce desired cell wall thickness, honeycomb density and mechanical properties. Blocks are then cured. Horizontal slices are sawed from the honeycomb block to required thickness. Alternatively, slices of the desired thickness may be sawed from the HOBE and then expanded and dipped to produce desired properties.

There are two ways to increase honeycomb core density: increase sheet or web thickness and/or increase resin coatings. Extremely thick webs (> 4 to 5 mils thickness) are difficult to expand. For a given web thickness there is a limit to which additional dippings only increase weight and do not add significant properties.

Corrugation (Figure 5-10) is typically used to produce higher density honeycomb cores, as well as sinusoidal and corrugated cores. Densities of 12 to 55 pcf are produced from this process using aluminum foil web material. For hexagonal honeycomb, web material is fed through corrugating rolls. The corrugated web material is then cut into sheets. Adhesive may be applied before or after corrugation. Sheets are stacked to form corrugated blocks. The honeycomb blocks are sliced into required thickness honeycomb sheets.

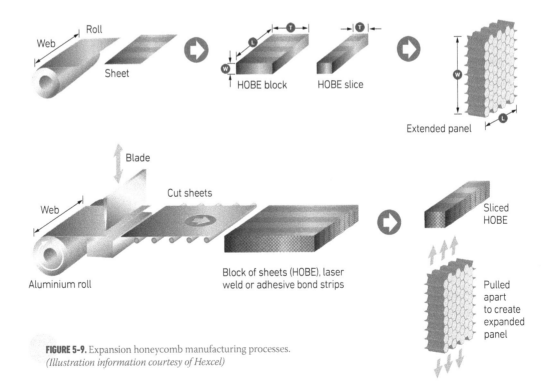

FIGURE 5-9. Expansion honeycomb manufacturing processes.
(Illustration information courtesy of Hexcel)

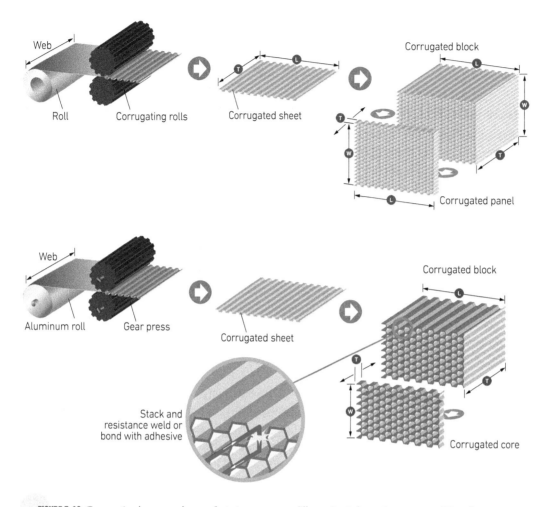

FIGURE 5-10. Corrugation honeycomb manufacturing processes. *(Illustration information courtesy of Hexcel)*

For sinusoidal honeycomb (Figure 5-11), two separate rolls of web material are used. One is fed through corrugating rolls, and the other is fed through a roll that applies adhesive. The two are adhered together by passing through a nip roll. This bonded sheet product is called single face. The single face now passes through slitting blades that produce multiple narrow bands of single face. These narrow bands are then laid side-by-side and squeezed together to form line boards. One outer surface of the line boards is coated with adhesive. The line boards are then stacked on a combining table to form large sheets of corrugated core. The thickness of the core is defined by the width of the slit bands of single face: wider bands produce thicker core, narrower bands produce thinner sheets of core.

❭ HONEYCOMB MATERIALS

Honeycomb can be made from a wide variety of materials, including: cellulose papers such as kraft paper and cardboard; more advanced papers such as those made from Nomex® meta-aramid and Kevlar® para-aramid fibers; fabrics from fiberglass, aramid and carbon fibers; foils and metal sheet; and plastics.

Carbon-carbon honeycomb is made by starting with epoxy-impregnated carbon fiber fabric, and heating it to 3,000°F. This carbonizes both the reinforcement and matrix resin, producing a carbonized-fiber-reinforced, carbonized-matrix-composite honeycomb core.

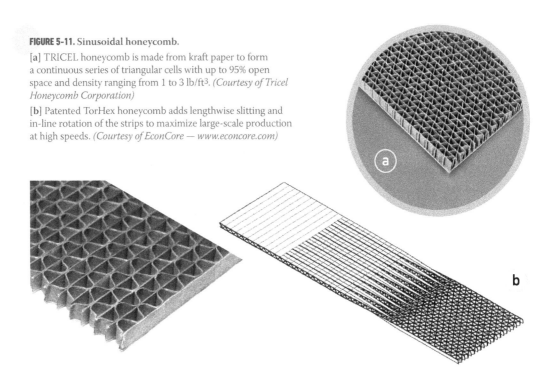

FIGURE 5-11. Sinusoidal honeycomb.

[a] TRICEL honeycomb is made from kraft paper to form a continuous series of triangular cells with up to 95% open space and density ranging from 1 to 3 lb/ft³. *(Courtesy of Tricel Honeycomb Corporation)*

[b] Patented TorHex honeycomb adds lengthwise slitting and in-line rotation of the strips to maximize large-scale production at high speeds. *(Courtesy of EconCore — www.econcore.com)*

TABLE 5.2 Honeycomb Core Materials

Material		Cell Shape	Product Name	Manufacturer
Cellulose papers	Kraft paper	Sinusoidal/ corrugated	Tricel	Tricel Honeycomb
			Paper honeycomb	ATL Composites
Advanced papers	Nomex®	Hexagonal	ACH, APH	Aerocell
			ECA, ECA-I	Euro-Composites
			HRH-10, HRH-78, HRH-310	Hexcel
			Gillcore HA, Gillcore HD	M.C. Gill
			PN1, PN2	Plascore
		OX	ACX, APX	Aerocell
	N36 Kevlar® paper	Hexagonal	KCH	Aerocell
			ECK-core	Euro-Composites
			HRH-36	Hexcel
			PK-2	Plascore
Fabrics	Fiberglass	Hexagonal	HDC, HFT, HRP, HRH-327	Hexcel
			UGF	Ultracor
			Glass fiber honeycomb (multiple resins available)	Euro-Composites
	Kevlar®	Hexagonal	HRH-49	Hexcel
			UKF	Ultracor
	Quartz	Hexagonal	UQF	Ultracor
	Carbon	Hexagonal	UCF (PAN or pitch fabric, cyanate ester or epoxy resin)	Ultracor

(continued)

TABLE 5.2 Honeycomb Core Materials *(continued)*

Material		Cell Shape	Product Name	Manufacturer
Metals	Aluminum	Hexagonal	ECM, ECM-P	Euro-composites
			ACG, CR-III, CR-PAA	Hexcel
			Dura-Core II, PAA Core	M.C. Gill
			PCGA-3003, PCGA-5056, PAMG-5002	Plascore
		Corrugated	Rigicell	Hexcel
			Higrid	M.C. Gill
	Stainless steel	Hexagonal	SSH	Plascore
Plastics	PP, PE, PS, PC, PET, PA, ABS, PPS, PEI	Hexagonal	ThermHex	EconCore
	Polyurethane	Hexagonal	TPU	Hexcel
	Polypropylene	Hexagonal	Air-Comb	Polyumac
			Nidaplast	Composite Essential Materials
		Tubular	PP	Plascore
			PP Tubus Core	Tubus Bauer
	Polycarbonate	Tubular	PC2	Plascore
			PC Tubus Core	Tubus Bauer
Other	Carbon-Carbon		PCC, UCC	Ultracor

Other Core Types

NOTE: Text in this section is adapted from "Engineered to Innovate" by Ginger Gardiner, High Performance Composites Magazine (Sept. 2006), reproduced by permission; © 2006 Gardner Publications, Inc., Cincinnati, Ohio.

❱ TRUSS CORES AND Z-AXIS REINFORCED FOAM

X-COR and K-COR

Albany Engineered Composites produces X-COR™ and K-COR™ pin-reinforced, closed-cell foam core sandwich materials. Both use pultruded pins arranged in the core to act as a shear- and compressive-load carrying truss system, while the foam provides buckling support. The pin material choice, reinforcement density, location, and angle of insertion may all be tailored for specific types of loading and end uses. (Figure 5-12)

Both materials are thermoformable, can be processed using the same cure techniques used for honeycomb-cored structures and are described as lightweight, durable and ballistic resistant. Applications have included the tail cone for the UH60M *Black Hawk* helicopter and spacecraft structures.

X-COR™—the pultruded rods extend from the foam core into the face skins during panel manufacture. Thus, the face skins are mechanically fastened and need no adhesive bond. Suitable for gentle contours or thick (>1.27 mm/0.050 inch) face sheets. Example construction: PMI foam with 0.28 to 0.5 mm/0.011 to 0.020 inches carbon fiber/epoxy rods.

K-COR™—the pultruded rods are splayed and captured between the core and face skins, which are bonded to the core. It is suitable for complex contours.

SAERfoam®

SAERTEX produces SAERfoam® structural core material combining lightweight polyurethane (PU), polyethylene (PE) or polyisocyanurate (PIR) foam with glass fiber reinforcement. Strength, weight and cost can be adjusted by tailoring the density and type of foam as well as the density and alignment of the 3D glass reinforcement. Reported to provide five times the shear stiffness of PVC foam and weight savings in comparison to balsa, SAERfoam® is compatible with almost all composite resins and can be supplied with cut grid patterns for conformability and flow channels for resin infusion. (Figure 5-13)

SAERfoam® I—fiber reinforcement at 90° provides isotropic properties plus excellent compressive strength and impact/delamination resistance.

SAERfoam® X—fiber reinforcement at 45° in one direction provides resistance to directional bending and shear stresses plus impact resistance.

SAERfoam® O—fiber reinforcement at 45° in two directions provides orthotropic properties plus excellent resistance to bending stresses in several directions as well as shear and thrust loads.

Seriforge 3D Stitched Foam Panels

Seriforge uses its automated net-shape preforming process to create foam-cored panel preforms with z-axis reinforcement for resin infusion. The process stitches together continuous carbon fiber through both face skins and foam core without knots, tension or loops in the stitch

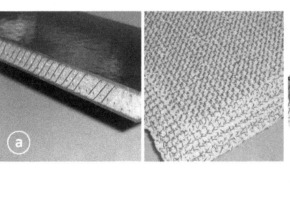

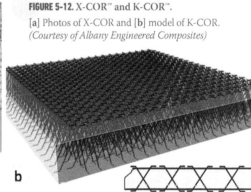

FIGURE 5-12. X-COR™ and K-COR™.
[a] Photos of X-COR and [b] model of K-COR.
(Courtesy of Albany Engineered Composites)

FIGURE 5-13. SAERfoam.®
SAERfoam® [a] processed as resin-infused composite panel and [b] supplied as conformable, fiber-reinforced core material.
(Courtesy of SAERTEX)

fiber, nor distortions in the dry fabrics. Therefore no disruptions are created to affect the face skins' in-plane properties. This can help prevent damage, for example, where foam core tapers and the two skins become one. Stitch fiber and spacing can also be tailored, placed only where needed or with regular spacing down to 3mm. Measured improvements over un-stitched sandwich panels include 90% greater shear strength, 200% greater flatwise tensile strength, and 12.6 times greater flatwise compressive strength. (Figure 5-14)

❱ NON-WOVEN CORES

Non-woven cores are used in vacuum infusion processing to provide channels for resin distribution as well as bulk within the infused laminate.

Bulking Materials vs. Print Barriers

The non-woven materials described here are generally referred to as bulking materials versus print barrier materials. The same manufacturers typically produce both types of materials; however, print barriers are not used as cores. They are instead thin layers placed behind the gelcoat in boat lamination to prevent the weave of the fabric reinforcement in a composite laminate from "printing" or showing through to the exterior of the gelcoat.

For example, Lantor produces Coremat® which is a flexible bulker mat and print blocker used in hand lay-up and spray-up processes. Lantor also produces Soric®, a thin, flexible, structural core material that also acts as an infusion flow medium. (Figure 5-15)

Examples of applications are wide-ranging (*see* Table 5.3). Soric XF3 has replaced honeycomb in the dorsal cover on the U.S. Air Force T-38 *Talon* jet trainer, when the part was transitioned from autoclaved cured prepreg and honeycomb core to a vacuum infusion of dry reinforcements and non-woven core with vinyl ester resin. Spherecore has been successfully used in Olympic medal-winning composite kayaks and canoes. The Atvantage CSC and CIC materials are used in the interiors of private and business jet aircraft.

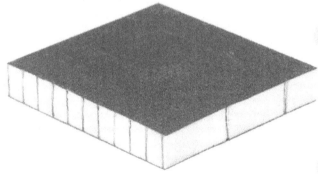

FIGURE 5-14. Seriforge 3D stitched foam panels. *(Courtesy of Seriforge)*

FIGURE 5-15. Non-woven core materials. Soric XF3 combines a non-woven core and resin infusion flow medium into one material. *(Courtesy of Lantor BV)*

TABLE 5.3 Non-woven Core Materials

Tradename	Manufacturer	Form	Fiber
Spherecore SBC	Spheretex	Stitch-bonded web (resembles felt)	Fiberglass/thermoplastic microballoons (note that carbon, aramid and other fibers can be used)
Soric	Lantor	Non-woven polyester mat with stamped honeycomb pattern	Polyester/stamped "cells" filled with microspheres
Atvantage CSC and CIC	National Non-wovens	Laid at specified angles and then needled for z-reinforcement	Discontinuous 3-6 inch (76-152 mm) aramid fibers or aramid fiber blends

❱ 3D PRINTED CORES

Composite sandwich construction may also be achieved by 3-D printing the core using thermoset or thermoplastic polymer that is unreinforced, filled with short fiber or reinforced with continuous fiber. This approach is being used to produce end-use composite parts as well as molds for manufacturing composites, including tools that can withstand autoclave cure. The examples in Figure 5-16 illustrate several different techniques, as follows.

Modified infill pattern—Infill is a repetitive structure used to take up space inside an otherwise empty 3D print. It allows printing over empty space with reliability and provides stability to the printed structure. The infill pattern can be printed with a higher or lower density, which affects the part's mass, strength and buoyancy, as well as the material used, print time and cost. In Figure 5-16 below, a cross-section of a brake lever (view [a]) created with a Markforged printer uses continuous carbon fiber filament to reinforce Onyx nylon and chopped fiber thermoplastic, to fabricate a structural infill. The autoclave-capable molding tool printed by Stratasys using unreinforced Ultem 1010 polyetherimide (PEI) high-temperature polymer (view [c]) is another example of infill patterns used as structural cores.

Core or mandrel for layup or fiber deposition—The "Superior" lower limb prosthetic (Figure 5-16 [b]) was produced by Moi Composites using its Continuous Fiber Manufacturing (CFM) process to print a continuous glass fiber/epoxy resin core, which was then wrapped in carbon fiber fabric using hand layup and epoxy resin and autoclave cured at 110°C and 2 bar of pressure for several hours. The final part met reduced weight and deflection requirements while enabling an attractive, customized design and significantly cutting production time and cost as opposed to alternative methods.

Surfacing infill patterns to produce molds—Polynt-Reichhold, in partnership with Cincinnati Inc. and TruDesign, have developed a high-build, room temperature spray coating into which final dimensions and details are machined to produce high-quality molds for producing composites. This enables molds 3D-printed with mostly lattice-type core and less solid structure, again reducing print material, time and cost, as well as weight. (Figure 5-16 [d])

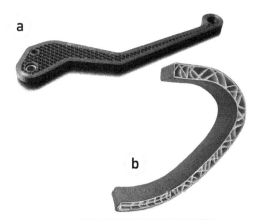

FIGURE 5-16. 3D printed cores. *(Photos courtesy of [a] Markforged, [b] Moi Composites, [c] Stratasys, and [d] CompositesWorld)*

Design and Analysis

Sandwich structures are used to provide a combination of light weight, high strength and high stiffness. Two examples of effective sandwich structures include building walls and space mirrors:

- When used as a wall, honeycomb-cored building panels act as a column and will carry loads up to ten times that of conventional frame construction.
- The benefit of using honeycomb core in the construction of large mirrors for space is that the deformation of the honeycomb under gravity is similar to that of a solid blank of the same dimensions, but the honeycomb has only one-fifth the mass of a solid blank.

Cored construction achieves this by acting like an I-beam, with the face skins acting like the flanges and the core functioning as the web. (Figure 5-17)

I-beams are based on the design principle that, in bending, the largest part of the load is carried near the extreme fibers (outer surfaces) of the beam, and very small bending stresses are developed near the neutral axis, as shown in Figure 5-18 below. Notice that the section modulus (stiffness for a given cross-sectional shape) of a rectangular beam is .167 times its height, while an I-beam is more than twice as stiff for the same height. The ideal beam cross-section is where the area is equally split between the top and bottom surfaces, with no material in the middle.

Thus, the top and bottom face skins of a cored sandwich carry most of the load, including in-plane bending, tension and compression. A general rule-of-thumb for lightweight sandwich panels is for face skins to comprise 60–67% of the total panel weight. Many aircraft flooring panels have been optimized for weight savings and have 50% of their weight in the facings and 50% in the core and adhesive.

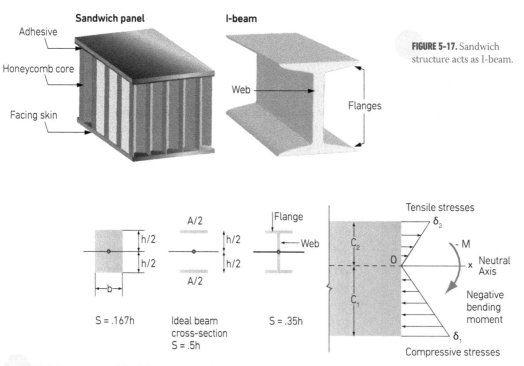

FIGURE 5-17. Sandwich structure acts as I-beam.

FIGURE 5-18. Section modulus (*S*) and load distribution of beams.

The core reacts to shear loads and resists out-of-plane compression (e.g., impact) while providing continuous, wide-span support for the face panels. Honeycomb cores have directional properties; thus, careful attention is required to make sure that this is accounted for during design and that the core is placed within the structure correctly to optimize overall structural performance.

〉GENERAL DESIGN CRITERIA

Some basic criteria to consider when designing sandwich structures include:

Skin compression failure

- Facings should be thick enough to handle the tensile, compressive and shear stresses caused by the design load.

- Core should be strong enough to withstand the shear stresses caused by the design loads. The adhesive must also have enough strength to carry shear stress into the core.

- Core should be thick enough and have sufficient shear modulus to prevent overall buckling of the sandwich under load, and also to prevent crimping.

- Compressive modulus of the core and of the facings should be sufficient to prevent wrinkling of the faces under design load.

- Compressive strength of the core must be sufficient to resist crushing by design loads that are normal to the panel facings and also by any compressive stresses induced through flexure.

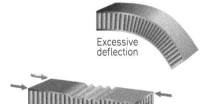

Excessive deflection

- Sandwich structure should have enough flexural and shear rigidity to prevent excessive deflections under design load.

- The cell size of honeycomb core must be small enough to prevent buckling within the cells.

〉GENERAL DESIGN GUIDELINES

1. Define loading conditions, including uniform distributed loads, end loading, point loading and possible impact loads.
2. Define panel type (e.g., cantilever, simply supported). Note that fully supported can only be considered when the supporting structure acts to resist deflection under the applied loads.
3. Define physical/space constraints including weight limit, thickness limit, deflection limit and factor of safety.

4. Preliminary calculations:
   - Assume skin material, skin thickness and panel thickness. Ignore core for now.
   - Calculate stiffness and deflection (ignoring shear deflection).
   - Calculate face skin stress and core shear stress.
5. Optimize design. Modify thickness and material selection for both skin and core, if necessary, to achieve desired performance.
6. Detailed calculations:
   - Stiffness.
   - Deflection, including shear deflection.
   - Face skin stress.
   - Core shear stress.
   - Check for the failure modes listed in "Design Criteria" (above), where each is applicable.

❱ KEY PROPERTY TESTS FOR SANDWICH PANELS

Long Beam Flex [a]

This is the standard test for determining the load bearing capability of a sandwich panel. It tests the face skins—which should fail before the core—and verifies how much weight the panel will support as well as how much deflection it will experience.

Core Shear

This test reduces the span from the flex test to 15 to 30 times the panel thickness in order to determine where the core will fail before the face skins.

Flatwise Compression [b]

This measures the strength of the core in resisting compressive loads.

Plate Shear [c]

This measures the shear strength and shear modulus of the core.

Flatwise Tension [d]

This test measures the strength of the adhesive bond between the face skin and core, and is useful in determining surface preparation for bonding and prepreg adhesion.

Smoke, Toxic Emissions and Heat Release

Most sandwich panels are used in applications that have some type of smoke, emissions and/or heat release requirements. The test methods and requirements vary greatly by industry and regulatory body. (*See* Chapter 10, Page 290, "Fire, Smoke and Toxicity (FST) Requirements and Heat Release Testing.")

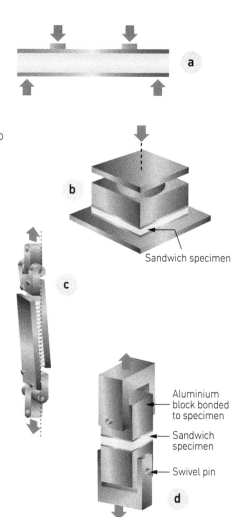

Sandwich specimen

Aluminium block bonded to specimen

Sandwich specimen

Swivel pin

Fabrication

If cored structures are being fabricated without using an autoclave, it is usually advisable to vacuum-bag the core using an appropriate adhesive or core bonding putty. Adding chopped or milled fibers to bonding putty with foam core can improve the core-to-skin bond, similar to how the reinforcing fibers in SAERfoam® and X-COR™ tie together face skins and foam core. Core manufacturers typically provide core bonding guidelines and recommended adhesives.

If using an autoclave, selecting the proper adhesive is vital. For example, a film adhesive developed to provide a good fillet between the face skin and honeycomb cell wall is mandatory when bonding prepreg face skins to honeycomb core.

Cored panels typically need to be sealed or closed off at edges and fastener holes. A variety of edge closure designs are possible, depending on aesthetic and attachment requirements. Core may be routed out at the edges and filled with potting compound or replaced with a more solid material or specially fabricated fitting. Fabricated "z"-sections or "u"-sections may also be bonded to the exterior of the panel edge. Core may also be machined to provide a tapered edge; however, this requires close tolerance work during bonding. Fastener holes typically need to be filled with potting compound or otherwise sealed to prevent moisture ingression. (Figure 5-19)

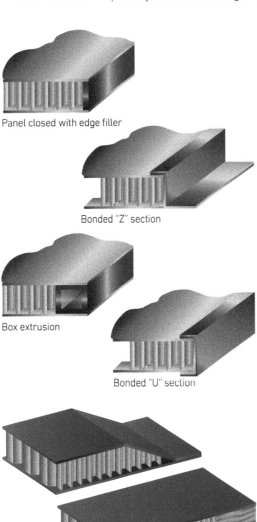

Panel closed with edge filler

Bonded "Z" section

Box extrusion

Bonded "U" section

FIGURE 5-19. Sandwich panel edge closure designs. *(Courtesy of Hexcel)*

In-Service Use

Two of the most serious considerations for in-service use of cored structures are impact and moisture absorption. These factors should be considered during the design stage, as careful selection of materials and design can prevent costly maintenance issues and even structural failures.

Aircraft manufacturers have learned to think about such issues as where service vehicles will contact the aircraft (e.g., baggage carts impacting doorways) and the frequency and magnitude of point loading on interior flooring (e.g., doorways and aisles are obvious, but the small strip between seats where shoe heels repeatedly land is equally problematic).

Core has been used successfully in marine structures for over thirty years, so the issue of moisture is not so much environment, as making sure to plan for in-service conditions. Boat builders typically use core bonding putties and vacuum bagging of core materials, but also carefully select the density and type of core depending on whether it is being used in a below waterline, forward section of a hull subjected to slamming from waves, or in a flat deck structure subjected to high heat and a range of compressive and shear loads. Aircraft manufacturers have learned that very thin skins over honeycomb core can result in early failures in structures due to moisture ingress and freeze-thaw cycling, if the proper film adhesives and exterior sealing systems are not utilized.

NOTE: Honeycomb illustrations in this chapter, including those on pages 142–144 and 151–153, have been recreated from originals that are copyright Hexcel Corporation, and the information in them is reproduced courtesy of Hexcel (all rights reserved).

6

Basic Design Considerations

CONTENTS

Composite Structural Design

❯ DESIGN CONSIDERATIONS

Manufacturability. All designers must ask (and answer) the following question: Can the part be manufactured as designed, with the chosen materials and forms, per the prescribed molding process? In addition to providing the laminate design, composite designers must also fully understand the constituent materials and molding processes to avoid excessive labor or manifestation of costly defects produced in manufacturing. Because of this, the following design considerations should be heeded.

Tight radii or sharp corners. With sheet metal, there is a very strict requirement as to bend radius allowables for each material, dependent upon material type and thickness. For similar reasons, radii in a composite part design should be scrutinized based upon material type (fiber and form) and thickness. For example, a 1/8-inch-thick prepreg carbon fiber fabric laminate molded into a 1/16-inch radius would result in a negative radius value on the inside surface and would certainly lead to defects in the laminate during processing.

Additionally, continuous fibers may be broken if molded into or over tight corners, resulting in reduced structural properties in the laminate. Joggle steps of 90° are discouraged in the design of composite parts, and are instead being replaced with 45° joggles that allow for a transition of fibers without the risk of breakage. (Figure 6-1)

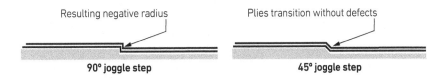

Resulting negative radius · Plies transition without defects

90° joggle step · **45° joggle step**

FIGURE 6-1. 90° vs. 45° joggle steps.

Laminate springback. In designing radial shapes or parts with flanges or angles, it is necessary to consider the amount of **springback** (a.k.a. "spring-in") of the part that will occur when molded. This is especially important if angle tolerances are ≤ ±2°. Springback of an angle is caused by several factors ranging from fiber type, fiber form, orientation of fibers, in-process compaction of layers, and the resulting resin volume in the radii.

Bridging of plies through an inside corner produces stress on the fibers when compacted during processing. Bridging also provides for a low-pressure area along the radius, resulting in excess resin accumulation and a greater amount of localized shrinkage after processing. This, along with the increased stress on the fibers contribute to the springback effect. Radial springback is caused by stress induced on inner plies during compaction in a similar manner to that in a corner. This is often intensified in cocured honeycomb core sandwich structures, as the plies tend to push into the core cells during processing, causing additional stress on the fibers.

Once the springback dimension has been calculated for a given laminate or angle, it is often mitigated with a correction to the mold design, to allow for a dimensionally accurate part. Often, designing the part with larger radii in the first place will alleviate many of the contributing factors to excessive laminate springback. (Figures 6-2, 6-3)

Dimensions and tolerances. With composite part design, it is common practice to allow for generous bilateral tolerances on ply drop-offs and core edge locations, core ramp angles, ply-splice joints, and other minimally critical locations of features within the laminate. Unless stringent control of any of these features is mandatory to the form, fit, or function of the laminate, the use of tight tolerances may be costly to achieve, or even unattainable. Location of such features may also be difficult to inspect after processing; therefore, dimensioning from fixed datum features is strongly encouraged.

Mold draft angles. Often considered to be a tool design issue, mold draft angles related to the configuration of a composite part should be considered at the design stage; e.g., a tall cylindrical shaped composite part with outward flanges on one end and closed on the other end. This would be difficult to demold if the vertical walls were designed at a 90° angle. Adding a 0.5° to 1.5° draft angle will greatly enhance part removal after the molding operation.

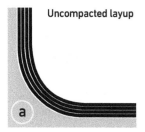
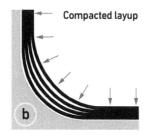

FIGURE 6-2. Fiber bridging adds to springback of laminates.

[a] The uncompacted materials are in a relaxed state.

[b] By compacting, the fibers are placed in tension and add stress to the laminate. This also allows plies to bridge and create a low-pressure area that fills with resin. The combination of resin shrinkage during cure and residual fiber tension after cure contribute to the springback condition.

FIGURE 6-3. Laminate and radial springback.

For the reasons discussed in the text, though the laminate in both of these illustrations was applied to the mold, once cured they deviated from the mold due to springback (also called "spring-in").

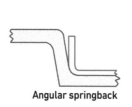
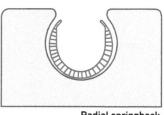

Angular springback

Radial springback

❱ MATRIX-DOMINATED PROPERTIES

The following are matrix-dominated properties:

- Compressive strength
- Off-axis bending stiffness
- Interlaminar shear strength
- Service temperature

Damage to the matrix system, delamination, porosity, or a low resin content will cause these properties to diminish rapidly. Matrix damage or degradation has a small effect on the (axial) tensile strength of a composite structure, yet it has a significant effect on compressive strength and shear properties. A high void volume (>1%) in a laminate will have similar effects.

❱ FIBER-DOMINATED PROPERTIES

Fiber-dominated properties include:

- Tensile strength
- Flexural modulus

The type of fiber reinforcement used, the fiber form that is selected, and the orientation arrangement of the fibers have a strong effect on the laminate performance.

❱ SOLID LAMINATE PANELS

Today's designers have a wide choice of material forms to choose from, including (fiber-placed) unidirectional tows, braided forms, spread tow fabrics, multi-axial fabrics, etc. Unidirectional forms yield the highest strength-to-weight ratios and the most efficient axial load transfer but typically require more layers to achieve quasi-isotropic, balanced, and symmetric properties (discussed later in this chapter). Fiber-placed unidirectional forms function much like unidirectional tape but may produce a higher fiber load per layer due to increased tow size, resulting in a greater layer thickness.

Solid laminate panels are typically specified for high flexural, tensile, and/or compressive loads in a structure; also where impact resistance or high damage tolerance is required. Solid laminates are often prescribed for heavily loaded structures such as aircraft wings and fuselage sections, spars for wind turbine blades, hulls for large marine vessels, etc. Many stiffener designs also utilize solid laminates and include *I-beam*, *blade* and *hat-section* stiffeners (*see* Chapter 5, page 136), named for their shapes, and preferably molded as an integral part of the structure. (Secondarily-bonded or fastened stiffener structures tend to be less efficient and more labor intensive to build.)

The specific performance properties of these structures are dependent upon the fiber type, fiber form, matrix resin selection, and ply orientation scheme of the laminate. Non-woven fiber "mat" materials are not at all dependent on orientation, and are quasi-isotropic by their nature; however, they cannot match the strength or stiffness of oriented continuous or long-fiber reinforced laminates.

Stiffened solid-laminate structures tend to be more durable and damage tolerant than sandwich panels, which are discussed below. (Figure 6-4)

❯ SANDWICH CORE PANELS

"Sandwich" structures rely on relatively thin skins bonded to a lightweight but relatively thick core material. Compared to solid laminates, they offer much higher bending stiffness-to-weight ratios. They are primarily used in applications in which high stiffness-to-weight is required, including aircraft interior panels, boat decks and hulls, rotor blades in helicopters and wind turbines, landing gear doors and control surfaces in aircraft, structural insulated panels (SIPs) for building and construction, skis and board-based sporting goods, and floors in aircraft, railcars, ships, trucks and cars. (Figure 6-5)

The primary disadvantage of sandwich structures is the reduced damage tolerance and durability compared to solid laminates. Thin skins can often be scratched or punctured. In addition, water or other liquids can ingress into honeycomb core cells and become trapped. These liquids can be difficult to locate and remove and can also cause disbonding of the skins

FIGURE 6-4. Stringer-reinforced solid laminate composite structures.

[a] CFRP hat stringer-stiffened fuselage panel representative of Airbus A350 construction, and [b] foam-cored, glass-fabric-skinned stringer structural grid for a power boat, prior to resin infusion.

(Photos courtesy of [a] Ginger Gardiner, CompositesWorld magazine and [b] Compsys)

FIGURE 6-5. Cored (sandwich) composite structures.

[a] Balsa wood core in one half of a resin-infused, glass fiber composite shell for a wind turbine rotor blade, and [b] foam-cored CFRP spare wheel well made using RTM.

(Photos courtesy of [a] Covestro Polymers (China) Co., Ltd, and [b] SGL Carbon)

when the liquid freezes and expands, or when subjected to temperatures above the boiling point during service or repairs.

More recently, thinner foam cores with added toughness or z-direction reinforcement have been developed to offer increased damage tolerance. (*See* Chapter 5 for more about sandwich construction.)

Fiber Orientation

❯ PLY ORIENTATION AND STANDARD ORIENTATION SYMBOL

Laminate performance is almost totally dependent upon fiber type and orientation. This must be closely controlled at the engineering, manufacturing, and repair levels. Because of this, there is an internationally recognized system for controlling fiber orientation on engineering drawings and all subsequent manufacturing processes. This involves the use of an orientation symbol and a specified location on the structure where the fiber orientation is prescribed and controlled. The fiber angle callouts are tabulated in a ply layup table and coordinated to the ply orientation symbol on the drawing.

The ISO recognized standard for use of this symbol states that the counter-clockwise (CCW) symbol (or a variant of the symbol), as shown in Figure 6-6, will be used on the main views of the composite panel drawing to indicate the relative fiber (or warp yarn in a fabric) orientation. The 0° axis is shown parallel to the primary load direction of the structure or global assembly.

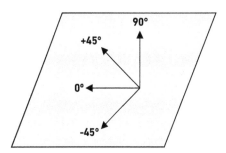

The clockwise (CW) symbol is shown from the engineering-design standpoint, whereas the fiber orientation is viewed from the outside of the structure, or from the tool surface, looking in. This perspective is often used in repairs.

In addition, the CCW symbol is always shown on the inside or bag-side surface of the part, as looking towards the tool surface, as it would be fabricated, precisely where it is applicable on the structure. The fiber orientation is controlled and inspected at this exact location in manufacturing. This is especially critical on contoured parts, as the fiber direction may change drastically as the fibers are draped over the contoured shape. The location of the symbol is less critical on flat panels.

Multiple symbols may be used on large or complex parts if necessary. Coordination of the applicable symbol/location with the ply layup table (or tables) is necessary when multiple symbols are used.

As a rule, the fiber orientation tolerance will be specified in the general drawing notes and coincide with the position of this symbol (e.g., ± 2° to ± 5° tolerance).

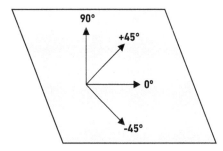

The counter clockwise (CCW) symbol is shown from the manufacturing point of view, whereas the fiber orientation is viewed from the inside of the panel looking toward the tool surface. This would be the case during layup.

FIGURE 6-6. Ply orientation convention symbols.

❱ PLY LAYUP TABLE

The **ply layup table** is used on the composite panel drawing to define the ply layup sequence, ply numbers, part numbers, orientation requirements, materials, splice control, and other pertinent information about the panel layup. An example is shown in Table 6.1.

TABLE 6.1 Ply Layup Table

Sequence	Ply Number	Orientation	Material	Splice Control	Revision
010	P1	+45	1 ▷	6 ▷	-
020	P2	-45	1 ▷	6 ▷	-
030	P3	0	1 ▷	6 ▷	-
040	P4	90	1 ▷	6 ▷	-
050	P5-P8	0	3 ▷	9 ▷	-
060	P9	90	1 ▷	6 ▷	-
070	P10	0	1 ▷	6 ▷	-
080	P11	-45	1 ▷	6 ▷	-
090	P12	+45	1 ▷	6 ▷	-

❱ COMMON LAYUP TERMS AND CONDITIONS

Symmetry

A *symmetric* laminate is one in which the ply axial orientations are symmetric about the middle (or mid-plane) of the laminate when viewed in cross-section. Another way of looking at this is to describe the ply orientations of the laminate as being a mirror image from the mid-plane.

Laminate symmetry is necessary to equalize thermal stress within the laminate and to maintain dimensional stability during elevated temperature processing or when exposed to changes in temperature while in service. Thermal stress within a symmetric laminate is equally offset when the same fiber type, material form, and axial orientation of each laminate layer (lamina) is paired at the same distance from the mid-plane, on each side of the mid-plane in the layup scheme. An asymmetric laminate can twist or warp from unequal thermal stress. (Figures 6-7, 6-8, 6-9)

Achieving a symmetric laminate is more challenging with harness-satin fabrics and stitched-biaxial or multiaxial forms versus plain or twill fabrics. A thorough understanding of how the fiber/fabric forms can affect laminate symmetry is necessary when designing laminate structures that will maintain shape when processed or exposed to changes in temperature in service. Figure 6-10 (page 162) shows how a single ply of 4-harness satin fabric is inherently asymmetrical compared with the inherently symmetrical plain weave fabric on the left.

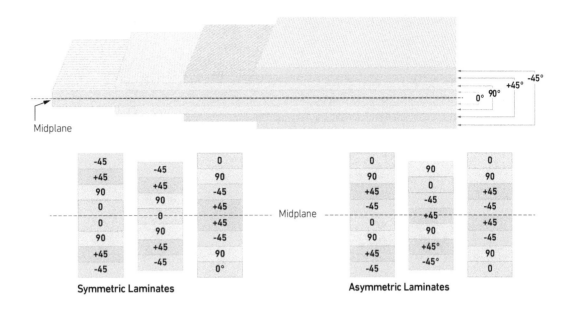

FIGURE 6-7. Laminate symmetry.

The laminate in the top drawing, and those on bottom left, are symmetric about the midplane, with the ply orientations above the midplane mirrored below the midplane. Those in the bottom right drawing are not symmetric (i.e., asymmetric).

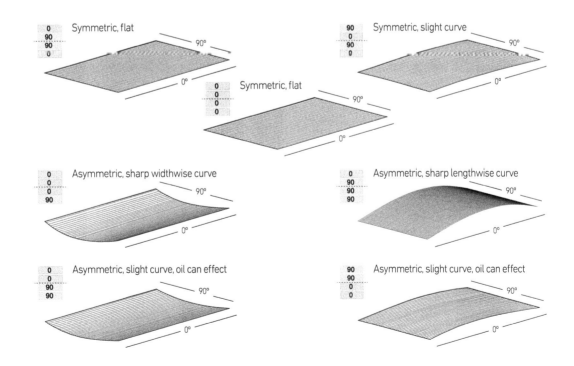

FIGURE 6-8. Effects of symmetry and asymmetry in 0° and 90° laminates.

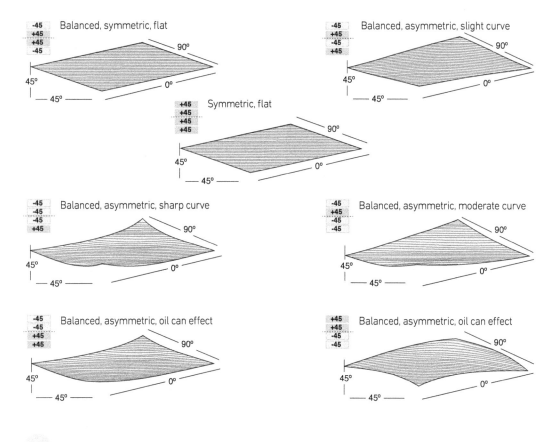

FIGURE 6-9. Effects of symmetry and asymmetry in +45° and -45° laminates.

FIGURE 6-10. Symmetry of plain vs. harness-satin fabrics.

Balance

A *balanced* laminate contains an equal number of plus and minus angled plies. This arrangement helps avoid twisting under applied loads. A laminate that is both balanced and symmetric will have an equal number of plus and minus angled plies on each side of the laminate, mid-plane.

Quasi-isotropic

A material that is isotropic has the same properties in all directions. **Quasi-isotropic** simply refers to having similar "in-plane" properties within a fiber-reinforced composite laminate. A quasi-isotropic composite laminate has either discontinuous or continuous fiber in a random orientation, or has continuous (long) fibers that are oriented in specific axial directions such that equal strength is developed all around the laminate when loaded in flexure, or in edgewise tension or compression.

Typically, a laminate is quasi-isotropic when constructed with an equal number of plies having fiber angles of 0°, 90°, +45°, and -45°. Quasi-isotropic properties can also be achieved with fewer fiber angles (e.g., 0°, +60° and -60°), or with additional fiber angles (e.g., 0°, +22.5°, +45°, +67.5°, 90°, -67.5°, -45° and -22.5°).

Note that the 0°/+60°/-60° example cannot be achieved with woven bidirectional fabrics, and in the other example more plies would be required to accomplish this configuration. Thus, the industry standard is to use 0°, 90°, +45° and -45° angles unless otherwise specified, which is compatible with unidirectional and bidirectional fabric forms (e.g., 0°/90° often called biaxial, and +45°/-45° called double bias). Modern forms such as the QISO™ fabric can produce 0°/+60°/-60° orientations with a single (flat-braided) ply. (Bidirectional and QISO™ fabrics are described in Chapter 3.) *See* Figure 6-11.

Nesting vs. Stacking

"Nesting" refers to the placement of harness-satin woven fabric plies so that the dominant (warp or fill) fiber direction at the surface of one ply aligns in the same direction as the interfacial surface of the adjacent ply. The fiber/yarn orientation at this interface is the same and thus the plies are "nested" together, providing intimate, on-axis load transfer between layers.

"Stacking" refers to the placement of (harness satin fabric) plies in the layup without regard for up/down positioning of the warp face. Plain weave and twill weave fabrics are always stacked, as they have equal warp and fill yarn exposure on both faces of the fabric and therefore nesting is not possible with those forms. *See* Figure 6-12 for examples of nesting versus stacking of harness satin weave fabrics. Note that stacked layers of the same harness satin fabric and axial orientation will produce an asymmetric laminate.

Quasi-isotopic, balanced, and symmetric laminates

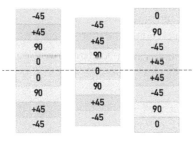
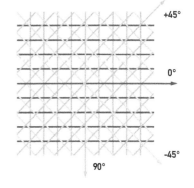

FIGURE 6-11. Quasi-isotropic laminates.

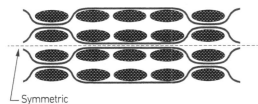
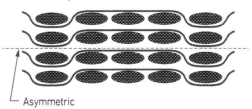

FIGURE 6-12. Nested vs. stacked harness-satin layers.

Rules for Nesting

- If the two plies in question will have the same warp fiber orientation, orient them warp face to warp face, or fill face to fill face to achieve a nested interface.
- If the two plies in question have warp fiber orientations 90° apart, orient them warp face to fill face and they will nest.
- Nesting does not apply to plain or twill weave fabrics, as they have equal fiber exposure on both faces.
- If the two plies in question are oriented at angles other than the same or 90° to each other, nesting is not possible; e.g., a +45 or -45 angled ply cannot nest with a 0° or 90° ply.

Considerations with Harness-Satin Weave Fabrics

There must always be an even number of harness-satin plies to achieve laminate symmetry. An uneven number would leave the laminate with a half-ply asymmetric condition at best. For example, style #1581 (or #7781) glass is an asymmetric harness satin fabric in which the yarn count is 57 warp yarns to 54 fill yarns per inch within the weave pattern. Thus, the warp direction of the fabric has a greater fiber count and is slightly stronger than the fill direction of the fabric. This fabric cannot be laid-up 0/90 or ±45 without recognition of the effect on symmetry and structural loading.

With symmetric (equal yarn size and count) harness satin fabrics, attention to the warp and fill face orientation is required to achieve laminate symmetry.

Considerations with Symmetric Fabrics

With a symmetric plain or twill weave fabric, the warp and fill faces are identical to each other and have the same number of identical yarns running in both directions. Therefore 0° or 90° plies are essentially the same thing and are called 0/90s. The same goes for +45 and -45 plies, which are also called ±45s. Balance can automatically be achieved, since each ±45° ply has both plus and minus 45° fiber axial directions, and therefore is inherently balanced.

- Symmetry in terms of ply orientation sequence from the mid-plane is still required to make a dimensionally stable laminate, although it can be achieved with fewer plies.
- A layup would still have an equal number of 0/90 and ±45° woven fabric plies to qualify as quasi-isotropic. The exception would be if a 0 or 90° ply is designated as the mid-plane ply.

Considerations with Asymmetric Fabrics

A good example of an asymmetric harness satin fabric is style #1581 (or #7781) glass, in which the yarn count for these fabrics is 57 warp yarns to 54 fill yarns per inch within the weave pattern. Thus, the warp direction of the fabric has a greater fiber count and is slightly stronger than the fill direction of the fabric. The fabric cannot be laid-up 0/90 or ±45 without recognition of the effect on symmetry and structural loading.

Considerations with Unidirectional Tape

Since each unidirectional tape layer has all fiber running in the same direction, careful consideration of layer placement and orientation is critical to the performance of a laminate. Care must be taken to insure minimal gaps without overlaps at edge-butt splices. Uni-tape will "nest" at each interface between **faying plies** (i.e., adjoining plies) oriented in the same direction. The rule of thumb is to limit the number of adjacent plies oriented on the same direction to five at a time, in order to minimize fiber splaying (i.e., spreading or slanting) in the resulting laminate.

❯ PLY ORIENTATION SHORTHAND CODE

Shorthand codes are used during engineering analysis or for other non-drawing uses. They are a way to condense a long descriptive representation of a multi-layer laminate into as few callouts as possible.

General Laminate Description

- Each laminate shorthand code is enclosed in a set of brackets.
- Each lamina is identified by its ply fiber angle callout. No degree symbol character is used within the shorthand brackets.
- Lamina are listed in sequence starting from the tool surface or the surface indicated by a leader arrow.
- Each lamina is separated by a forward slash (*see* Figure 6-13).
- Repeating groups of plies or special entities within a laminate can be placed in parentheses. For example, multiple laminae of the same angle are put in parentheses with the number of repetitions indicated by a subscript numeral after the closing parenthesis (as shown in Figure 6-15 on next page).
- Symmetric laminates with an even number of plies are represented by listing all plies on one side of the mid-plane enclosed in brackets, followed by a subscript S.
- Symmetric laminates with an odd number of plies are listed with a line over the centerline ply to indicate that it is the mid-plane ply, enclosed in brackets, followed by a subscript *S* (as shown at right in Figure 6-13).

Unidirectional Laminae Designation

- +45 indicates that the fiber is oriented in the +45 direction.
- -45 indicates that the fiber is oriented in the -45 direction.
- ±45 indicates two unidirectional plies in sequence starting with a +45 followed by a -45.
- ∓45 indicates two unidirectional plies in sequence starting with a -45 followed by a +45.

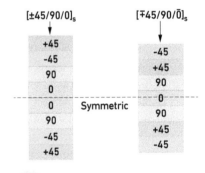

FIGURE 6-13.
Ply shorthand—unidirectional laminate.

Fabric Laminae

- Fabric plies are identified with a capital *F* following the ply angle callout. (Figure 6-14)
- The angle value represents the direction of the warp fibers.
- ±45F indicates a woven fabric is oriented at either a +45 or -45 direction.
- +45F indicates that the warp fiber is oriented in the +45 direction.
- -45F indicates that the warp fiber is oriented in the -45 direction.
- (0 or 90F) indicates a fabric ply oriented at either 0 or 90 degrees.

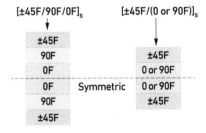

FIGURE 6-14.
Ply shorthand—bidirectional fabric.

Hybrid Laminates

- Individual plies of different materials located in a laminate are identified by a subscript fiber type and/or fabric code following the ply angle in order to identify the type of material.

- When numerous plies of the same material type are grouped together, the material subscript is shown on only the first and last ply in the group. Everything in between is assumed to be the same material.

The real power of shorthand code can be realized when identifying cross-section areas of very thick laminates with a great number of plies. For example: $[\pm45F_2/(+45/-45/90/0)_3/(90/0)_2]_{2S}$ —which describes a 72-ply laminate made with 4 plies of fabric and 68 plies of unidirectional tape.

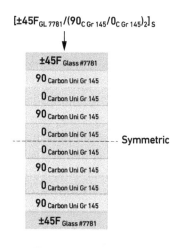

$[\pm45F_{GL\ 7781}/(90_{C\ Gr\ 145}/0_{C\ Gr\ 145})_2]_S$

±45F Glass #7781	
90 Carbon Uni Gr 145	
0 Carbon Uni Gr 145	
90 Carbon Uni Gr 145	
0 Carbon Uni Gr 145	Symmetric
0 Carbon Uni Gr 145	
90 Carbon Uni Gr 145	
0 Carbon Uni Gr 145	
90 Carbon Uni Gr 145	
±45F Glass #7781	

FIGURE 6-15.
Ply shorthand—hybrid laminate.

Fiber to Resin Ratio

For advanced composite laminates, there is a narrow range of functional fiber to resin ratios which will produce the best strength-to-weight ratio in the end product. With too little resin, compressive and shear properties drop off. With too much resin (assuming all the fiber is still in the composite), the laminate's weight and stiffness go up without any significant increase in strength, which lowers the strength-to-weight ratio.

With woven fabric forms in the laminate, the target fiber to resin percentage is approximately 60% fibers to 40% resin, by volume. In the shop environment, a "by weight" percentage is usually easier to calculate. However, it should be noted that a 60/40 weight percentage does not necessarily equal a 60/40 volume fraction for all fiber/resin combinations.

There will be many variations of this fiber/resin relationship, depending on the type of fiber, form, and the type and amount of resin used. See the example in Table 6.2 below. Unidirectional tape forms typically require a lower resin content, as there is not much area between the fibers that need to be filled with resin like there is with a woven form. Resin ratios in the low to mid 30% range, by volume, yield the best results (e.g., 31% resin/69% fiber).

TABLE 6.2 Calculating Fiber to Resin Ratio

By weight	By volume	
Wet-layup example. 60% fibers to 40% resin by weight.	This side illustrates just how different the actual fiber-volume could be if the weight ratio is used to make the laminate.	
Calculating Resin Weight fabric weight ÷ .6 (60%) = total weight total weight - fabric weight = resin weight	Fiber Weight Composite Weight Resin Weight	= 90 g = 150 g = 60 g
Example: 90 g fabric ÷ .6 = 150 g total weight 150 - 90 = 60 g resin weight	Fiber Density Resin Density Fiber Volume	= 2.65 g/cc = 1.24 g/cc = 33.96 cc

TABLE 6.2 Calculating Fiber to Resin Ratio *(continued)*

By weight	By volume
Calculating Actual Ratio actual total weight - fabric weight = resin weight resin weight ÷ total weight = resin %	Fiber/Resin Weight Ratio = 1.5 Fiber/Resin Volume Ratio = 0.7
Example: 150 g actual weight - 90 g fabric = 60 g resin 60 ÷ 150 = .40 (or 40%) resin	Void Volume = 1.0% Fiber Volume = 40.2% Resin Volume = 58.8%
Therefore, panel's fiber to resin by weight ratio is 60:40, if all 60 grams of resin is retained in the laminate.	Therefore, excluding void volume, the panel's fiber to resin volume ratio is about 40:60, just the opposite of the "by weight" ratio shown in the previous example.

Service Life Considerations

❱ TEMPERATURE AND MOISTURE

Hot-Wet

The wet T_g of a composite matrix resin dictates the maximum service temperature properties of the combined matrix/fiber structure and can be examined via thermal analysis and destructive coupon testing.

Cold and Cryogenic

All materials are inherently more brittle at low temperatures. Composite structures are unique in that they are made up of (fiber and matrix) constituent materials that have very different properties from one another. This difference in both mechanical and thermal properties, primarily in a polymer matrix, present challenges for designers in selecting the right resin to pair with a given fiber for cold or cryogenic temperature service.

The ability of a matrix resin to withstand very cold (-65°F/-54°C) or cryogenic (-423°F/-253°C) temperatures without microcracking is tied to the resin chemistry itself and the degree of cure. Most resins such as epoxy and polyurethane perform well at very cold temperatures but may not be as durable at cryogenic temperatures. High-performance chemistries are typically selected for this range. For example, cyanate ester (CE) resins, having a low coefficient of thermal expansion (CTE), tend to be more resistant to microcracking at lower temperatures than epoxy resins which have a higher CTE.

There is evidence that increasing fracture toughness with polymer modifiers or nano-particles may increase resistance to microcracking and thus improve laminate properties at very cold or cryogenic temperatures. Much research in this area promises to provide more solutions for future designs.

❱ ENVIRONMENTAL EFFECTS ON CURED COMPOSITE STRUCTURES

Chemical and Corrosion Resistance[1]

Composites are used in corrosion-resistant applications in industries including chemical processing, desalination, food and beverage, mining, oil and gas and wastewater treatment. There are specific examples of glass fiber-reinforced polymer pipes and ducting operating nonstop in harsh chemical environments for more than 25 years. A composite's corrosion resistance is determined by the resin and reinforcement used, as well as the part design.

Resistance of Resins to Corrosion/Chemicals

There are a number of resin systems that provide long-term resistance to a wide variety of chemicals. However, not all are resistant to the whole spectrum of chemical attack. For example, chlorendic acid polyester resin offers particular resistance to wet chlorine gas and oxidizing acids such as nitric acid, sulfuric acid, and chromic acid—but it is not very effective in caustic environments. Bisphenol A epoxy vinyl ester resins were developed to overcome cracking issues with earlier corrosion-resistant unsaturated polyesters. Providing higher-toughness, these vinyl esters remain the primary choice for corrosion-resistant resins today.

Isophthalic and terephthalic unsaturated polyester resins have also been developed for corrosion-resistant applications (*see* Table 6-3) with *flexibilized* and *high-crosslink density* versions available as well.

TABLE 6.3 Corrosion-resistant Resins

AOC Vipel®	Ashland	DSM	Interplastic CoREZYN®	Polynt-Reichhold
Bisphenol A Epoxy Vinyl Ester				
F013	Hetron® 9221	Atlac® 430	VE8300	Dion 9100
F010	Hetron® 942/35	Atlac® 5200	VE8100	Dion 9102
F007	Derakane® 411		VE8360	Epovia RF 1001DMV
	Derakane® 441-400			
High-Crosslink Density Vinyl Ester				
F080	Hetron® 980		VE8710	
F083			VE8470	
			VE8770	
Epoxy Novolac Vinyl Esters				
F085	Hetron® 970	Atlac® 590	VE8730	Dion 9400
F086	Derakane® 470		VE8440	Dion 9480
High-Elongation Vinyl Esters				
F017	Derakane® 8084		VE8550	Dion 9085
			VE8510	Dion 9500
			VE8515	Epovia KF3202L-00
Terephthalic polyester				
F774		Atlac® 490	77-AQ-201	Dion 490
			75-AA-450	Dion 495
Chlorendic Anhydride Polyesters				
K190-B	Hetron® 72			Dion 797
	Hetron® 92			
	Hetron® 197			

---

1 Text in this section adapted from "Corrosion Resistance" section of the American Composites Manufacturers Association (ACMA) website, http://compositeslab.com/benefits-of-composites/corrosion-resistance/. Also from "Different FRP Resin Chemistries for Different Chemical Environments," Don Kelley, Jim Graham and Thom Johnson, Ashland Performance Materials.

TABLE 6.3 Corrosion-resistant Resins *(continued)*

AOC Vipel®	Ashland	DSM	Interplastic CoREZYN®	Polynt-Reichhold
Isophthalic Polyester				
F701	Aropol™ 7241 Aropol™ 7242	Synolite® 1717 Synolite® 266 Palatal® A410	75-AQ-001 75-AQ-010 75-AQ-011	Dion 6631
Flexibilized Isophthalic				
F737 F738	Aropol™ 7334 Aropol™ 7530 Aropol™ 7532		75AQ610	Dion 6246
Urethane-modified Vinyl Ester				
				Dion 9800
Bisphenol A Fumarate Polyesters				
F282	Hetron® 700			31734-01 (Dion 6694)
Flexibilized Bisphenol A Fumarate Polyesters				
	Hetron® 800			

Epoxy resin is the most common composite matrix used in the aerospace industry, and there are a variety of epoxy formulations that, in general, offer excellent resistance to most chemicals. There are some exceptions, as noted in Table 6.4. Thus, it is important to check each formulation's chemical resistance with the resin supplier.

Thermoplastic polymers are gaining use in aerospace composites, having long been used in non-aerospace applications for their chemical resistance and other properties. Polyetheretherketone (PEEK), polyetherketoneketone (PEKK) and polyphenylene sulfide (PPS) are semi-crystalline thermoplastic polymers used in aerospace applications that offer excellent resistance to chemicals, including strong acids and Skydrol®. Polyetherimide (PEI) is an *amorphous thermoplastic* used in aerospace applications that offers very good chemical resistance, but not quite as high as PEEK, PEKK and PPS.

TABLE 6.4 Chemical Resistance of Epoxy Resins

Type of Chemical	Cured Epoxy Resistance
Gasoline, jet fuel, kerosene, diesel fuel, liquid propane	Excellent
Other Considerations: Can penetrate porous skins in sandwich construction and saturate core materials	
Military hydraulic fluids (e.g., 5606 and 83282)	Excellent
Skydrol® brand hydraulic fluid	Not recommended
Other Considerations: Some types of Skydrol® can weaken some types of epoxy	
Deicing/anti-icing fluids	Excellent
Other Considerations: Fair resistance to ethylene glycol	
Weak acids (e.g., boric acid, 20% acetic acid, maleic acid)	Excellent
Strong acids (e.g., fluosilicic acid, formic acid)	Good to Fair
Other Considerations: Not recommended: nitric acid	

(continued)

TABLE 6.4 Chemical Resistance of Epoxy Resins *(continued)*

Type of Chemical	Cured Epoxy Resistance
Alkalis (e.g., ammonia, borax, soaps/detergents)	Excellent
Oxidizing agents (e.g., hydrogen peroxide)	Fair
Solvents	Some Excellent (alcohol, heptane, naphtha) Some Good (hexane) Some Fair (ethyl acetate, methyl ethyl ketone) Not recommended: methylene chloride

Other Considerations: Methylene chloride paint strippers cannot be used on composites. For alternatives, see Ch. 13.

Source: "Epoxy – Chemical Resistance" published by Engineering Toolbox, https://www.engineeringtoolbox.com/chemical-resistance-epoxy-d_786.html

Resistance of Fibers to Corrosion/Chemicals

In general, fibers are less resistant to chemical attack than resins, so it is important that every fiber in a composite is encapsulated and impregnated by resin. For composites that must resist long exposure to chemicals, it is standard practice to use a resin-rich surface (70–90% resin by weight) often achieved by using a nonwoven veil as fiber reinforcement in the surface plies. Thermoplastic veils, films and liners may also be used, such as polyvinylidene fluoride (PVDF) liners common in oil and gas tubulars.

E-glass fibers may be weakened by acids, alkalis and other chemicals. For this reason, C-glass and E-CR glass fibers have been developed to provide increased chemical and corrosion resistance and AR glass fiber has been formulated for alkali resistance, enabling use in concrete structures.

Resistance of Core Materials to Corrosion/Chemicals

Chemicals that penetrate sandwich skins and reach core materials may have a damaging effect on uncoated paper and metal foil honeycombs. Note that some foam cores may be dissolved by certain chemicals including ordinary detergents. This effect varies substantially depending on the type of foam. Each core material's chemical resistance should be checked with the material supplier.

Resistance to Radiation

Ultraviolet (UV)

Exposure to ultraviolet (UV) radiation can cause discoloration and (micro-surface) degradation in a composite matrix, which can affect the quality of that composite's interface for secondary adhesive bonding or painting. This is due to UV attack on the carbon bonds in the exposed polymer chains, creating free radicals that further react with oxygen, causing surface pitting, etc. Moisture (humidity) in the environment tends to accelerate this process. Therefore, limiting both UV intensity and time of exposure is warranted in manufacturing. Composites subject to long-term UV exposure are also often protected with UV-resistant coatings.

Infrared (IR)

Infrared radiation also transfers heat and can cause a problem in composites due to the high temperatures generated. One example of this is the direct infrared radiation received by the underside of wings and tail surfaces on aircraft parked on a black asphalt ramp on a hot summer day. Temperatures of 225°F/107°C have been measured in such conditions, while air temperatures were approximately 120°F/49°C. If the T_g of the resin is not sufficient, the composite structures may be damaged or weakened. However, painting the structures white on all lower surfaces will reflect away most of the infrared radiation.

GALVANIC CORROSION

Galvanic corrosion occurs when two materials from opposite locations on the **galvanic scale** come in direct contact with each other (*see* Table 6.5). It is aggravated by the presence of moisture. Dissimilar metal corrosion problems are well known. However, carbon fiber composites also exhibit this galvanic effect with many metals. Though the carbon fiber composite itself is unaffected, corrosion in the metal can be significant and occur fairly rapidly.

To avoid corrosion, carbon fiber composites must be isolated from metal components with a suitable dielectric material. This can be accomplished with a fiberglass layer or adhesive with or without a carrier at the interface. The protection of aluminum hard points, inserts, and other interfaces with carbon fiber automotive structures is typically accomplished by anodizing their surfaces combined with a corrosion resistant (CR) primer and a thick non-conductive polymer or elastomeric barrier between the components.

Mechanical fasteners must also be selected with this problem in mind. Severe fastener corrosion is not uncommon in carbon fiber composites. Cadmium-plated fasteners, for example, can corrode rather quickly in such structures. There are multiple solutions, however, including:

- Bonded fasteners with an isolation layer in addition to adhesive;
- Corrosion-inhibiting coating and "wet" installation of fasteners;
- Fasteners made from low galvanic potential metals such as titanium and 300 series 18-8 stainless steel.

TABLE 6.5 Corrosion Potential of Various Metals with Carbon Fibers

Material	Galvanic Scale	
Magnesium	12	ANODIC
Zinc	11	
Aluminum 7075 Clad	10	
Aluminum 2024 Clad	9	
Aluminum 7075-T6	9	
Cadmium	8	
Aluminum 2024-T4	7	
Wrought Steel	6	
Cast Steel	6	
Lead	4	
Tin	4	
Manganese Bronze	3	
Brass	2	
Aluminum Bronze	2	
Copper	2	
Nickel	1	
Inconel	1	
300 Series 18-8 CRES	0	
Titanium	0	
Monel	0	
Silver	0	
Carbon Fiber	0	CATHODIC

VIBRATION AND NOISE

Vibration Damping

While viscoelastic elastomers have long been used to isolate structures from vibration and noise, they do so at the sacrifice of added weight. Recent advancements in composite damping systems do not necessarily isolate the structure from the vibration source like conventional elastomeric dampers, but instead diminish the magnitude of vibration within the laminate structure itself.

The idea is to combine low modulus (e.g., thermoplastic or natural fibers) and high modulus (e.g., carbon fibers) either in comingled tow or yarn forms, or in a hybrid laminate construction.

Lamina of differing fiber types may also be positioned within the laminate to help dampen vibration and mitigate resonant frequencies within the structure. (Figure 6-16)

Natural and Resonant Frequencies

Composite materials and the structures made from them can be tuned to have certain natural and resonant frequencies, and also to avoid certain natural frequencies so that they don't resonate while in service. This may be achieved in composites by altering their laminate stacking sequence and fiber orientations, the materials and forms of reinforcement used and the molded geometry of the structure.

Avoiding resonance is important for many applications, for example: drive shafts that rotate, space structures that must withstand launch loads, radar masts on ships and automotive suspension parts. Vibrational analyses are commonly used to integrate passive vibration damping into a composite structure's design. (Figure 6-17)

❱ DAMAGE TOLERANCE AND TOUGHNESS

Damage tolerance is the ability of a laminate or structure to maintain its original stiffness and strength after impact. Damage tolerance can be increased by enhancing the interlaminar fracture toughness of a laminate, which is the resistance to in-plane brittle failure (cracking) of the matrix, or disbonding at the fiber-to-matrix interface.

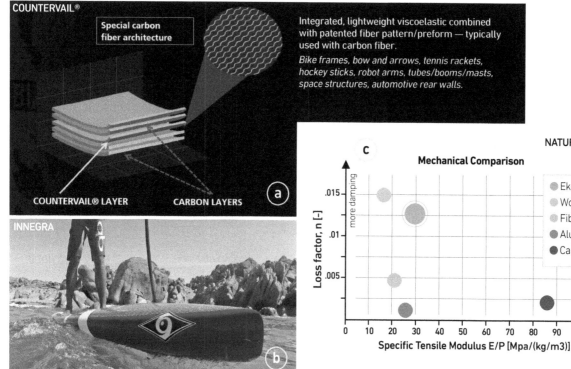

Fiber made from high modulus polypropylene (HMPP) which is viscoelastic; typically used in a hybrid construction with glass, carbon and/or aramid fiber.

Bike frames, snow/surf/paddle boards, bike seats, canoes/kayaks, oars/paddles, hockey sticks.

Made from the natural polymers cellulose, lignin and pectin, natural fibers are viscoelastic and can be used in hybrids; flax fiber laminates are most common.

Bike frames, snow/surf/paddle/skate boards, paddles, speakers, interiors, automotive interiors, automotive door panels.

FIGURE 6-16. Examples of vibration damping in composites. *(Photo and diagrams courtesy of: [a] Countervail®, [b] Innegra Technologies, [c] Lingrove)*

FIGURE 6-17. CFRP-damping composites for higher performance machine tools.
(Photos courtesy of CompoTech)
Vibration from accelerating parts, milling teeth and the harmonics of a cutting tool are the limiting factor in their performance. CompoTech integrates carbon fiber-reinforced composites into a wide range of machine tools, cutting tools, and tool holders for -75% mass and 12X vibration damping compared with steel, resulting in higher precision and performance in machining.

Matrix Toughening

One way to reduce brittle failure is to toughen the matrix. Modifying a thermoset matrix resin with elastomeric or thermoplastic microphase particles, nano-scale particles, or with graphene nano-platelets can greatly reduce in-plane brittle failure. Using an inherently tough thermoplastic matrix may also be an option. Other methods are shown below in Table 6.6. Many of the same ideas employed in designing for vibration apply to designing for damage tolerance.

TABLE 6.6 Methods for Increasing Damage Tolerance in Composites

Matrix toughening	• Thermoplastic matrix • Nanoparticles (e.g. nanosilica) • Carbon nanotubes and nanofibers	• Thermoplastic/elastomeric additives • Nanoplatelets of graphene
Toughened fibers	• Commingled reinforcements with higher elongation fibers (e.g. thermoplastic) • Hybrid reinforcements with higher elongation fibers (e.g. para-aramid, Innegra, etc.)	
Interleafing between plies	• Lower modulus layers • Rubber strips	• Film adhesives • Nanomaterial sheets
Strengthening fiber-matrix interface	• Modifying fiber surface chemistry • Fiber size/coating compatible with matrix • Nanomaterial applied to fiber surface • Nanofibers grown radially from fiber surface	
Thin plies /spread tow	• Thin plies of spread tow reinforcements • Combination of thin plies and conventional thick plies	

Interleaving

These methods may also act as crack-stoppers in the laminate. Another way to achieve this is to interleave nanomaterial sheets (*see* the NanoStitch discussion in "Vertically Aligned CNTs," Chapter 4) to increase interlaminar shear strength, which not only prevents delamination between plies but also cracks from propagating through plies.

Thin Plies/Spread Tow

As discussed previously in Chapter 3 "Thin Ply Bi-Angle Fabrics" (page 106), thin-ply and spread-tow reinforcements can significantly increase the stress a laminate can take before first-ply and last-ply failure. Research has shown that both 100% thin-ply laminates as well as those combined with conventional thick plies show delay or suppression of transverse cracking and delamination as well as increased fatigue strength, even with holes in the laminate.[2]

Fiber-to-Matrix Bond

Strengthening the fiber-to-matrix bond can be achieved by adjusting the surface chemistry on the fiber to best enhance wetting and adhesion to a given resin. For example, fiber size and/or nanomaterial coatings have been developed specifically for carbon fiber in thermoplastic matrix materials. Current research also suggests that adding multi-walled carbon nanotubes (MWCNTs) to the matrix can achieve nano-mechanical links to carbon fiber reinforcements, improving fracture toughness. Carbon nanotubes may also be grown radially from the fiber surface, known as "fuzzy fibers," which act both as links between fiber and matrix and as crack-stoppers in the matrix.

Void Content

Regardless of matrix formulation or surface treatment of the fiber, fracture toughness decreases with increased void volume fraction. For this reason, minimizing **void content** during processing is paramount to achieving toughness.

❱ ELECTRICAL CONDUCTIVITY

Composites are often designed to either conduct electricity or act as an insulator. As always, the choice of fiber reinforcement and matrix plays an important role. Glass, ceramic, aramid, polymer and natural fibers are not good electrical conductors. In fact, glass fiber composites are often used as electrical insulators.

While carbon fiber is a much better electrical conductor than most other composite rein-forcements, it is still not as effective as metals. Hence, the need for specific measures to deal with lightning strikes in composite structures, as discussed below (pages 181–186).

Although plastics and polymers are not inherently conductive, they can be made so by adding conductive fillers (e.g., carbon black, carbon nanotubes, metal flakes and nanoparticles) and conductive fibers, including carbon fiber and metal-coated fibers. Conductive adhesives, for example, have been used for decades. In general, the more filler added, the more conductive the composite, but fillers may also affect mechanical properties. Also, continuous fiber-rein-forced composites are generally more conductive than filled plastics.

The electrical conductivity of composites can be tailored to provide parts that dissipate static electricity, providing electrostatic discharge (ESD) shielding, or to be more conductive for electromagnetic interference (EMI) shielding. Composites may be even more conductive to transfer electrical charge out of structures quickly and efficiently, or just enough to provide resistive heating for deicing and self-heated molds.

Note that carbon nanotubes (CNTs) and graphene offer the promise of very high electrical conductivity—orders of magnitude higher than copper. However, for CNTs, this depends upon their structure. The properties they show in use range from semiconductors to very good conductors. CNTs are being developed in antistatic coatings for composites, as well as elec-trodes, flexible and printable circuits, ESD and EMI shielding, and electrical cables for aircraft and satellites. (Figures 6-18, 6-19)

---

2 "Strength and fracture of thin ply laminate composites: experiments and analysis," R. Amacher, J. Botsis, J. Cugnoni, G. Frossard, Th. Gmür, and S. Kohler, École Polytechnique Fédérale de Lausanne (EFPL), Comptest 2017, Leuven, Belgium; https://www.mtm.kuleuven. be/Onderzoek/Composites/comptest2017/Day1_keynote_Botsis. Also, "Toward aerospace grade thin-ply composites," C. Dransfeld, R. Amacher, J. Cugnoni and J. Botsis, June 2016, presented *17th European Conference on Composite Materials* (ECCM17, Munich, Germany); www.researchgate.net/publication/304659240_TOWARD_AEROSPACE_GRADE_THIN-PLY_COMPOSITES

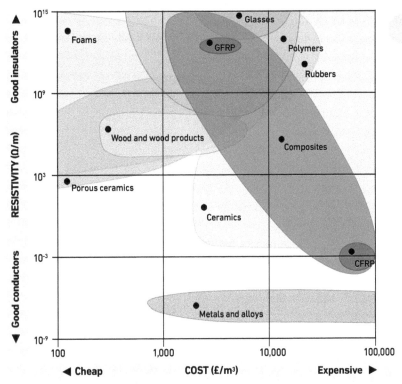

FIGURE 6-18. Electrical resistivity of composites vs. other materials, and surface resistivity spectrum.[3]

3 Source: University of Cambridge, Dept. of Mechanics, Materials and Design, http://www-materials.eng.cam.ac.uk/mpsite/interactive_charts/resistivity-cost/basic.html

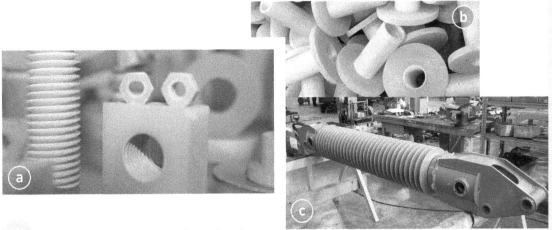

FIGURE 6-19. Glass fiber composites as electrical insulators.

[a] Dielettrika Ligure uses glass fiber-reinforced composites with epoxy, silicone, polyester or phenolic matrix resins as insulative components for applications that can withstand temperatures of 220°C and higher. [b] The sleeves were used to insulate structures in Qatar's Doha metro train system. [c] This segment for a France Elevateur lift bucket is a glass fiber composite tube filament wound by CompoTech (beneath a corrugated rubber cover with steel end fittings) that prevents conductivity in wet weather permitting crews to safely maintain electrical lines.
(Photos courtesy of Dielettrika Ligure, CompoTech)

❯ THERMAL CONDUCTIVITY

Thermal conductivity in composites is quite similar to electrical conductivity. Most fibers and resins are not good thermal conductors. Carbon fiber, again, is the exception, but its thermal conductivity is not as good as metals. The thermal conductivity of polymers has traditionally been enhanced by the addition of thermally conductive fillers including graphite, carbon black, carbon fibers, ceramic or metal particles. In general, higher concentrations of fillers produce higher thermal conductivity. However, these fillers may also negatively affect mechanical properties.

Thermal conductivity of composites is highest along the fiber. This is also true for carbon nanotubes. For this reason, in-plane heat transfer along the fibers in a laminate is more efficient than out-of-plane. For applications that require heat dissipation, this is a challenge. One approach to overcoming this has been the use of 3D woven textiles that orient significant amounts of fiber in the out-of-plane direction.

Composites as Thermal Breaks

One advantage to composites as poor thermal conductors is their ability to act as thermal breaks in building structures. Window profiles, support beams and connectors in buildings made from metal conduct heat and cold in and out of buildings, reducing energy efficiency and increasing operational costs. The can also act as sources for condensation, resulting in mold growth. Composites have gained use as a means to prevent these thermal breaks in a building's envelope and structures.

Coefficient of Thermal Expansion

A related issue is the expansion in resins, fibers and composites as a result of thermal loads. It is important to understand this behavior, especially in applications with large temperature differences and where composites are in contact or joined with metals or other materials which may have very different thermal expansions. Differences in the coefficients of thermal expansion (CTE) for metals and composites may be used to help demold a composite part from a metal tool or could produce serious dimensional tolerance issues. They can also create stress in structures where the differing materials are joined and lead to premature failure. Note that both carbon fiber and aramid fiber have a negative linear CTE (*see* Table 6.8).

TABLE 6.7 Thermal Conductivity of Composite Reinforcements and Resins vs. Other Materials

Polymer	Thermal Conductivity (W/mK)
Polyethylene	0.28–0.50
High density polyethylene	0.38–0.58
Epoxy resin	0.17–0.29
Polypropylene	0.12–0.24
Phenol resin	0.24–.029
Carbon fiber – PAN	1–100
Alumina (Al_2O_3 ceramic)	39
Steel	52
Aluminum	247
Gold	315
Copper	400
Silver	420
Carbon fiber – Pitch	1,000
Diamond	2,450
Graphene nanoplatelets	3,000
Multi-walled carbon nanotubes	3,000

SOURCE: "Thermal conductivity of polymeric composites: A review," I.A. Tsekmes, R. Kochetov, P.H.F. Morshuis, and J.J Smit, Delft University of Technology, 2013 *IEEE International Conference on Solid Dielectrics*, Bologna, Italy.

TABLE 6.8 Coefficient of thermal expansion for composite reinforcement fibers

	Linear CTE (10^{-6}/K)
PAN carbon fiber	-0.5–0.1
KEVLAR 49	-5– -2
E-glass	4.7
Silicon carbide	5.7
Aluminum oxide	7.5
Boron	8.3
Steel	12
Aluminum	24

❭ FIRE RESISTANCE

Composites are used in many fire-resistant and high-temperature applications in aircraft, ships, oil and gas platforms, rail cars and the interiors and exteriors of buildings. However, composites must be specially designed for these applications. Table 6.9 shows the general categories for measuring fire resistance in composites (and other materials). These also form the basis for the fire tests used by various industries, including:

- UL 94
- ASTM E 84 and NFPA 285, 286 for buildings
- EN 45545 for rail
- FAA 14 CFR 25.853 OSU test for aircraft
- IMO Fire Test Procedures (FTP) Code 2010 for ships

For more details on fire testing, *see* Chapter 10.

TABLE 6.9 Fire Resistance

Ignitability	How readily a material ignites or resists ignition
Self-extinguishing	How rapidly or easily a material can extinguish flame
Flame Spread	How rapidly a fire spreads across a surface
Burn Through	How rapidly a fire burns through a material – how long it can withstand direct flame
Heat Release	How much heat is released and how quickly
Smoke Generation	Amount and rate of smoke generation
Smoke Toxicity	Toxicity of smoke produced

Inorganic vs. Organic

Most polymer matrices and organic fibers decompose at high temperatures. Note that *inorganic* matrices and fibers (e.g., ceramic, carbon, basalt and glass) do not burn and many can withstand very high temperatures. However, 30–100% of *organic* fibers and polymers will be volatized (disbursed into vapor) by very high temperatures, if left uninsulated or untreated, and may also release flammable gases and/or toxic smoke. Important exceptions for organic polymer fibers are KEVLAR para-aramid and NOMEX meta-aramid fibers, which are both used in fire protection products.

The best-performing organic resins for fire resistant composites have historically been phenolics (phenolformaldehyde) and furan/furfuryl alcohol-based systems. Even though phenolic and furan resins will burn, they burn slowly and have excellent flame retardance, good heat resistance and low smoke/toxic gas emission. Their resin chemistry also produced char, a carbon residue that remains after burning, which acts as thermal insulation and a barrier to flame propagation. The major drawback of phenolic and furan resins is that, historically, they have been brittle and difficult to process.

Inorganic polymers that resist fire often contain silicon-nitrogen, boron-nitrogen, and phosphorus-nitrogen monomers. Many of the polymers shown in Table 6.10 tend to be expensive and/or difficult to process. Some are also brittle or sensitive to notching and impact damage.

Thus, fire-resistant composites typically use more common resins and fibers and apply two main approaches:
- Improve the fire resistance of fibers and/or polymers by chemistry, additives, etc.
- Apply coatings to the composite for insulation from fire and heat.

TABLE 6.10 Fire-Resistant Inorganic and Organic Polymers*

Polymer	Description	Thermal Stability
INORGANIC		
Polysialate	Also called geopolymers, contain silicon, aluminum and oxygen in the backbone (e.g., aluminosilicates)	1300-1400°C/2372-2552°F
Polysiloxane	Polydimethylsiloxane (PDMS, a.k.a., silicone rubber) is the most common	300-350°C/572-662°F
Polysilane	Contains silicon only in the backbone – used to modify composites and to make fiber	300°C/572°F
ORGANIC		
Polybenzoxazole (PBO)	Used mostly as fiber, film and coating	650°C/1202°F
Polybenzthiazole (PBZT)	Used mostly as film	650°C/1202°F
Polybenzimidazole (PBI)	Class of heterocyclic thermoplastic	500°C/1300°F
Polyimide (PI)	Can be thermoset or thermoplastic	Continuously to 260°C/500°F Intermittently to 480°C/900°F
Melamine	Nitrogen-rich crystalline compound used in Formica	400°C/752F°
Polyarylate	Type of aromatic polyester	400°C/752F°
Polyamideimide (PAI)	Can be thermoset or thermoplastic	350°C/662F°

*These properties have been compiled from multiple sources, using various materials forms and tests, and thus are meant for general reference only and should be verified using actual material supplier data.

Flame Retardant (FR) Treatment of Fibers[4]

Flammable polymer and natural fibers may be treated with inexpensive, non-durable flame retardants. More costly durable flame retardants are not needed since fibers will remain encased in a matrix. Non-durable FR materials include inorganic salts such as borax/boric acid mixtures and ammonium salts of strong acids. These have been tested on cellulose natural fibers with an addition of 10–15% solids showing good results. (Figure 6-20)

FIGURE 6-20. Fire-resistant carbon fiber composites.
FR 10 materials comprise chopped carbon fiber and inorganic resin formed into 2mm-thick panels or net-shape preforms that are lightweight (<3 kg/m²) and tested to show 7-hour resistance to 1,500°C flame. *(Photo courtesy of CFP Composites)*

4 "Flame Retardant Polymer Composites," Mahadev Bar, R. Alagirusamy, and Apurba Das, Indian Institute of Technology Delhi; *Fibers and Polymers* 2015, Vol.16, No.4, 705-717. www.researchgate.net/publication/279173778

FR Additives to Resins

There are three different ways of adding flame retardance to resins:

- Incorporate an FR compound into the polymeric backbone.
- Mix FR compounds, particulates and/or nanomaterials into the resin.
- Add an intumescent system to the matrix.

These may be used independently or as a synergistic combination—i.e., one FR lowers heat release while the next reduces smoke and the third produces char, etc.

Polymer Backbone Synthesis

The inorganic polymers listed in Table 6.10 can be blended into a resin matrix or synthesized into the backbone of organic polymers, as can inorganic monomers. This approach has been used successfully in development work with polypropylene, polyethylene, epoxy, polyvinyl, polyester, polyamide and polyurethane resins.

Another approach has been to blend and cocure unsaturated polyester and vinyl ester resins with an appropriate char-forming resin such as phenolic, melamine-formaldehyde or furan. Initial evaluations show promise for reduced heat release and fire propagation rates.[5]

Additives Mixed into Resin

The most widely used FR additives to resins, however, are compounds based on halogens, phosphorus, nitrogen and silicon. Nanoparticles as FR additives are also being developed. The most effective FR have been those based on the halogens *chlorine* and *bromine*. However, new environmental directives such as **REACH** restrict their use because they are toxic to organisms and the environment. This has led to the development of non-halogenated FR compounds, including:

- Calcium carbonate
- Alumina trihydrate (ATH)
- Other metal hydroxides, such as magnesium and phosphorus-based additives
- Melamine

While these mineral fillers do not release toxic gases, the high concentrations required to significantly improve fire retardance can significantly degrade a composite's mechanical properties. They can also make it difficult to process using resin infusion and resin transfer molding (RTM), common in the marine industry and also with polyester and vinylester resins.

Nanoclays have also been shown to provide high FR performance at a low cost. Nanoclays promote formation of char. Also, because of their very small particulate size and ability to disperse at a sub-micron scale, lower amounts of nanomaterials are needed compared to macro-scale additives. Amounts as low as 5–10% by weight, when uniformly dispersed in a resin system, can reduce the peak heat release by 70%.[6] Initial work on graphene nanoplatelets (GNPs) and CNTs have also shown positive results.

It is important to understand how each FR additive works, how to use it effectively in a given matrix resin and composite, and how to combine it with other FR methods. Notable as well is the fact that with both macro-additives and nanoparticles, homogeneous dispersion in the polymer matrix is a key factor in obtaining sufficient FR improvement.[7] (Figure 6-21)

5 "Fire Performance Evaluation of Different Resins for Potential Application in Fire Resistant Structural Marine Composites," Baljinder Kandola and Latha Krishnan, Institute for Materials Research and Innovation, University of Bolton, UK; *Fire Safety Science-Proceedings Of The Eleventh International Symposium* pp. 769-780; 2014 International Association for Fire Safety Science, DOI: 10.3801/IAFSS.FSS.11-769. Also "Development of vinyl ester resins with improved flame retardant properties for structural marine applications," Baljinder Kandola, John Ebdon and Chen Zhou, *Reactive and Functional Polymers* 129, August 2017.

6 Ibid., Kandola and Krishnan.

7 "Composites and fire: developments and new trends in flame-retardant additives," Belén Redondo, Composites Department of AIMPLAS, Plastics Technology Centre, 2018.

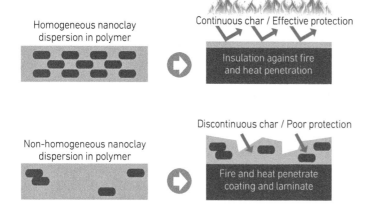

FIGURE 6-21. Dispersion of FR additives determines performance.[8]

Intumescents

Intumescent FR additives undergo a thermal degradation process when heated to produce a thermally stable, foamed, multicellular residue called "intumescent char." This char acts as a physical barrier to the transfer of flame and heat into the composite.[9] These systems can be used in many polymers such as polyamides, polyesters, polyolefins, epoxies and styrenes.

TABLE 6.11 Intumescent Products for Fire-Resistant Composites

Supplier, Product	Description	Applications
Zoltek PX30 Scrim Fabrics	Carbon fiber leno weave fabric high-temperature treated to >99% carbon with an intumescent coating	Oil and gas, jet fuel exposure applications
Technical Fibre Products (TFP) TECNOFIRE®	Veil made from mineral fibers, exfoliating graphite and ATH. Expands up to 35 times its original thickness in temperatures >190°C.	Buses, trains, bridges, building interiors and facades. Gap filling fire protection for doors and other building applications.
Fire Protection Developments FabShield	Coatings for fiber reinforced plastics	
Firefree 88	Water-based, non-toxic, thin-film intumescent fire retardant and resistant paint that can be applied to spray PU foam, fiberglass, carbon fiber, plastics and other composite materials	Building and transportation applications
SAERTEX LEO® System SAERTEX LEO® Coated Fabric SAERcore LEO®	Low-viscosity, unfilled matrix, an intumescent protection layer and fire-retardant multiaxial non-crimp fabrics. Smoke emission reduced by a factor >50	Floors in Deutsche Bahn ICE trains, sundeck and exterior walls for 110-m river cruise ship saving 45% weight vs. steel and aluminum

8   Source: "Flame Retardant Nanocomposites," Christophe Swistak, Valentin Chapuis, Alexandre Durussel, École Polytechnique Fédérale de Lausanne, Switzerland, published online September 9, 2009. https://www.slideshare.net/vchapuis/composites-flameretardant
9   "Flame retardants: Intumescent systems," Giovanni Camino, *Plastics Additives* 297-306, 1998. Part of the *Polymer Science and Technology Series* (POLS, volume 1), Springer Science+Business Media, Dordrecht. https://link.springer.com/chapter/10.1007/978-94-011-5862-6_33

FR Coatings for Composites

These coatings can be intumescent, producing char that protects the composite, or non-intumescent. Intumescent coatings have been used for decades on wood and metal structures. A variety of products are available for use with composites. Non-intumescent FR coatings include ceramic coatings that impart very high temperature resistance, and inorganic coatings that resist ignition and may also char.

❯ RADIOLUCENCE AND BIOMEDICAL

Composites are used in a wide variety of medical applications including tables for radiology procedures (e.g., X-ray, MRI, etc.) surgical instruments and orthopedic implants. Just as in aerospace, composites' lightweight, high stiffness and strength, and chemical/corrosion resistance make them attractive alternatives to metal and plastics. Composites are also *radiolucent*—that is, transparent to X-rays. This is because they absorb very low levels of radiant energy, which minimizes signal attenuation. This allows doctors to see through surgical tools or orthopedic implants with an x-ray or fluoroscope (real-time x-ray) during surgery. Stainless steel or titanium products are *radiopaque*, appearing white in x-rays.

A primary benefit of using carbon fiber reinforced composites in knee and hip replacement hardware is the ability to tailor the design to best match the stiffness of the adjacent bone. This decreases the localized stress on the joint, adding flexibility, which prevents atrophy of the surrounding bone. This results in an extended service life of the prosthetic implant. Recent research also suggests composite implants stimulate faster tissue formation for quicker recovery. Epoxy and polyetheretherketone (PEEK) resins have been used successfully in such implants and have successfully passed biocompatibility and safety tests, receiving Federal Drug Administration (FDA) approval.

❯ LIGHTNING STRIKE PROTECTION (LSP)

NOTE: Text in this section is adapted from "Lightning Strike Protection for Composite Structures" by Ginger Gardiner, High Performance Composites Magazine (July 2006), reproduced by permission; © 2006, Gardner Publications, Inc., Cincinnati, Ohio.

Unlike metal structures, parts made from composites do not readily conduct away the extreme electrical currents and electromagnetic forces generated by a lightning strike. Composite materials are either not conductive at all (e.g. fiberglass) or are significantly less conductive than metals (e.g. carbon fiber). When lightning strikes an unprotected composite structure, up to 200,000 amps of electrical current seeks the path of least resistance to ground, usually finding any metal available. Along the way, it can cause a significant amount of damage. It may vaporize metal control cables, weld hinges on control surfaces and explode fuel vapors within fuel tanks If current arcs through gaps around fasteners.

These direct effects also typically include vaporization of resin in the immediate strike area and possible burn-through of the laminate. Indirect effects occur when magnetic fields and electrical potential differences in the structure induce transient voltages, which can damage and even destroy any electronics that have not been shielded. For these reasons, lightning strike protection (LSP) is a significant concern, especially in vulnerable composite structures such as sailboat masts, aircraft, and wind turbines.

Lightning strikes usually occur in Zone 1 but can occasionally occur in Zones 2 and 3. (*See* Figure 6-22 and Table 6.12) Lightning usually attaches to the airplane in Zone 1 and departs the aircraft from a different area in Zone 1. Components that are likely to receive a lightning strike include: radomes, nacelles, wing tips, horizontal stabilizer tips, elevators, vertical fin tips, leading edge flaps, trailing edge flap track fairings, landing gear, and various air data sensors such as pitot tubes and air temperature probes.

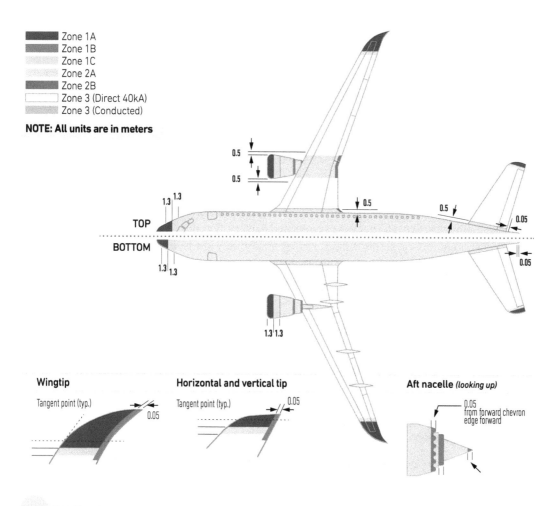

FIGURE 6-22. Lightning zones in aircraft (*see* Table 6.12).

TABLE 6.12 Aircraft lightning zones as defined by SAE Recommended Practices 5414

Zone	Description	Definition
1A	First return stroke zone	Surfaces where a first return is likely during lightning channel attachment with a *low* expectation of flash hang-on.
1B	First return stroke zone with a long hang-on	Surfaces where a first return is likely during lightning channel attachment with a *high* expectation of flash hang-on.
1C	Transition zone for first return stroke	Surfaces where a first return stroke of reduced amplitude is likely during lightning channel attachment with a low expectation of flash hang-on.
2A	Swept stroke zone	Surfaces where a first return of reduced amplitude is likely during lightning channel attachment with a low expectation of flash hang-on.
2B	Swept stroke zone with long hang-on	Surfaces into which a lightning channel carry subsequent return stroke is likely to be swept with a *high* expectation of flash hang-on.
3	Strike locations other than Zone 1 and Zone 2	Those surfaces not in Zone 1A, 1B, 1C, 2A, or 2B, where any attachment of the lightning channel is unlikely, and those portions of the airplane that lie beneath or between the other zones and/or conduct a substantial amount of electrical current between direct or swept stroke attachment points.

LSP Design Fundamentals

LSP strategies have three goals:

1. Provide adequate conductive paths so that lightning current remains on the structure's exterior.
2. Eliminate gaps in the conductive path to prevent arcing at attachment points and possible ignition of fuel vapors.
3. Protect wiring, cables and sensitive equipment from damaging surges or transients through careful grounding, EMF shielding and application of surge suppression devices where necessary.

Traditionally, conductive paths in composite structures have been established in one of the following ways: (1) bonding aluminum foil to the structure as the outermost ply; (2) bonding aluminum or copper mesh (expanded foil) to the structure either as the outermost ply or embedded within a syntactic surfacing film; or (3) incorporating strands of conductive material into the laminate.

All these approaches require connecting the conductive pathways to the rest of the structure in order to give the current an ample number of routes to safely exit to ground. This is typically achieved by using metal bonding strips (i.e., electrical bonding) to connect the conductive surface layer to an internal "ground plane," which is usually a metal component such as an aircraft engine or wind turbine motor, a metal conduit pipe, a metal base plate in a sailboat, etc. Because lightning strikes can conduct through metal fasteners in composite structures, it may be desirable to prevent arcing or sparking between them by sleeving countersunk holes with conductive sleeves that contact the primary LSP, or encapsulating fastener nuts or sleeves with plastic caps or polysulfide coatings.

LSP Materials

The need for protection of composite structures has prompted development of a number of specialized LSP materials.

Metal Mesh/Foil Products

For external surface protection, thin metal alloys and metallized fibers have been developed into woven and non-woven screens and expanded foils. These mesh-like products enable lightning current to quickly travel across the structure's surface, reducing its focus. For example, Dexmet supplies a large variety of conductive metal products for aircraft, including aluminum, copper, phosphorous bronze, titanium and other materials and can modify any design to meet precise customer needs, working with them to test and evaluate LSP designs. Dexmet can also supply some LSP in 48" widths, which can reduce labor cost during application. (Figure 6-23)

FIGURE 6-23. Microgrid® expanded metal LSP. *(Photos courtesy of Dexmet Corporation)*

Aluminum wire was one of the first LSP materials used, interwoven with carbon fiber as part of the laminate. However, using aluminum with carbon fiber risked galvanic corrosion, so copper wires were then tried, but are three times as heavy as aluminum. As the aircraft industry began using more fiberglass composites in aircraft, it investigated foils and then expanded foils, which can be cocured with the laminate's exterior ply.

Coated fibers (nickel or copper electrodeposited onto carbon and other fibers) are also used but perform much better as EMF shielding than as direct lightning strike protection. Hollingsworth & Vose has developed a non-woven veil from nickel-coated carbon fiber. It is flexible, lightweight and has performed well in lightning strike testing, with no damage to any fibers below the LSP.

Nanomaterial LSP Products

Veelo VEIL seamless carbon nanotube (CNT) sheet and film meets Zone 1A LSP requirements for aircraft and provides broadband EMI shielding. Designed to replace metals, it is 50–75% lighter than expanded copper foil. The CNT sheet reportedly offers 3–100 times lower electrical resistance versus metallized nonwovens and does not require corrosion protection materials, thus simplifying layups. It is conformable to complex curvature and available in continuous rolls and prepreg for both autoclave and out-of-autoclave processing.

Nanocomp Technologies manufactures Miralon® CNT nonwoven sheet, in standard areal densities of 12 g/m^2 and 25 g/m^2, available up to 54 inches wide and 8 feet long. Sheets can be slit into tape as narrow as ¼ inch, and both tapes and sheets can be seamed to deliver rolls up to 1,000 feet in length. Miralon® has been used as ESD and EMI shielding in aerospace applications.

Work to develop LSP products using graphene has shown initial success, including a lab-scale thin, flexible coating that reduced damage in carbon fiber composite laminates by up to 96% in simulated lightning strike testing. EMI shielding results also showed great potential and work continues to develop commercial products using this technology.

LSP Prepregs

"All-in-one" LSP prepregs contain pre-embedded woven or non-woven metal meshes and are applied first-down in layups. By combining the LSP and external prepreg surface layers into one ply, these products may reduce kitting and manufacturing costs.

Cytec-Solvay produces SURFACE MASTER 905® with embedded mesh, which is drapeable like a film adhesive. SURFACE MASTER 905® contains enough resin to permit surface sanding without damaging the embedded metal screen. It is available in several areal densities and foil thicknesses for weight optimization.

Henkel Hysol also produces a prepreg surface film that goes by the brand name SynSkin®. This material is available with the expanded foil in different densities to meet specific structural applications.

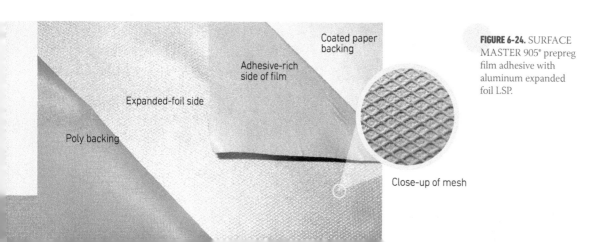

Coated paper backing

Adhesive-rich side of film

Expanded-foil side

Poly backing

Close-up of mesh

FIGURE 6-24. SURFACE MASTER 905® prepreg film adhesive with aluminum expanded foil LSP.

Post-Manufacture External Products

Integument Technologies produces polymer-based, peel-and-stick appliqués, which can be installed on composite surfaces after construction. Lightning Diversion Systems makes a thin **conformal shield** product for composite exterior surfaces, which protects against both direct and indirect lightning damage, is lightweight and produces a smooth finish. Vought Aircraft and Kaman Aerospace have used the product, which is more expensive than machined metal scrims but also can be stretched, providing enhanced conformability to structures with complex curvature.

LSP Issues

Paint Thickness

If primer and paint coatings are too thick, this will cause a significant dielectric barrier and the lightning energy will not readily conduct through to the copper mesh or other LSP material. This can be problematic to the function of the LSP layer and requires careful application and verification of coating thickness within designated specifications put forward in the design. (Figure 6-25)

After a lightning strike, repairs must be done properly to regain the conductive path. Traditional methods of repairing LSP-enabled composite structures using metal mesh and film adhesive eliminate surface porosity. The optimum surface is obtained by using separate layers of mesh and surfacing film, applied so that the film overlaps the mesh and provides an adequate fillet around the repair edges. Compared to previous methods, this approach replaces the outermost layer of film adhesive with surfacing film to bond and seal the mesh-to-mesh interface, so that the repair patch mesh is bonded and sealed to the original surface mesh, resulting in a successfully restored conductive path.

LSP For Wind Turbine Blades

While expanded foil mesh is the material of choice in the aerospace industry and may be applicable to large CFRP blades of the future, it is not currently used for protection of GFRP wind turbine blades. Instead, lightning protection for wind blades is more about *attracting* strikes than avoiding them. These blades are often designed with conductive metal air termination receptors that route from the blade surface through the blade to an internal metal cable or "down conductor" that transfers the lightning energy through the center of the blade to the hub, tower, and eventually to the ground.

The design typically includes x-number of receptors spaced along the maximum chord line of the blade, along the forward and aft facing surfaces, starting at the tip and at selected locations along the span. The tip and forward facing receptors on the blade are more susceptible to lightning attachment; however, aft facing receptors are also utilized for best protection. The number of receptors used in each blade design is determined based upon blade size and calculated air-termination requirements. (*See* Figures 6-26, 6-27 on next page)

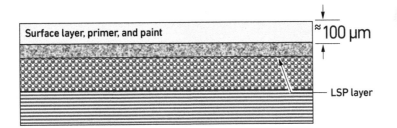

FIGURE 6-25. Primer and paint thickness must be controlled to maximize effectiveness of the LSP layer.

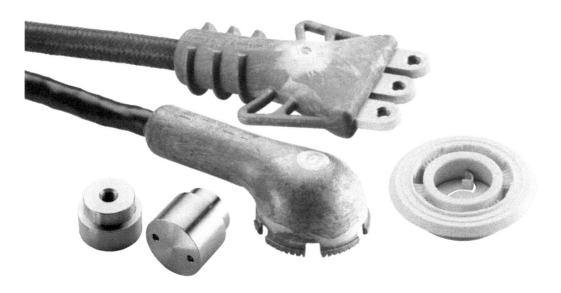

FIGURE 6-26. Lightning strike air termination receptors.

These lightning receptors by nVent ERICO are designed to protect the turbine blades and attract the lightning strike to a preferred attachment point. The system is designed to help minimize the probability of streamer initiation inside of the blade where the receptors serve as the strike termination device and earthing conductor connection point. *(Photos courtesy of nVent)*

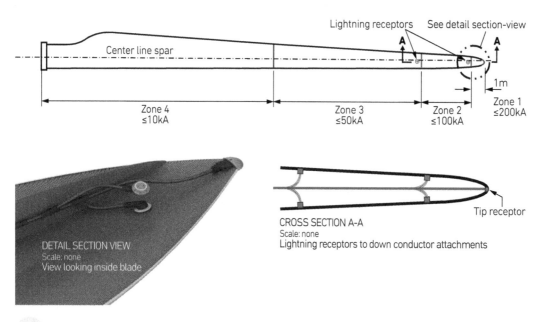

FIGURE 6-27. Wind-blade lightning strike receptor locations. *(Inset photo courtesy of nVent)*

Molding Methods and Practices

7

CONTENTS

Overview of Molding Methods and Practices

There are many different molding methods and techniques employed in the manufacture of advanced composite parts. The most common classification is by the type of fiber used (short or long/continuous) and the type of resin used (thermoset or thermoplastic). A third classification is *open* versus *closed* molding. (Figure 7-1, next page)

Open molding includes applying liquid resin to dry reinforcements in a mold open to the atmosphere (hand layup), spraying a mixture of resin and fiber onto an open mold (spray-up), or winding a reinforcement impregnated with wet resin onto a mandrel (filament winding). These laminates are most often cured at room temperature. Note that with hand layup, the laminates may be vacuum-bagged after the resin is applied. This is rare, however, with fabricators choosing instead to infuse resin into an already vacuum-bagged dry laminate stack (vacuum assisted resin infusion), which is a closed molding process.

Alternatively, **closed molding** encases dry reinforcements in a vacuum bag or closed set of molds before infusing or injecting resin. Prepreg methods are inherently closed molding because they require vacuum bagging or pressure in a closed set of molds before heating, in order to flow and cure the B-staged thermoset resin without voids. Most thermoplastic methods are also closed molding. However, the matrix is not cured (crosslinked) but instead heated to flow and infiltrate the reinforcement, and then cooled to shape.

The choice of molding method is typically driven by the materials being used, the complexity of the shape, and production rate requirements. Each method involves some layup (or placement) of fiber reinforced forms, thermoset prepreg or liquid resins, thermoplastic prepreg or polymers, and a heating/forming process or cure cycle. (Figure 7-2)

Molding liquid resin and dry fibers often demands a different approach than molding pre-impregnated (prepreg) materials, although some parts of these processes are similar. The

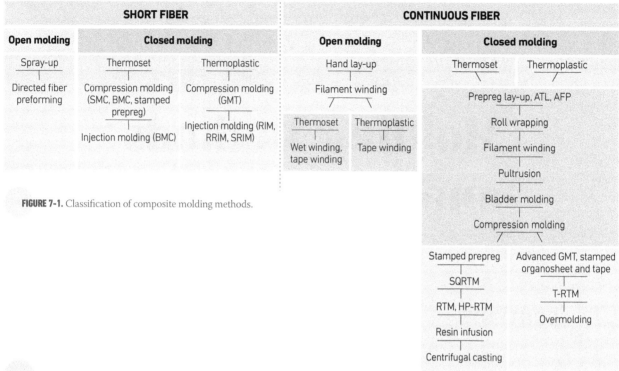

FIGURE 7-1. Classification of composite molding methods.

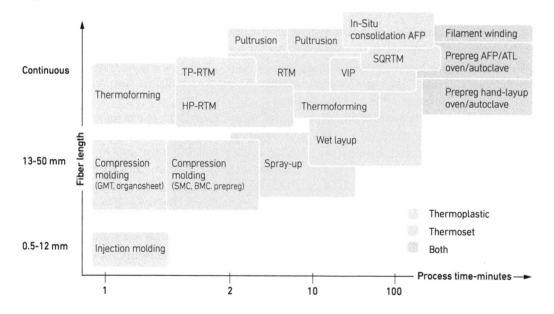

FIGURE 7-2. Comparison of fiber length and production molding methods.

primary difference involves the mixing and distribution of A-staged resin throughout the fibers versus having the resin already pre-impregnated in the fiber/fabric layers. (*See* Chapter 8, **Liquid Resin Molding Methods and Practices**.)

Prepregs are used in a variety of different and sometimes unique molding processes ranging from contact molding in a single-sided mold with a vacuum bag and oven or autoclave process, to closed molding processes that often involve a press.

With increased computer speed, robotics, and enhanced machine interaction, methods have evolved from previously labor-intensive hand-placed processes to more streamlined automated processes. Pultrusion, filament winding, automated tape laying, and fiber place-ment technologies are a few of the more common processes that have evolved since the 1960s. For example, today's 8-axis fiber-placement process was spawned from yesterday's 2-axis filament winding technology. Many new hybrid processes continue to develop as the industry finds more innovative ways to efficiently produce fiber-reinforced plastic parts for rate production.

This chapter touches on fundamental procedures such as application of mold release and vacuum bagging, as well as advanced processes such as prepreg layup, automated fiber place-ment (AFP), automated tape layup (ATL), and various press molding processes and technologies.

Semi-Permanent Mold Release Agents

Release agents are used for preventing parts from sticking to the mold or fixture, regardless of the process. High performance release systems are prescribed for rate production processes. Much like Teflon® coating a pan, semi-permanent mold release systems are in fact cured polymers that are bonded to the mold surface. These materials are not wax! They are designed to minimize or eliminate transfer to the part.

A typical semi-permanent system is made up of two different polymer coatings:

1. Tool sealer coat to fill the micro-porous sites on the tool surface (reducing mechanical attachment).
2. Low-energy release agent to prevent chemical attachment (*see* Figure 7-3).

These materials work best when applied in very thin layers. Application procedures gener-ally require that the materials be sprayed or wiped sparingly onto the tool surface. The desired result is to have the least amount of release agent on the surface that will do the job. This reduces the possibility that the release polymer will transfer to the part surface during the molding operation. Worse yet, the release polymer can act like an adhesive when it is over-applied or under-cured. This sometimes results in the part sticking to the tool and perhaps damaging both the part and the tool upon demolding. Thus, proper application procedures are crucial for good performance.

A semi-permanent release agent is designed to provide multiple releases. Typically, 2–10 parts can be molded from a properly treated surface prior to re-application of the release agent. Some end-users choose to re-apply the release coat every molding cycle. While this is thought to ensure a good release, it adds to quicker build-up and possible transfer problems in production.

After a certain number of applications, it is recommended that the release agent be stripped back to the sealer coat and reapplied. This is necessary for long-term release performance without release transfer. Most manufacturers will set up regular maintenance intervals

FIGURE 7-3. Semi-permanent release systems.

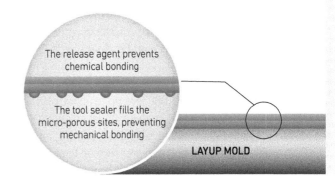

The release agent prevents chemical bonding

The tool sealer fills the micro-porous sites, preventing mechanical bonding

LAYUP MOLD

between 20 and 50 application cycles, to strip and reapply the release system depending on the tooling and the material system used.

Semi-permanent release systems are available in both solvent and solvent-free (water-based) formulas. Many manufacturers are choosing to use the solvent-free, non-VOC formulas in lieu of the old solvent-based materials to reduce health and safety risks and other environmental concerns that solvents cause.

Vacuum Bagging

Vacuum bagging is a fundamental part of manufacturing composites across a range of different molding methods. Understanding how vacuum works, how each layer under the bag performs, and why it is important to the end-result, is paramount to making good composite laminates.

The primary function of a vacuum bag is to provide compaction pressure and consolidation of the plies within the laminate. This is done by creating a pressure differential between the inside and the outside of the bag using a vacuum pump, a hose, and a sealed film membrane (bag film), and results in down-force on the bag equal to the ambient atmospheric pressure.

The secondary function of a vacuum bag is to facilitate resin movement, removal of excess resin, and extraction of air and gas from the laminate during processing. The route for extraction is via the bleeder/breather sequence, during vacuum debulking, especially for **out of autoclave (OoA)** or **vacuum bag only (VBO)** processes. Both the compaction pressure provided by a vacuum bag and the extraction of air and volatiles are crucial to obtaining a good quality laminate.

In an autoclave process, much higher pressures are available (>10 bar/145 PSIG). The vacuum bag is still a necessary part of the process because it acts like an intimate membrane that separates the vessel pressure from the laminate. This is analogous to having a highly conformable internal tool surface, compacting against the backside of the laminate in a large press molding operation. The resultant product is uniformly compacted throughout the vacuum bag. It should be noted that the vacuum bag schedule for autoclave processing is different than OoA or VBO; there is typically no bleeder system used for autoclave processing.

❱ UNDERSTANDING VACUUM AS ATMOSPHERIC PRESSURE

Planet Earth is surrounded by gaseous matter that forms our atmosphere. This extends over 62 miles/100 km above the earth and is held to the earth by gravity. Being a gas, the atmosphere has weight, and that weight is normally measured in pounds per square inch (psi) or kilograms per square centimeter (kg/cm^2).

If you were to take a square inch column of the air extending to the outer atmosphere, its standard recognized weight (pressure exerted on the earth) at sea level would be 14.7 lbs. This is called atmospheric pressure. Pressure measured that is greater than atmospheric pressure is referred to as gauge pressure (PSIG). Pressures below this are referred to as vacuum and are measured in inches (or mm) of mercury (Hg).

This same square inch column of air weighing 14.7 psi is equal to a square inch column of mercury 29.92 inches high. (In other words, 29.92 inches^3 of mercury would weigh 14.7 lbs on a scale.)

Atmospheric pressure decreases at higher elevations. The average loss is approximately 1 in/Hg (.46 psi) per 1,000 feet, up to about 13,500 feet. Beyond that, the atmospheric pressure decreases by approximately 50% at an altitude of about 3.5 miles (5.6 km). This pressure drop is roughly exponential, so each doubling in altitude after this point results in a decrease of about half the measured pressure up to about 62 miles (100 km). Obviously, most composite processing occurs below 10,000 feet. (Figure 7-4)

❭ VACUUM BAGGING REQUIREMENTS

To obtain a good vacuum, certain requirements must be met:

- The vacuum bag and the tooling must not leak (i.e., must be airtight and hold a vacuum).
- Leaks must be minimized within (gauge leak test) specifications or preferably, eliminated.
- The vacuum pump and plumbing must be sufficient to accommodate volume flow.
- The vacuum hoses must be large enough to provide adequate volume flow.
- A vacuum gauge should be used to verify that minimum vacuum or leak rate is achieved.

The breather and bleeder layers are an important factor in achieving good vacuum. The basic rules are:

1. Have a continuous breather layer under the bag to maintain a path for air and volatile extraction.
2. Prevent resin from filling the breather layer to ensure a breather path.
3. Use separator film to filter resin movement into the bleeder, or block movement into breather layers.

If both the bleeder and breather layers fill up with resin, the dynamics beneath the vacuum bag change from atmospheric pressure to a **hydrostatic** state (equilibrium pressure). When this happens, it is no longer possible to maintain pressure (down force) on the layup inside the bag.

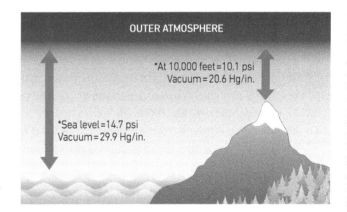

OUTER ATMOSPHERE

*At 10,000 feet = 10.1 psi
Vacuum = 20.6 Hg/in.

*Sea level = 14.7 psi
Vacuum = 29.9 Hg/in.

FIGURE 7-4. Vacuum vs. atmospheric pressure.

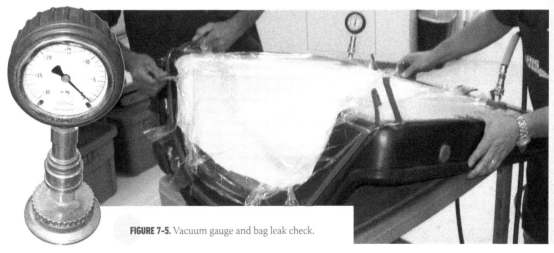

FIGURE 7-5. Vacuum gauge and bag leak check.

For this reason, a separator layer is used between the bleeder and breather layers so that excess resin being pulled into the bleeder layer does not move up and saturate the breather layer. However, the bleeder and breather layers must make contact around the edges or through small, evenly spaced perforations so that air and volatiles have a continuous path for extraction out of the laminate before, or during cure.

VACUUM BAG SCHEDULE AND FUNCTION

Bleeder, Breather, and Vacuum Bag Schedule

The vacuum bag, bleeder, and breather schedule is made up of a series of different materials that are applied to the uncured composite laminate prior to curing. Each layer in sequence is designed to function in a specific manner and in unison with the adjacent layers to achieve proper resin content and low void volume in the finished laminate. One schedule is not applicable to all conditions, especially for OoA or VBO processes. (Figure 7-6)

There are many different options to consider when designing a vacuum bag schedule for a certain process. For example, a bleeder layer may be necessary in an OoA or VBO process to promote movement of resin, air and gas. A bleeder layer is not required for an autoclave process where several atmospheres of pressure are applied to the vessel/laminate and no resin bleed is desired.

Because there are many variables involved in choosing the right vacuum bag schedule for a given process, it is recommended that small-scale testing be done to evaluate the proper arrangement prior to attempting a large-scale layup.

Release Film or Peel Ply

The first layer that goes against the uncured laminate is either a release film or a peel ply layer that is used as to release the laminate from subsequent bleeder or breather layers. This layer can be nonporous or porous material depending on whether resin bleed is necessary. Often a porous peel ply is used when you want resin bleed, or a peel ply surface texture is required. Perforated release film is used when a controlled resin bleed is desired. The diameter and the spacing of the holes can vary depending on the amount of resin flow desired. (Figure 7-7)

A nonporous peel ply commonly known as TFNP, or Teflon® coated fabric or a solid film such as fluorinated ethylene propylene (FEP) film is used when no resin bleed is required. This layer usually extends beyond the edge of the layup and can be sealed and secured with flash breaker (FB) tape as needed.

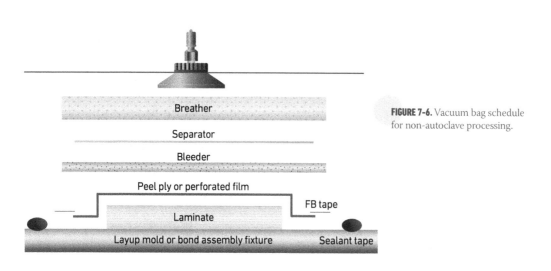

FIGURE 7-6. Vacuum bag schedule for non-autoclave processing.

Bleeder Layer

The bleeder layer is used to absorb resin from the laminate either through a porous peel ply or a perforated release film as described above. The bleeder layer is usually a non-woven synthetic fiber material that comes in a variety of different thicknesses and/or weights that range from between 2 oz/yard^2 to 20 oz/yard^2. Glass fabric can also be used as a bleeder material and is often prescribed for processing repairs. (*See* Chapter 13, **Repair of Composite Structures** for more information.) Multiple layers can also be utilized for heavy resin bleed requirements. This layer usually extends beyond the edge of the layup and is secured in place with FB tape as required.

Separator or Film Layer

The separator layer is used between the bleeder layer and the subsequent breather layer to restrict or prevent resin flow. This is usually a solid or perforated release film that extends to the edge of the layup but stops slightly inside the edge of the bleeder layer, to allow a gas path to the vacuum ports.

Breather Yarns

Intermittent glass yarns are often deployed as a breather link from the layup, extending out beyond a solid film layer that overlaps the edge of the layup at a prescribed distance. This is then sealed 100% around the periphery with FB tape so that resin does not exceed a designated "flash zone." The yarns allow gas to escape without moving resin beyond the seal.

These yarns are usually spaced 10 to 24 inches apart, depending upon the size of the part and the volatile content of the prepreg. They are typically utilized in autoclave or press molding processes where elevated pressure is applied during the cure process.

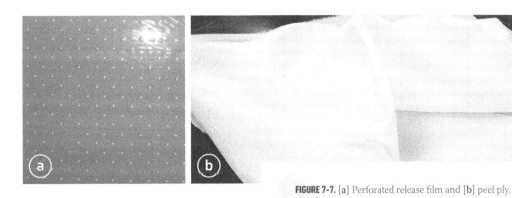

FIGURE 7-7. [a] Perforated release film and [b] peel ply.

FIGURE 7-8. Breather materials.

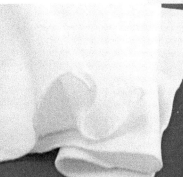

Breather Layer

The breather layer is used to maintain a breather path throughout the bag to the vacuum source, so that air and volatiles can escape, and uniform atmospheric pressure is applied to the laminate. Typically, synthetic fiber materials and/or heavy fiberglass fabric is used for this purpose, depending on the pressure to be exerted on the bag during processing. (Figure 7-8)

The breather layer extends past the edges of the layup and contacts the bleeder ply around the separator film. The vacuum ports are connected to the breather layer. It is especially important that adequate breather material be used for autoclave processing at pressure.

Flash Breaker Tape

Flash breaker tape is an adhesive backed polyester, Nylon®, or Kapton® film material used as a multipurpose tape to secure ancillary bagging materials, mask-off mold surfaces, and seal film layers at the edges against the mold surface to stop (break) resin movement at a "flash" zone around the layup. The tape comes with a pressure sensitive cured rubber (or acrylic) adhesive that is designed to peel off a surface after exposure to elevated temperature and pressure.

Bag Film and Sealant Tape

The bag film is used as the vacuum membrane that is sealed at the edges to either the mold surface or to itself (if an envelope bag is used). A rubberized sealant tape is used to provide the seal around the periphery. The bag film layer is generally much larger than the area being bagged as extra material is required to form pleats at all the inside corners and about the periphery of the bag as required to prevent bridging. (*See* "Pleats," below.) Bagging films are made of Nylon®, Kapton® or other hybrid film materials, each of which are used for different applications and temperatures. (Figure 7-9)

Vacuum Ports

A machined or cast metal fitting that connects the vacuum bag to the vacuum source. The port is connected to the vacuum source with a hose. Both the hoses and ports typically incorporate quick-disconnect fittings that allow the hoses to be removed from the ports without losing vacuum on the bag. Multiple vacuum ports are recommended for all vacuum bag process for best results.

It is important to have enough ports to provide for adequate volatile extraction during processing. A simple rule of thumb is to have a minimum of two ports for even the smallest

FIGURE 7-9. Vacuum bag film and sealant tape.

vacuum bag, adding one (minimum) extra port for each additional 9 square feet of area. One way to think about this is to consider that the ports are vents and you want all trapped air and gas to reach the vents during the molding process. (Figure 7-10)

Pleats

Pleats are strategically placed anywhere that additional vacuum bagging material is needed to accommodate for changes in configuration, at the base of vertical walls (inside corners), at the base of core ramps, and other intricate areas that require extra bag film materials. Often, small pleats will be place around the periphery of a flat bag surface just to alleviate shear stress on the sealant tape during processing.

Figure 7-11 illustrates the four-corner vacuum bagging method that uses a measured amount of bag film to cover the part area, including extra material for pleats. This approach ensures that all bag film is used, and pleats are located where needed.

Vacuum Debulk

Intermittent vacuum debulking (compacting) of prepreg materials during layup is prescribed as part of the manufacturing process. This is typically done after the first ply goes down in the mold and after every 2–5 plies thereafter, depending on the material form and complexity of the part being laid up. The idea is to tightly compact the plies and remove air and volatiles from the plies at specified intervals during the layup process, in order to minimize voids prior to the curing process and thus maximize mechanical properties.

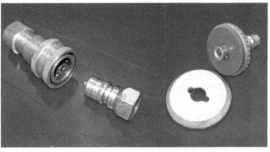

FIGURE 7-10. Vacuum ports and quick disconnect fittings.

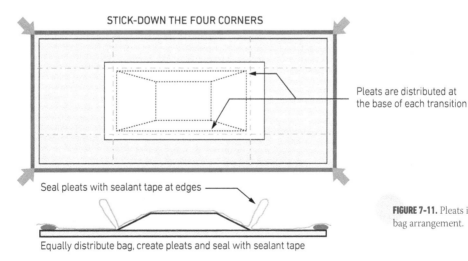

STICK-DOWN THE FOUR CORNERS

Pleats are distributed at the base of each transition

Seal pleats with sealant tape at edges

Equally distribute bag, create pleats and seal with sealant tape

FIGURE 7-11. Pleats in vacuum bag arrangement.

A vacuum table can be used for debulking flat or (supported) gently contoured shapes of a size in which the layup can be transported to and from the table. Many of these tables have heaters and can be used to perform elevated temperature debulks (hot debulks). (Figure 7-12)

Hot Drape Forming

Hot drape forming (HDF) is essentially a more industrialized form of hot vacuum debulking, which enables dozens of layups to be consolidated simultaneously. It also potentially enables the debulking of many plies at once. For example, FBM Composites claims its HDF process can debulk 50 unidirectional plies at a time, compared to the traditional lengthy and multi-step process of placing 2–5 plies, debulking, placing another 2–5 plies, debulking, etc.

The process involves placing a stack of pre-laid (thermoset) prepreg materials "draped" over a mandrel while heating the layup so that it is pliable, and then vacuum-forming it to shape. Note that the heat applied is only to soften the prepreg and should not advance any reaction in the resin matrix. Multiple layups on their mandrels may be placed into the HDF equipment. Thus, the process aids rate production and is well-suited for molding channels and other continuous, or near-continuous tapered or multi-tapered cross-section shapes that lend themselves to this method.

The prepreg plies layups may be produced using hand layup or more automated methods such as automated tape layup (ATL) and automated fiber placement (AFP) (*see* page 199). The prepreg plies may be placed directly onto a mold release-treated mandrel. Alternatively, they may be placed on top of a layer of release film against a flat tool. For the latter, the stack is removed from the flat tool and placed on release-treated mandrel with the release film side up and a breather on top. The mandrel with layup is then placed in an HDF box where it is heated and simultaneously vacuum-formed against the mandrel. The mandrel with layup is then removed from the HDF box and vacuum-bagged for autoclave or oven cure. (Figures 7-13, 7-14)

Hand Layup – Prepreg

Prepreg hand layup is one of the most widely used methods of fabricating advanced composite parts and structures. Prepreg materials allow for very precise resin to fiber ratios, along with more accurate control of fiber orientation during layup than with wet layup.

 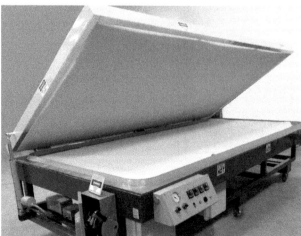

FIGURE 7-12. Heated vacuum debulk table.
(Photos courtesy of BriskHeat)

❯ COMPACTION METHODS

Prepregs are typically processed by laying the prepreg on a mold, vacuum-bagging the layup and then curing in either an oven or an autoclave. However, some flat or gently shaped prepreg parts are may also be press-molded in a heated press without the aid of a vacuum bag. Other prepreg layup methods use a cavity mold (bladder molding) or set of matched molds (e.g., compression molding, SQRTM). See later sections in this chapter on "SQRTM" (page 216) and "Bladder Molding" (page 219).

TABLE 7.1 Prepreg Molding Using Matched Molds

Prepreg molding method	Tooling	Pressure
Prepreg stamping or press-molding	Matched molds	Press only
Bladder molding	Matched molds, bladder	Clamped mold set and air to inflate bladder
SQRTM	Matched molds	Press-clamped mold set and liquid resin

FIGURE 7-13. The HDF process.

Layup stack located on mandrel

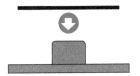 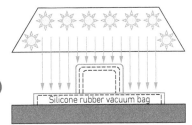 

Laminate is heated and vacuum formed

Layup is formed on mandrel and ready to bag and cure

Silicone rubber vacuum bag

FIGURE 7-14. Hot drape forming in process.
Prepreg layups are de-bulked prior using an HDF machine from Electrotherm Industry (Migdal HaEmek, Israel).

❭ CLEAN ROOM

Prepreg layup is usually done in an environmentally-controlled clean room where the temperature and humidity are regulated and monitored. While this is not required for some manufacturers, it is commonplace in the aerospace industry. Such strict controls are important as these materials require careful processing to achieve the desired structural properties and in-service performance. Such controls are also required by regulatory authorities (e.g., FAA, EASA, NASA, ESA) for use on primary and secondary aircraft and spacecraft structures.

❭ CUTTING AND LASER PROJECTION

Since prepreg materials are already impregnated with resin, there is little to no mess involved with the material **kitting** (cutting to size) or the layup process. While manual kitting and layup is typical with prepregs, the materials also lend themselves to automated table cutting systems that are increasingly being utilized in high-rate production facilities.

Both the automated cutting table and the hand layup process benefit today from laser projection systems, which utilize lasers to guide operators through each task. To minimize scrap, cutting tables use sophisticated software to "nest" ply patterns so that they fit with minimal space between each pattern. This typically results in a very complex configuration of pieces left on the cutting table. A laser system eliminates the cumbersome job of manually determining the order in which pieces should be removed from the table. It automatically points operators to each piece sequentially, so that they are "kitted" in the order in which they will be laid up. Then during layup, the laser projector outlines the position in which the ply is to be placed. The precision of these laser projection systems helps to ensure highly accurate material positioning. (Figure 7-15)

❭ VACUUM DEBULKS AND TOOLING

While prepreg hand layup is perceived to be less labor intensive than **wet layup**, the reverse can be true when fabricating parts with complex geometries. In this case it may be necessary to layup, vacuum bag, and debulk each ply individually, in sequence, to get the best results.

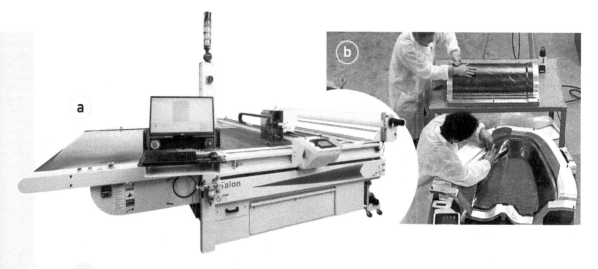

FIGURE 7-15. Automated kit cutting and prepreg layup with laser projection.
[a] Typical prepreg cutting table and prepreg hand layup of CFRP prepreg on metal molds.
[b] The operators are using a projected laser pattern to locate the plies on two tools at the same time.
(Photos courtesy of Eastman Machine Company and Aligned Vision)[1]

1   Photos copyright and used with permission of Eastman Machine Company, Buffalo NY, and Aligned Vision, Chelmsford, MA.

These intermittent vacuum debulks are part of the normal prepreg layup process. Without these steps, *bridging* in the inside corners and wrinkling of the fibers on the outside corners may occur.

The tooling (molds) used for prepreg manufacturing is normally of better quality and materials than those used for wet layup processes. This is primarily because they are used in an elevated temperature environment and must maintain vacuum and/or pressure integrity through multiple thermal cycles. Tools of this type can be costlier up front and expensive to maintain. (*See* Chapter 9, **Introduction to Tooling**.)

Automated Tape Laying and Automated Fiber Placement

Automated tape layup (ATL) and automated fiber placement (AFP) are methods used to apply pre-impregnated continuous fiber reinforcements to a mold-tool or mandrel. They are true additive manufacturing (AM) processes and share a number of characteristics. Both methods:

- use computer-controlled equipment with a multi-axis placement head on a gantry or robot;
- operate at high speeds;
- are integrating automated, inline inspection;
- minimize waste compared to traditional fabric-based processes; and
- require a significant capital investment.

ATL was patented in 1973 by Goldsworthy Engineering, Inc. The original vision was to place 2" to 3"-wide tapes using a gantry system, but that quickly evolved into placing multiple rolls of tape slit into widths of ½", ¼" or ⅛". Today, ATL uses slit unidirectional prepreg tape (1.5 to 12 inches wide) fed from a spool incorporated into the tape laying head and up to 32 rolls of tape can be placed simultaneously.

AFP developed out of ATL and was originally distinguished by placing impregnated carbon fiber *tows* versus *tapes*. Today, AFP uses up to 32 separately collimated pre-impregnated tows or narrow tapes (0.125" to 0.5"-wide) fed from one or more creels on or near the head (*see* Figure 7-16).

FIGURE 7-16. Diagram of AFP/ATL head.

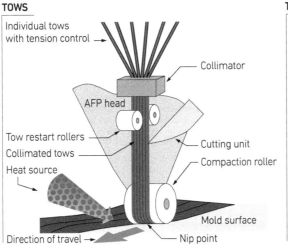

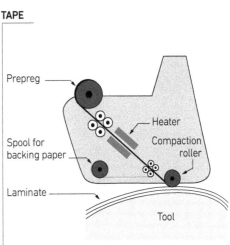

Although the two terms are often interchangeable, there have been several traditional differences:

TABLE 7.2 Comparing ATL and AFP

ATL	AFP
More often gantry-based	Both robot and gantry-based
Slower than gantry-based AFP: ~2,400 in/min	Gantry-based is faster: up to 4,000 in/min
Multi-row tape capability for wide width laydown	Can lay down single or multiple tows, or narrow tapes
Better suited for simpler, flatter shapes	Better able to produce complex curves and geometries

The term AFP is now gaining favor as it implies greater freedom and flexibility in the placement process and also seems able to contain all of the variations in both fiber formats and process technologies being developed.

Shown in Figure 7-17, CHARGER™ tape layers achieve up to 25-degree contours and up to 60 m/min (2,362 ipm) layup rate.

FIGURE 7-17. ATL machine for flat and gentle contours. *(Photos courtesy of Fives Cincinnati, Hebron, KY)*

FIGURE 7-18. AFP machine for complex shapes and high flexibility.

[a] With +/-95 degrees of A-axis travel, this AFP system enables complex lamination of a one-piece aircraft wing spar, applying fiber to both vertical and horizontal surfaces at speeds up to 4,000 ipm. It can round a corner in 0.8 seconds while the C-axis turns the head 180 degrees in 0.75 seconds. *(Photo courtesy of Electroimpact)*

[b] TORRESFIBERLAYUP is an AFP system that produces complex-shaped composite layups at a speed of 60 m/min (2,500 ipm). Its modular design can apply 12, 16, 24 or 32 tows in 1/2, 1/4 or 1/8 inch width, and can accommodate different architectures—gantry type, column type, and with or without a head stock/tail stock system for revolution parts. *(Photo courtesy of MTorres)*

The ATL tape-laying head includes one or more spools of tape, a winder with winder guides, a heater, a compaction shoe, a position sensor and a tape cutter. AFP machine heads typically include a heater system and compaction roller (or "nip-roller"), as well as tow/tape placement guides and cutting mechanisms. For both processes, the head may be located on a gantry suspended above the tool that enables movement over the mold surface, or it may be located on the end of a multi-axis articulating robotic armature that moves around the tool or mandrel. (Figures 7-18)

For both ATL and AFP, the position of the head is computer-controlled, enabling very precise placement of the prepreg tow or tape. The computer software used to control the machine is programmed using numerical data from part design and analysis. This input defines the lay down pattern for the part, comprised of multiple courses, with a course being one pass of material of any length at any angle. After the required number of plies have been applied, the part and tool/mandrel are vacuum-bagged and placed in an autoclave or oven for curing.

Newer machines can switch from ATL to AFP with interchangeable heads. This allows for flexibility for a wider range of capabilities within a single work cell. (Figure 7-19)

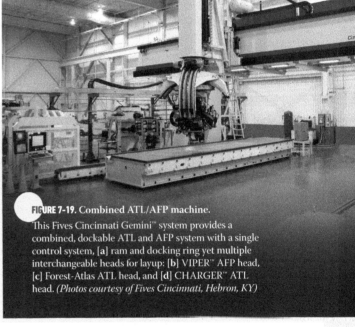

FIGURE 7-19. Combined ATL/AFP machine.
This Fives Cincinnati Gemini™ system provides a combined, dockable ATL and AFP system with a single control system, [a] ram and docking ring yet multiple interchangeable heads for layup: [b] VIPER™ AFP head, [c] Forest-Atlas ATL head, and [d] CHARGER™ ATL head. *(Photos courtesy of Fives Cincinnati, Hebron, KY)*

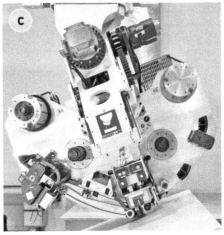

Advantages of both ATL and AFP include high layup rates for large parts, resulting in lower labor costs and fewer human errors. Because the layup process is automated, parts made with these processes are more uniform and consistent. Because the placement heads automatically cut material at the end of each pass, both processes deliver large reductions in material waste.

The computer-controlled precision of automated tape and fiber placement processes make for a high degree of accuracy and repeatability in manufacturing. Many machines have cameras mounted along the gantry or in the head to record the layup and detect defects in-process. This greatly enhances the quality assurance of these parts.

Automated inspection with laser technology further enhances layup accuracy. Newer laser systems may house both laser projectors and high-resolution cameras. The projector displays the correct fiber orientation, ply edge and other key inspection information. Images then captured by the high-resolution camera may be sent to an inspector remotely or may be analyzed by a computer algorithm to detect any errors. (Figure 7-20)

❱ MATERIALS

AFP and ATL have traditionally used carbon fiber epoxy or BMI prepreg materials; however, thermoplastic (TP) prepregs, including PEEK, PPS, PEI, PA, PP, PE, PFA and others have been adapted, mostly via AFP technologies. The key to using thermoplastics is having a heat source at the head to melt the TP and promote controlled crystallization of the resin as it cools. The process is compatible with virtually all fiber reinforcements.

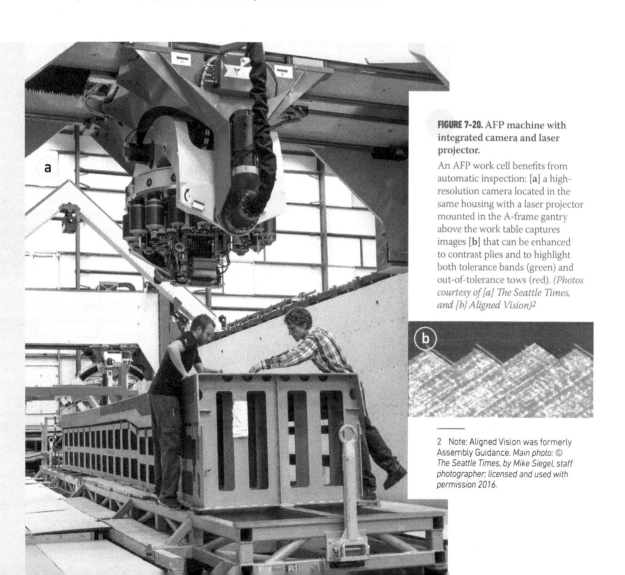

FIGURE 7-20. AFP machine with integrated camera and laser projector.

An AFP work cell benefits from automatic inspection: [a] a high-resolution camera located in the same housing with a laser projector mounted in the A-frame gantry above the work table captures images [b] that can be enhanced to contrast plies and to highlight both tolerance bands (green) and out-of-tolerance tows (red). *(Photos courtesy of [a] The Seattle Times, and [b] Aligned Vision)*[2]

2   Note: Aligned Vision was formerly Assembly Guidance. *Main photo: © The Seattle Times, by Mike Siegel, staff photographer; licensed and used with permission 2016.*

❯ PARAMETERS

ATL delivery heads include a heating system that warms the prepreg tape to tack levels just ahead of the compaction shoe/roller. This is especially useful for BMI and other dry resin systems that lack the necessary tack for tape-to-tape adhesion. Heated tape temperatures range from 100°F to 160°F. The head also provides "automatic debulking" via the compaction force typical delivery heads apply, which can range from 60 to 293 pounds of force (lbf) across a 6-inch wide tape and as high as 601 lbf across 12-inch wide tapes.

Today's AFP equipment has the ability to perform a warm debulk "on the fly," as the fiber placement head heats the materials at the nip point of the compaction roller, reducing the viscosity of the matrix resin and compacting the materials together simultaneously during laydown. In the case of thermoplastic composite placement, the materials are not only tacked together with heat, but are compacted and bonded instantaneously via an **in-situ consolidation (ISC)** process. This eliminates the need for a post-process autoclave cycle.

Heating is done just prior to the nip-roller with a hot gas torch (HGT), infrared (IR) lamps, or a laser heating system (LHS). An IR camera mounted nearby is used to monitor the temperature at the roller. The HGT method was developed in the late 1980s by Automated Dynamics[3] (now Trelleborg) for use on thermoplastic prepreg tow materials in early AFP development. HGT uses superheated nitrogen aimed at the nip-point. IR lamps serve the purpose for heating thermoset materials, as the temperature requirements are not as high as those required to melt TPs. (Figure 7-21)

---

3 Automated Dynamics, Niskayuna, NY, became part of Trelleborg Group, Trelleborg, Sweden.

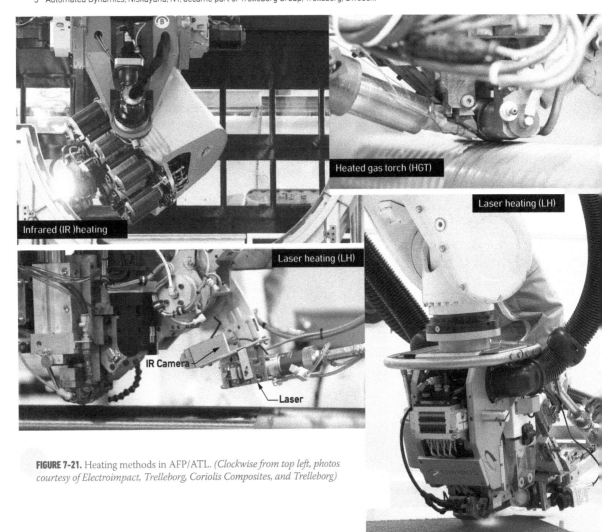

FIGURE 7-21. Heating methods in AFP/ATL. *(Clockwise from top left, photos courtesy of Electroimpact, Trelleborg, Coriolis Composites, and Trelleborg)*

Unlike HGT heating, an LHS provides very precise and uniform heating along the width of collimated (accurately aligned) tows or tape, at the nip-point, up to 10 times faster than the HGT. This allows for a much faster lay-down rate and for true in-situ consolidation of TP composites.

) IN-SITU CONSOLIDATION (ISC) OF THERMOPLASTIC COMPOSITES

When using AFP or ATL to produce TP composites, the prepreg tape or towpreg may be placed and subsequently cured in an oven or autoclave the same as when using thermoset prepreg, or it may be heated, consolidated and cooled in-situ ("in place") with no further processing.[4] This is possible because thermoplastic matrix materials do not require crosslinking. They do, however, require controlled heating, cooling and pressure to produce the high mechanical properties for load-bearing structures. This results from achieving a consolidated laminate with less than 2% voids; or when using semi-crystalline materials (PEEK, PEKK, PAEK, PPS), achieving the necessary crystallinity (which is usually > 35%).

In-situ consolidated TP composite parts were first developed in the 1980s by Accudyne Systems and Automated Dynamics (now Trelleborg). By the early 1990s, Automated Dynamics was producing in-situ consolidated cylindrical parts every day. Note that the hoop stress created during winding thermoplastic composite cylinders consolidates the laminate, achieving a void content of < 2%, so that no additional compaction is needed. In the 1990s, Accudyne Systems made carbon fiber/PEEK flat panels at a rate of 3 m/min with voids less than 2%. Trelleborg reports that it has made in-situ consolidated TP composite parts since 1993 ranging from 500,000 parts per year as small as 2 grams, to multiple large parts per year up to 40 feet long and using 2,500 lbs of composites.

In 2012, Trelleborg produced a full-size, CF/PEEK fuselage with lightning strike protection and stiffeners integrated into the AFP skin using one-step ISC (*see* Figure 7-22). It has used hot gas torch heating for ISC, but now typically uses a laser because it is three times faster. However, laser heaters require safety equipment and procedures while the system is running.

Recently, Heraeus Noblelight has developed the humm3® pulsed light technology for AFP of thermoset, dry fiber and thermoplastic materials that does not have laser safety issues, nor precludes technicians from being in the work cell during operation. The process temperature required for ISC varies by matrix: 340°C for low-melt (LM) PAEK, 375° for PEKK, and 385° for PEEK.

---

4   Text in this section taken from "Consolidating thermoplastic composite aerostructures in place, Part 1" and "Part 2", *CompositesWorld magazine*, February and March, 2018.

FIGURE 7-22. In-situ consolidated thermoplastic fuselage demonstrator.

Trelleborg produced this carbon fiber/PEEK helicopter fuselage demonstrator in 2012 using one-step ISC and AFP. *(Photo courtesy of Trelleborg)*

However, ISC requires more time during AFP than placing a laminate that will be consolidated afterward—for example, 60–100 mm/s for ISC, versus 600–700 mm/s for straight AFP and then post-autoclave. This post-autoclaving is basically a heated vacuum-bagged debulk, where cooling is controlled to produce the required crystallinity in the matrix. Proponents of using this second step claim it is a short cycle: heat for 30 minutes to reach melt/process temperature, hold for additional minutes (not hours) and then cool down—a fraction of current thermoset prepreg autoclave cycles.

ISC supporters argue that overall part cost is competitive because the longer AFP time is offset by elimination of the vacuum bagging, part handling and duration of the second step, as well as additional tooling required. Also, with ISC, it is possible to produce a finished part with integrated stringers, ribs and bulkheads without adhesive or fasteners by placing these prefabricated pieces in the tooling and AFP placing on top of them. (Figure 7-23)

Though it is possible to integrate stiffeners without using ISC, the tooling needed for the second consolidation step becomes more complicated. Proponents of ISC also claim that newer, higher-quality TP tapes that are flatter, with more consistent thickness will improve ISC speed, as will lower process temperature materials such as low-melt PAEK.

Current CF/PEEK tape   **Potential flatter tape**

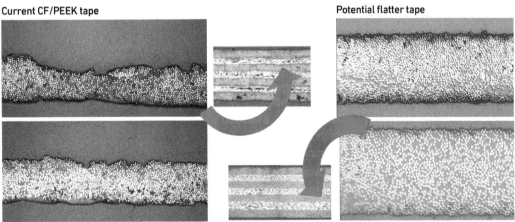

FIGURE 7-23. In-situ consolidation (ISC) of thermoplastic composite structures.
FIDAMC produced this carbon fiber/PEEK integrated stringer fuselage demonstrator panel by placing prefabricated omega stringers into the tool and AFP placing the skin laminate on top, effectively welding the stringer flanges to the skin during placement. The bottom diagram shows how higher quality tape reduces voids in ISC laminates. *(Photos courtesy of FIDAMC)*

Oven and Autoclave Equipment

For advanced composite parts that require an elevated temperature cure, an oven or an autoclave (a cylindrical pressure vessel with heating and cooling capabilities) may be prescribed. The parts are generally laid-up, vacuum bagged, and placed in the oven or autoclave for processing.

Both ovens and autoclaves are fitted with vacuum plumbing and thermocouple connections. Autoclaves typically have advanced vacuum/venting capabilities to allow the vacuum lines to be vented to the atmosphere or regulated during the cure process. In addition, there may be special tracks and racks for loading and unloading large parts or groups of smaller parts.

Temperature controllers range from single-setpoint manual devices to full-blown computer-aided control systems that may be necessary to control hundreds of sensors in the oven or autoclave, and on multiple tools and parts during the curing process. Both ovens and autoclaves have heaters and fans that provide convection heating in a manner that uniformly distributes the heat throughout the vessel. Autoclaves also have water cooling capabilities where cool water is distributed via plumbing within the vessel for aid in cooling the part during the cool-down portion of the cure cycle.

Ovens are normally rectangular and usually have simple hinge/latch mechanisms because they have no internal pressurization capabilities. Autoclaves are cylindrical in shape and have dome-shaped doors that have a rotating lock-ring or door-locking design to provide clamping, sealing, and safety functions during pressurization cycles. (Figure 7-24)

Autoclaves are typically pressurized with nitrogen to multiple atmospheres of pressure inside the vessel, providing additional compaction force on a composite laminate during the curing process. This allows for excellent consolidation of the materials within the vacuum bag and converts residual gas back into liquid (resin), minimizing the void content of the laminate.

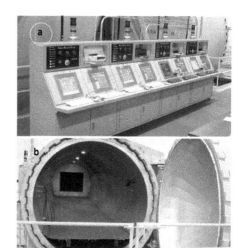

FIGURE 7-24. Autoclave and controls. [a] Multi-vessel control panel with large production autoclaves in background, and [b] small-diameter production vessel. *(Photos courtesy of American Autoclave Co.)*

Thermoforming

Thermoforming[5] is one of the oldest and simplest plastics forming processes. Baby rattles and teething rings were thermoformed in the 1890s, using cellulose-based plastics. The process saw major growth in the 1930s with the development of the first roll-fed thermoforming machines in Europe.

In the thermoforming process, heat and pressure are used to transform flat sheet thermoplastics (unreinforced or reinforced) into a desired three-dimensional shape. The sheet is preheated using one of three methods:

1. *Conduction* via contact heating panels or rods;
2. *Convection* heating, using ovens which circulate hot air;
3. *Radiant heating* achieved with infrared heaters.

The preheated sheet is then transferred to a temperature-controlled, pre-heated mold and conformed to its surface until cooled. The final part is trimmed from the sheet, and the trim can be reground, mixed with virgin material and reprocessed. This ability to recycle trim waste is typically most viable for unreinforced and some chopped glass-reinforced thermoplastics.

---

5 Note: This section includes text from "The Thermoforming Process," April 2006 *Composites Technology* (© 2006, Gardner Publications, Inc.), used with permission.

There are numerous variations of thermoforming, distinguished primarily by the method used to conform the sheet to the mold.

Sheet bending uses a folding machine or jig to produce a gently angled or bent surface.

Drape forming, where the preheated sheet is stretched down over typically a male mold using either gravity alone or application of vacuum.

Vacuum forming uses negative pressure (vacuum) to conform the material to the mold. *See* Figure 7-25.

Pressure forming uses positive pressure (compressed air or press equipment) to conform material to the mold.

Diaphragm forming is a lower pressure process suitable for simple geometries, which uses air pressure to compress a flexible diaphragm against the sheet and conform it to the mold. *See* Figure 7-26.

Rubber forming, or rubber block stamping is a high-pressure process suitable for more complex geometries. It uses air pressure to compress an upper molding surface made from rubber to conform the sheet against the lower metal mold surface.

Hydroforming uses a pressurized fluid to conform the material. A thick rubber, flexible diaphragm, typically much larger than the part being formed, is attached to the upper platen of a press and supported from behind by a fluid medium, usually hydraulic fluid. The preheated composite material is placed on a lower mold surface, usually made from machined metal, which is attached to the lower platen. During compression, the fluid behind the diaphragm applies pressures as high as 69 MPa (~10,000 psi) for large systems, and produces more complex shapes in short cycle times. *See* Figure 7-27.

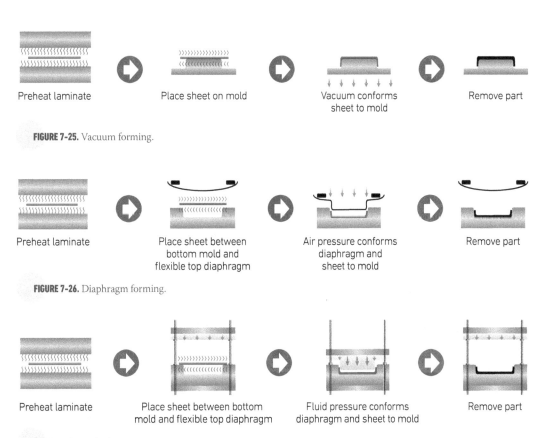

Preheat laminate   Place sheet on mold   Vacuum conforms sheet to mold   Remove part

FIGURE 7-25. Vacuum forming.

Preheat laminate   Place sheet between bottom mold and flexible top diaphragm   Air pressure conforms diaphragm and sheet to mold   Remove part

FIGURE 7-26. Diaphragm forming.

Preheat laminate   Place sheet between bottom mold and flexible top diaphragm   Fluid pressure conforms diaphragm and sheet to mold   Remove part

FIGURE 7-27. Hydroforming.

Matched die forming, or matched metal die stamping uses higher pressures and matched upper and lower metal molds. The mold is typically maintained below the thermoplastic melt temperature so that the part is formed and cooled at the same time. Cycle times range from 30 seconds to 15 minutes, with mold temperatures between room temperature and 350°F (25–177°C) and pressures from 100–1,500 psi. *See* Figure 7-28.

Twin-sheet forming positions two preheated sheets between two female molds with matching perimeters or contact surfaces. The twin sheets are drawn into the molds, using a combination of vacuum and air pressure, formed and then joined to make a single hollow part.

❯ THERMOFORMING MATERIALS

Materials commonly used in thermoforming include commingled thermoplastic fibers such as Twintex (fiberglass/polypropylene made by Saint-Gobain Vetrotex) and the glass, carbon and aramid products made by Comfil using a wide variety of thermoplastic polymers including PET, LPET, PP, PPS, PAEK, and PEEK.

There are also semipregs such as CETEX (made by Toray Advanced Composites) and TEPEX (by Bond-Laminates), which combine glass, aramid and/or carbon unidirectional or fabric reinforcements with a range of thermoplastic matrices (*see* Table 7-3). These materials may also be referred to as organosheet, which can describe either woven fabric laminates or cross-ply stacks of uni-tapes. Other suppliers include Porcher, Solvay and Teijin. In addition, there are suppliers that produce only unidirectional fiber-reinforced tapes and laminates.

Preheat laminate   Move sheet to die   Close die   Remove part

FIGURE 7-28. Matched die forming.

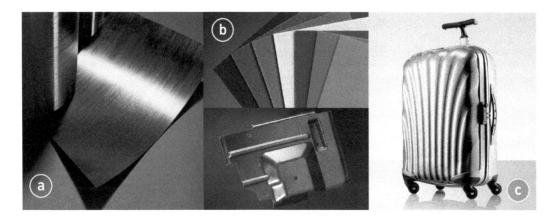

FIGURE 7-29. Thermoforming materials.
Reinforced thermoplastic materials for thermoforming include [a] Plytron unidirectional glass/polypropylene (PP) prepreg made by Gurit, and [b] CURV PP fiber reinforced PP made by Propex Fabrics, which is used for a variety of molded end-uses such as [c] the part shown here from Samsonite's "Cosmolite" suitcase. *(Photos courtesy Gurit, and Propex Fabrics)*

Another type of material is self-reinforced plastics (SRPs) such as Curv polypropylene fiber reinforced polypropylene made by Propex Fabrics. Note that thermoforming materials may also include thermoset prepregs (typically epoxy), specifically the fast-cure/snap-cure products formulated for press-molding. (Figure 7-29)

TABLE 7.3 Thermoforming Semipreg and Prepreg Materials

Product	Supplier	Description
Thermoplastic semipregs (tapes, fabric and woven or cross-ply laminate organosheets)		
CETEX	Toray	AF, CG, GF — PA6, PC, PC-ABS, PAEK, PEEK, PEKK, PEI, PES, PET, PP
Evolite™	Solvay	CF, GF — PA6, PA66, PPA, PPS, PVDF
Pipreg	Porcher Industries	AF, CF, GF — PEEK, PEKK, PPS, PEI, PC, TPU, PA12, PA6, PA66
Tenax®	Teijin	TPUD, TPWF (woven fabric), TPCL (consolidated laminate)
TEPEX	Bond-Laminates/Lanxess	CF, GF — PA6, PA66, PA12, PC, PP, PPS, TPU
	SGL Carbon	CF, GF — PA6, PA66, PP
Thermoplastic prepreg tapes		
Celstran® CFR-TP	Celanese	CF, GF — PA6, PA66, PBT, PP, PPS, TPU
Maezio™	Covestro	CF, GF — PC, PC-ABS, PC-PBT, TPU
Plytron	Gurit	GF — PP
Rilsan®	Arkema	CF — PA
Suprem™ T	Suprem SA	AF, CF, GF — PA12, PEEK, PES, PPS, TPI
UDMax™	SABIC	GPE 46-70, GPP 46-70
	CompTape	AF, CF, GF — PA6, PA66, PA12, PC-ABS, PE, PP
	Mitsui Chemicals Europe	CF — PP
Thermoset prepreg for press-curing (compression molding)		
CM10, CM11 OM14	Kordsa	Epoxy — under 3-min cure at 150°C Epoxy/vinyl ester — 1-2 min at 150°C
E732	Toray	Toughened epoxy — 4-min cure at 160°C
HexPly® M77	Hexcel	Epoxy — 2-min cure at 150°C
RP-570	PRF Composites	Epoxy — 4-min cure at 140°C
Smartcure™ SC160 SC110	Gurit	Epoxy — 5-min cure at 150° Epoxy — 15-min cure at 160° Epoxy — 20-min cure at 150°
SolvaLite™ 710-1 SolvaLite™ 730 SolvaLite™ 23	Solvay	CF, GF/epoxy — 3-min cure at 130-150°C CF/epoxy — 3-min cure at 150°C, a 1-min cure at 170°C GF/epoxy
Vestanat PP	Evonik	Polyurethane — 2-min cure at 140°C

❯ THERMOFORMING TAILORED BLANKS, OR PREFORMS

Thermoforming, in the form of matched die stamping, can be used for industrialized production of composite parts from prepregs and tapes. Though this type of production most commonly uses thermoplastic tape, it is also possible to use fast-cure thermoset prepreg. As a first step, the prepreg is stacked to form a tailored blank, which may include a mix of nonwoven, woven and unidirectional materials at various orientations. Industrialized production of these tailored blanks is commonly achieved using AFP and ATL.

There are numerous companies offering automated systems for rate production of preforms. Most use a rotary table and apply tape from a robot arm, gantry or cartridge system (e.g., the Fill "MULTILAYER"). The Broetje Automation STAXX 1700 and the Fill MULTILAYER systems can apply up to 16 tapes at a time, while Coriolis Composites can apply up to 32 tapes simultaneously. Compositence claims it can produce four 1.25m x 0.75m preforms in 60 seconds and deposit material at a rate of 250 kg/hr, while Dieffenbacher also creates four preforms simultaneously on its Fiberforge RELAY system yet touts a rate of 490 kg/hr. (Figure 7-30)

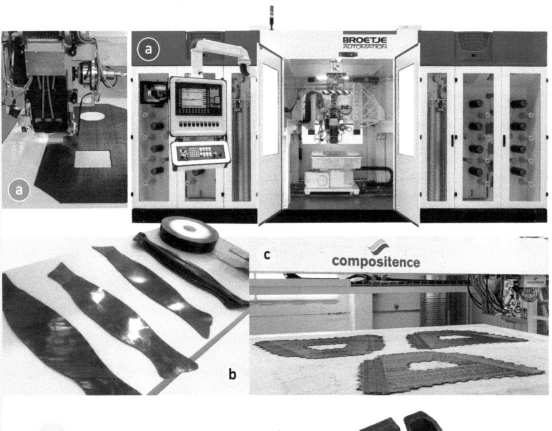

FIGURE 7-30. AFP and ATL systems for producing tailored blanks. *(Photos courtesy of [a] STAXX 1700 by Broetje Automation; [b] Van Wees UD and Crossply Technology, photo by Ginger Gardiner; [c] Compositence 2D DRY-Fiber Placement System; [d] multimaterials preform and "Part with QSP Process" by Cetim.)*

Other features shared by most of these systems include:

- Ability to handle different materials in a single preform.
- Ability to place local reinforcements and varying thicknesses.
- Automated spool changing system for uninterrupted production.
- Precise tape cutting.
- Heating method to tack and/or thermally bond tapes together.
- Production of 100,000 to 1 million parts/year can be supported with a single tailored blank and preforming line.

One of the main benefits offered by these systems is the ability to produce net-shaped preforms with almost no waste compared to cutting plies from roll goods (*see* Figure 7-31).

Once the tailored blanks have been completed, they may proceed directly to the stamping press or they may be preformed first. Either step requires preheating of the blank in an oven or infrared heating station. For preforming, the preheated blank will then be placed in a set of matched preforming tools, where heat and vacuum and/or pressure are applied to transform the 2D blank into a 3D-shaped, consolidated preform. (Figure 7-32)

FIGURE 7-31. Tailored blanks reduce waste.
Compositence claims use of its ATL/AFP system to create tailored blanks from prepreg tape cuts scrap waste by 25%, compared to cutting plies from fabric. *(Source: Fill Gesellschaft)*

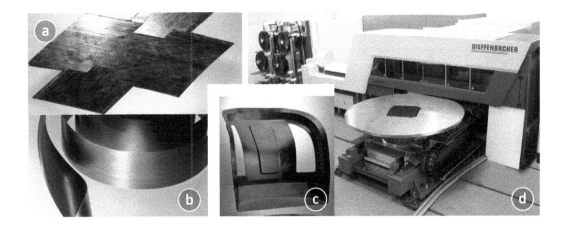

FIGURE 7-32. Thermoforming advanced composites from tailored blanks.
[d] Fiberforge RELAY Station that is used to [a] layup multiple plies of continuous unidirectional fiber-reinforced TP prepreg tape into [b] a tailored blank, which is then consolidated before thermoforming into a final composite structure like [c] this carbon fiber/polyamide 6 automotive seat frame. *(Photos courtesy of Dieffenbacher-Fiberforge)*

For an example of preforming process parameters, Dieffenbacher's Fiberforge tailored blank system uses temperatures of 300–800°F (150–425°C), depending on the thermoplastic polymer matrix used, and a typical pressure of 220 psi in a 450-ton press with a cycle time of 30–90 seconds. After consolidation/preforming, the final step is then molding which may include injection overmolding (*see* Chapter 8, page 221) in order to achieve intricate geometries and functionality on the part's surfaces.

❱ THERMOFORMING PARAMETERS

Thermoforming temperatures depend on the specific thermoplastic being used, but typically range from 300–800°F (149–425°C). Forming pressures can range widely. Low-pressure processes may require as little as 15 psi for vacuum forming to over 1,500 psi for high pressure matched metal die forming. Typical pressures are around 500 psi.

For low-pressure thermoforming materials, tooling does not have to be steel but, depending on the length of a production run, can be aluminum or machined epoxy or ceramic tooling board. Wood and even plaster tools can also be used for prototyping.

Compression Molding

Compression molding uses a heated matched mold similar to matched die forming, and has traditionally used thermoset matrix resins in bulk or sheet molding compounds (BMC or SMC) that liquefy and flow into the voids, producing complex composite shapes while holding a very high tolerance. The material properties are less than that of conventional composite parts because of the discontinuous fibers in the molding material. However, these materials have evolved to include unidirectional, woven and braided continuous fibers.

Compression molding also includes press molding of traditional aerospace prepregs as well as newer, snap-cure prepregs, developed for automotive applications. Molds can be made from steel, cast iron or aluminum and are built with integral heating and cooling for rapid turnaround and high-volume production.

The main differences between thermoforming and compression molding are those of pressure and material flow. Thermoforming has historically been a shaping process for plastic materials using heat and much lower pressures than are common with compression molding. When higher pressures are used in thermoforming, they simply help with surface finish, stamping fine detail into the parts, and forming certain complex geometries. The material to be shaped is generally the same size as the mold and is not expected to flow very much. In compression molding, however, a charge of material is placed in a tool, and with significant pressure and heat, is expected to flow throughout the mold and fill all cavities. (Figure 7-33, 7-34)

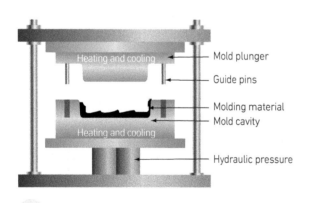

Mold plunger
Guide pins
Molding material
Mold cavity
Hydraulic pressure

FIGURE 7-33. Compression molding concept.

FIGURE 7-34. Compression molding method: [a] Install mold in press, [b] place material in mold, close mold and cure, [c] remove cured blank from mold, and [d] machine part to final dimensions. *(Source: Litespeed Racing, Los Angeles, CA)*

❭ COMPRESSION MOLDING MATERIALS

Thermosets

Originally, SMC and BMC materials were mostly made from polyester resin reinforced with chopped glass fiber. However, there are manufacturers of epoxy, cyanate ester, and BMI resin based bulk molding compounds used for manufacturing highly detailed structures with thin and thick sections, as well as molded-in features such as threads and inserts for satellite, aerospace and defense, commercial aircraft, industrial and sporting applications.

Carbon fiber sheet molding compound (CFSMC) was utilized by Lamborghini and Callaway Golf Co. in a joint effort to develop Forged Composite™ technology that was later used to manufacture the Lamborghini Diablo Octane body and tub, as well as a variety of composite drivers for Callaway. The material consists of chopped **carbon fiber tows** that are sandwiched between adhesive layers in a prepreg type format. The material is easily cut to near net size, layered or rolled up and compression molded to almost any shape. (Figure 7-35)

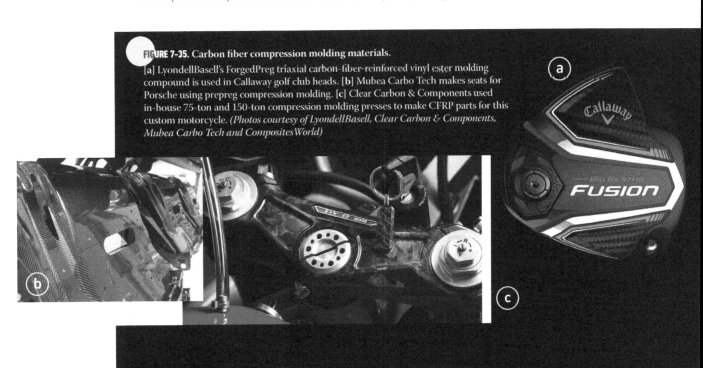

FIGURE 7-35. Carbon fiber compression molding materials.
[a] LyondellBasell's ForgedPreg triaxial carbon-fiber-reinforced vinyl ester molding compound is used in Callaway golf club heads. [b] Mubea Carbo Tech makes seats for Porsche using prepreg compression molding. [c] Clear Carbon & Components used in-house 75-ton and 150-ton compression molding presses to make CFRP parts for this custom motorcycle. *(Photos courtesy of LyondellBasell, Clear Carbon & Components, Mubea Carbo Tech and CompositesWorld)*

Thermoplastics

No longer confined to thermoset materials, compression molding is used to mold glass mat thermoplastic (GMT) and long fiber reinforced thermoplastic (LFRTP) materials into automotive hoods, fenders, scoops and spoilers. There are a wide variety of GMT thermoformable sheet materials made by companies such as Azdel, Quadrant Plastic Composites, Owens Corning, and FlexForm Technologies (which uses natural fibers such as kenaf, hemp, flax, jute or sisal, blended with thermoplastic matrix materials, such as polypropylene or polyester).

Prepregs

Autoclave may be the first method that comes to mind when processing prepreg, but actually compression molding—also called press molding, hot press molding, press curing and stamping— has been used with prepregs for decades. Press molding is listed as a processing method in guidelines from the late 1990s for products such as Advanced Composite Group (ACG) LTM (low temperature molding) epoxy prepregs, and Fiberite epoxy and cyanate ester prepregs. Since that time, most epoxy prepregs list press molding as a process method, as do many cyanate ester and bismaleimide prepregs. NASA published its optimized press molding for polyimide prepregs in 1991.

Though part quality is comparable to autoclave processing, due to the use of presses, this method is typically limited to small and medium-sized parts as well as relatively simple geometry when processing continuous fiber prepreg. Because matched molds are used, high surface quality is possible on both sides. Example parts include aircraft interior linings, automotive decorative parts, cowls and screens, industrial fittings and covers, glass fiber printed circuit board laminates, skis and prosthetics.

By 2010, prepreg press molding was an established method for producing composite automotive parts. In 2011, Mitsubishi Rayon Co., Ltd. promoted its prepreg compression molding (PCM) process and epoxy PCM material, using either unidirectional or woven carbon fiber, claiming up to 65% fiber content by weight and 2-minute cure at 150°C/302°F using a molding pressure of 3–10 MPa. It also claimed lower total manufacturing cost compared with autoclave for >200 parts per month (assuming a 1,200 mm × 700 mm part, 1.1 mm thickness and 120-min autoclave cure versus 10-min PCM cure). PCM was used for the deck-lid inner and outer panels of the 2014 Nissan GT-R supercar. (Figure 7-36)[6]

Compression molding of prepreg is also very suitable for hybridizing. For example, the Composite Technology Center (CTC, Stade, Germany) has developed a process for compression molding of aircraft parts—such as door frames—using continuous fiber prepreg with sheet

FIGURE 7-36. Compression molded prepreg.

[a] Huntsman Advanced Materials compression molded this 550 mm x 500 mm x 35 mm part demonstrating the 2.5D capability of its snap-cure epoxy in a continuous carbon fiber prepreg. [b] Mitsubishi Rayon created a net-shape blank from towpreg using automated fiber placement and then preformed and compression molded that into this liftgate inner panel, similar to its PCM part on the 2014 Nissan GT-R supercar. *(Photos courtesy of Huntsman Advanced Materials, Mitsubishi Rayon Co., Ltd. and CompositesWorld)*

6   Source: Peggy Malnati, "Lower cost, less waste: Inline prepreg production," March 2016 *CompositesWorld*.

molding compound (SMC) on the outer surfaces to provide more complex geometry, such as ribs, bosses and attachment points. Such hybrid SMC parts can be cured in 120–180 seconds at 145–155°C, achieving a low-waste, high material usage rate of 90%. (Figure 7-37)

Prepreg compression molding has also been combined with injection molding. In the OPTO-Light project completed by AZL Aachen (Aachen, Germany), unidirectional carbon fiber/epoxy tape was transformed into net-shape blanks using automated placement methods. These blanks were then compression molded and subsequently overmolded (*see* Chapter 8, page 260) with short glass fiber-reinforced PA6 thermoplastic to demonstrate a section of the BMW i3 floor structure, with molded-in clips for attachment. Dozens of such parts verified a 75-second cycle time using a single hybridized molding machine. Two different process routes have been demonstrated, a 3-step method using laser ablation as a pretreatment of the CF/epoxy surface prior to overmolding, and a shorter 2-step process where laser pretreatment is eliminated, instead using latent reactivity in the epoxy to bond with the overmolded PA6 composite. (Figure 7-38)

FIGURE 7-37. CTC Stade hybrid compression molded prepreg and SMC. *(Photos courtesy of and © CTC Stade, an Airbus Company)*

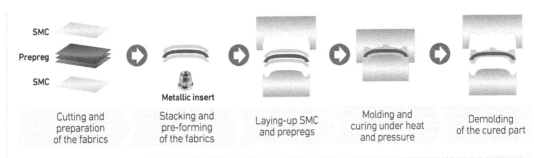

FIGURE 7-38. OPTO-Light hybrid compression molding and injection molding. *(Photo courtesy of AZL Aachen)*

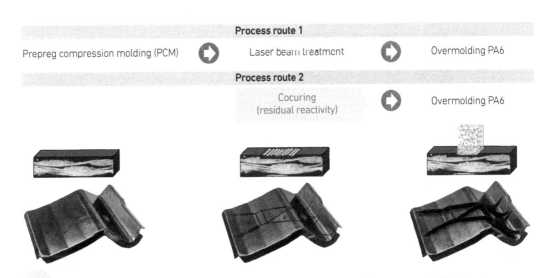

❯ PROCESS PARAMETERS

SMC is typically molded at pressures close to 1,000 psi, and sheet materials competitive with GMT can require pressures as high as 3,000 psi. There are also low-pressure SMC materials that only require 150 psi. These low-pressure molding compounds (LPMCs) require only a 700- to 750-ton press versus the 3,000-ton presses more common with traditional SMC and BMC materials.

It is also possible to use smaller presses. For example, as shown in Fig. 7-35 above, a pair of 75-ton and 150-ton presses are used by Clear Carbon & Components to make a wide variety of composite parts, including drone bodies, coiled springs, and fittings for carbon fiber furniture and yachts.

❯ CONTINUOUS COMPRESSION MOLDING

Continuous compression molding (CCM) is different from conventional compression molding in that the process is designed to make continuous cross-section stock such as flat laminates, I-beams, C-channels, Z-stringers, hat sections, etc. from fiber reinforced thermoplastic (TP) prepregs or dry forms stacked with TP films. Low void (<1%), highly consolidated parts can be made using the CCM method.

CCM processing allows almost endless manufacturing of laminated composite stock; the material is pulled across a heater and through the press by way of a feeder unit, which opens and closes intermittently as the heated materials progress through an automated platen or die-set located in the press. Pressures of up to 400 psi and temperatures of up to 800°F (427°C) enable the production of flat organo-sheets, or shaped stock that are cut to desired lengths. (Figure 7-39)

Same Qualified Resin Transfer Molding Process

The "same qualified resin transfer molding" (SQRTM) process was developed in the late 1990s by engineers at Radius Engineering (Salt Lake City, UT). This out-of-autoclave process allows net-shape manufacturing of complex, monocoque composite parts. The SQRTM process uses

FIGURE 7-39. Continuous compression molding.

Examples of possible shapes that can be molded via CCM. *(Photos courtesy of XELIS GmbH, a member of AVANCO Group)*

aerospace-qualified prepreg including standard toughened prepreg systems. SQRTM follows the qualified cure process and provides quality superior to autoclave-processed parts, and it also consumes less energy and allows faster cycle times than typical autoclave processing. (Figure 7-40)

To facilitate the SQRTM process, a platen press, located within a work cell and sized for the tooling it will support and cure, is equipped with upper and lower steel bolsters that have been ground flat on one side. The tool/part is placed within the press on the lower bolster plate, which sits on a series of air bags (like fire hoses) that, when inflated, force the plate up, providing uniform pressure against the tool and upper bolster plate. Both the upper and

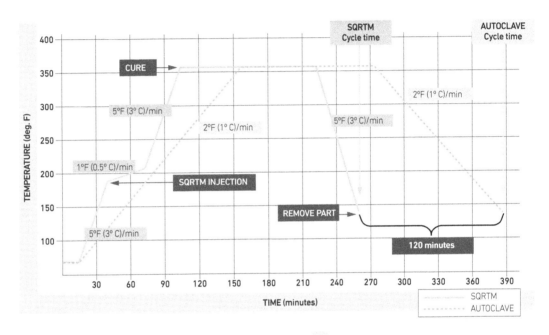

FIGURE 7-40. SQRTM vs. autoclave cure cycle.

Higher thermal conductivity with the press and tool *(bottom)* allow for faster heating and cooling; as a result, the cure cycle can be as much as two hours shorter than an autoclave cycle *(top)*. *(Photo/chart courtesy of Radius Engineering)*

lower bolsters are heated (electrically or via water) and cooled (water) through independent heating/cooling zones within the plates, allowing for precise temperature adjustments during cure. (Figure 7-41)

The way SQRTM works is that the prepreg plies are laid up on mandrels and assembled in a cavity mold, the mold is closed and installed in a press, high vacuum is drawn, and then liquid resin is injected into the tool. The injected resin is the same qualified resin as that used in the prepreg but is not intended to impregnate the material, only to provide uniform hydrostatic pressure around the edges of the part and throughout the preform. This compresses volatiles back into solution, thus preventing void formation during cure. (Figure 7-42)

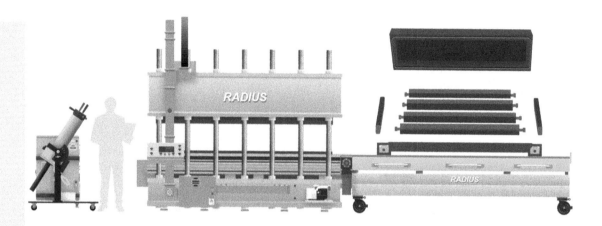

FIGURE 7-41. SQRTM work cell and manufacturing process.

From right to left: the internal mandrels with prepreg layups are assembled on the tool base and enclosed within the outer tool shell. The sequence is as follows: move tool into press, clamp and heat tool, prepare resin injector, connect resin and vacuum lines, inject resin, develop an internal hydrostatic pressure, and heat to cure temperature. Hold at cure temperature, then cool press, disconnect lines, unclamp press, move tool out of press and demold part. *(Source: Radius Engineering)*

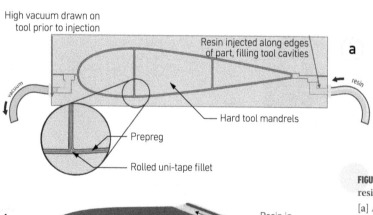

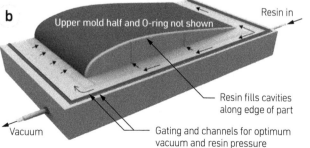

FIGURE 7-42. The SQRTM concept; injected resin provides fluid pressure on laminate.

[a] Air and gas are evacuated from the prepreg layup prior to injection, facilitated by precisely placed gating and channels within the tool. [b] Resin is injected and fills the cavities along the edges of the part, producing a uniform fluid pressure of approximately 100 psi (6.89 bar). The part is then heated and cured. *(Source: Radius Engineering)*

Bladder Molding

Bladder molding is used to make hollow parts with a good surface both inside and out. Prepreg and a silicone or natural rubber bladder are placed into a cavity mold. A press or other method is used to clamp the mold shut. Heat is applied to flow and cure the resin. Air pressure is used to inflate the bladder and push the laminate outward against the cavity mold. The part is consolidated and cured, after which the deflated bladder is removed. (Figure 7-43)

Note that bladder molding can also be used with a dry laminate stack (preform) and impregnated with resin using a resin infusion or resin transfer molding (RTM) process. For more details, *see* Chapter 8, **Liquid Resin Molding Methods and Practices.**

FIGURE 7-43. Bladder molding process. *(Photos courtesy of (top and left) Piercan, and (right) Heidi Dorworth)*

8

Liquid Resin Molding Methods and Practices

CONTENTS

Overview of Liquid Resin Molding

Applying liquid resin to fabric by hand—what we know as hand layup or wet layup—was one of the first composite molding methods, evolving out of dope and fabric used to make early aircraft wings. This is an open molding process, meaning the shaping and curing of the resin is achieved while it is open to the atmosphere/ambient environment. (The exception is when wet-layup is vacuum bagged and cured.)

Spraying, where chopped fiber and resin are mixed and applied to a mold using spray equipment, is another common open molding method, used mostly to produce shower and bath enclosures, spa tubs and some boats. Hand layup can be semi-automated via **impregnators**, or wet-pregging machines, which draw in fabric and apply precisely mixed resin to produce lengths of wet-out fabric that are then applied to a mold.

Another type of automation can be seen in **filament winding**, specifically **wet winding**, where equipment draws individual fiber tows through a resin bath and applies it to a rotating mandrel. This is also an open molding process.

The remaining liquid molding methods are closed molding processes. These include **liquid compression molding** (LCM, or wet pressing), which can be either a manual or automated form of layup hybridized with compression molding. That is, in this process people or robots mix and apply resin to fabric and then transfer the wet-out fabric into a heated press for cure and consolidation. Other methods covered in this chapter include:

- **Resin infusion** and **resin transfer molding (RTM)**, where liquid resin is infused or injected into a dry fiber perform.
- **Pultrusion**, a continuous process where dry reinforcements are pulled through a set of resin injection and heated die mechanisms.
- **Centrifugal casting**, a process used to make pipe where resin is injected into a centrifugally spinning mold, permeating fabric laid on the mold's interior surface.
- **Injection molding**, where a thermoplastic polymer reinforced with discontinuous fibers is injected into a mold, typically under high pressure.

TABLE 8.1 Comparison of Wet Layup vs. Infusion vs. Prepreg

Wet Layup	Vacuum Infusion	Prepreg Layup
Advantages		
• Inexpensive • Usually low-temperature • Special equipment needs minimal • Vacuum bag processing is an option but not always necessary	• Inexpensive • Usually low-temperature cure but elevated temp. cures possible • Less messy than wet layup • Containment of volatile organic compounds (VOCs) • Easy to control resin/fiber ratio accurately	• Little mess or smell • Lower exposure to health risks • Easy to control resin/fiber ratio accurately • Many high-performance matrix systems available • Easier to cut to shape than many dry materials • Long working time with most production prepregs
Disadvantages		
• Messy • Can be smelly, especially with polyester and vinyl ester resins • Greater exposure to health risks • Runaway exotherm is a possibility • Limited amount of time to finish the lay-up • Difficult to control resin/fiber ratio accurately	• Low wetting of fibers between layers • Poor cosmetic surfaces without surface coat resin • Runaway exotherm possibility in source buckets • Accurate cutting and placement of preforms is vital • Special materials and equipment required • Vacuum bagging is necessary	• More expensive than wet layup or infusion • Frozen storage required for most systems • Proper thawing technique critical to quality of finished part • Requires vacuum bagging and elevated-temperature curing, often requires autoclave cures • Sensitive to moisture, dirt, and oil contamination

Liquid molding methods range in cycle time from short (seconds for injection molding) to long (hours for low-pressure RTM of large, aerospace structures). They may use single-sided molds (hand layup, resin infusion, filament winding, centrifugal casting) or matched mold die-sets (pultrusion, RTM, injection molding).

Matrix materials may be thermoset or thermoplastic, with the latter sometimes relying on in-situ polymerization. This is when a thermoplastic monomer—which typically has a very low viscosity like water—is infused or injected into the fiber reinforcement and then converted into a polymer (long molecular chains), while the composite is being shaped and consolidated with heat and pressure during molding.

As noted in the previous chapter, the selection of which process to use depends on the reinforcement type, resin type, part-shape complexity, production volume and cost, among other factors.

Hand Layup – Wet Layup

Wet layup is the one of the simplest, least expensive, yet most labor-intensive methods of composite manufacturing because it involves manually wetting out the reinforcement fibers with resin. It remains a common method for applications like boats, kit-built small aircraft, canoes and other small-volume sporting goods.

Fiber reinforcement is typically in the form of continuous or chopped strand mat, woven fabric, non-crimp fabric and/or nonwoven materials. The resins used are normally (but not always) cured at room temperature and include polyester, vinyl ester and epoxy. (Figure 8-1)

❱ PROCESS AND PRINT-THROUGH

Wet layup open molding with polyester or vinyl ester resins often begin by coating the mold with a gelcoat, which is cured before lamination can begin. Next, a layer of filled resin called an interface coat or barrier coat may be applied. Alternatively, the first layer is the skin coat, which comprises a wet layer of laminating resin into which reinforcement is applied—typically nonwoven veils or chopped strand mat. A skin coat is used to prevent print-through, which is when the pattern of the heavier fiber reinforcement and/or core material in a laminate can be seen on the surface of the finished part due to resin shrinkage. The amount of shrinkage is affected by:

- **Type of resin**—polyesters can shrink up to 8%, vinyl esters are typically a little less (shrinkage for epoxies is up to 2%).
- **Fiber volume**—the more fiber, the less shrinkage.
- **Amount of exotherm**—high exotherm causes more shrinkage than low exotherm during resin reaction and cure.
- **Temperature cycling of finished part**—extremes in high and low temperatures can cause shrinkage (for example, leaving molds out in winter, black hulls in hot climates).

More wet resin is applied on top of the skin coat fiber reinforcement and worked into the fibers using a squeegee (*see* Figures 8-1 and 8-2), brush and/or roller. Successive layers of reinforcement and resin are applied, with the goal of completely impregnating every fiber with resin. Once all laminate plies have been laid and wet out, the laminate is left to cure.

The process is similar for wet layup of epoxy (or other) resin laminates that will be processed under a vacuum bag. The key differences are that a simple liquid resin coat is applied in lieu of a gel coat or surface coat and that the starting resin quantity is carefully measured in order to obtain the best results. The laminate may be vacuum bagged with a prescribed bleeder breather system and cured under vacuum. This results in a laminate with a higher fiber volume and low shrink that minimizes print-through.

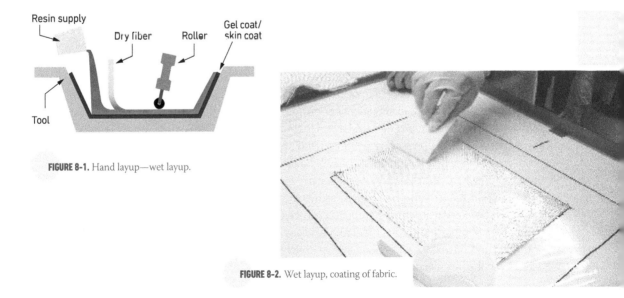

FIGURE 8-1. Hand layup—wet layup.

FIGURE 8-2. Wet layup, coating of fabric.

❱ QUALITY ISSUES

Variation in the wet layup technician skill level and techniques can result in significantly different laminate properties. For example, one worker may easily wet out the reinforcement using 1 oz/yd^2 of resin while another worker uses 4 oz/yd^2 of resin on the same amount of fabric. Thus, consistent resin content and resin-to-fiber ratio is not easy to achieve without careful material calculations and controls. Also, too much rolling can break fibers and bring air back into the laminate, causing porosity and voids. It is sometimes difficult to control fiber orientation in a wet layup, as the materials tend to slide around. These issues will decrease the laminate strength and overall part performance.

Shop conditions may also cause issues with consistent quality of parts. Decreased temperature usually means more time to work the resin but may increase resin viscosity, which makes wetting the fibers more difficult. Increased temperature usually lowers resin viscosity but also the working time before the resin gels. The process can be improved by controlling the ambient temperature in the work space and prescribing the necessary amount of resin that is used to wet out the fibers without producing a resin rich part.

To increase quality, the wet-out laminate can be vacuum bagged with a properly designed bleeder and breather schedule to achieve good compaction, excess resin extraction, and removal of trapped air and gas in the laminate during cure. The part is cured under vacuum using a heated mold, oven or other form of heating. To achieve full cure throughout the part, thermocouples or other sensors should be used to measure the laminate temperature and manage either manual or automatic control of the heating equipment. The best quality can be achieved using material state management (MSM) or cure state management, as explained in the text and illustrations about cure cycles in Chapter 2 (*see* pages 55–60).

Preforms

Preforms are fiber reinforcements that have been oriented and pre-shaped to give the finished composite structure the desired shape and load path requirements. They are most commonly used in liquid molding processes such as resin infusion and resin transfer molding (RTM). In these processes, preforms must be permeable to the injected resin, enabling good fiber encapsulation without dry areas or voids, in order to achieve higher properties.

Depending on the desired molding process and rate of production, making preforms (preforming) may involve manual methods, such as hand placement wherein a technician manipulates materials over a plug or inside a mold. In more sophisticated automated processes, robots cut and place materials into a preforming tool where heat and pressure are applied to create an easy-to-handle single piece. Shapes range from a simple contour to complex shapes with corners, local recesses and multiple thicknesses.

Preforming can be adapted to nearly any material type or form. *Fiber* preforms can be made from dry or prepreg chopped fiber, strand, tows, yarns, tapes, braids, fabrics, nonwoven materials or combinations thereof. Layers that make up a preform can be bound together with polymers, binders, 3D weaving, pinning, stitching, knitting, or other methods of attachment. (*See* Chapter 3 **Fiber Reinforcements**.)

Historically, preforming only applied to dry fabric materials for liquid molding processes. Today, preforming of dry fiber, as well as thermoset and thermoplastic prepregs are mainstream methods in composite production processes. Table 8.2 lists various technologies, suppliers of preforming equipment and services, as well as example composite parts. (Braiding, knitting and 3D textiles were discussed in Chapter 3.) The remaining methods are detailed briefly in the next section.

TABLE 8.2 Preforming Technologies and Suppliers

Technology	Companies	Examples
3D textiles	• Albany Engineered Composites • Sigmatex • Seriforge	• LEAP engine fan case, blades • One-piece nodal connectors • Brackets
Automated Tape/Fiber Placement	• Broetje-Automation (STAXX) • Cetim (Quilted Stratum Process) • Cevotec (Fiber Patch Placement) • Compositence • Coriolis • Dieffenbacher (Fiberforge) • Fill Austria (Multilayer) • Steeg & Hoffmeyer (F2 Compositor) • Van Wees	• Auto and aerospace parts, auto B pillars • Bumpers, door beams • Helmet, bike seat, truck chassis part • Auto, aero, industrial parts, seat shell • Auto roof, B-pillar, engine mount • Auto roof, load floor • Auto rear walls, chassis parts • Auto and aerospace parts • Auto door beam, B pillar
Braiding	• A&P Technologies • Herzog braiding machines • Munich Composites (Braidform) • Revolution Composites • SGL Carbon	• Sporting goods, GE engines • Dowty propellers, BMC bikes • Tail rotor driveshaft, bike parts • Robots, rockets, submarines, struts • BMW 7 Series roof rails
Continuous Forming	• Applus (Glide forming) • Broetje-Automation • COPRO (continuous roll forming)	• Aircraft stringers • Aircraft frames and stringers • Curved aircraft frames, stringers
Knitting	• Saint Gobain Performance Plastics	• Aircraft ducts
Stacked Woven or Noncrimp Fabric	• American GFM • CIKONI • Composite Alliance Corp. • Danobat • Dieffenbacher • Schmidt & Heinzemann • SAERTEX	• Auto front bumper beam • Auto rear trunklid, beams • Aircraft beams, stringers, seats • Large spans, aircraft wings, wind blades • Auto trunk lid inner shell • Auto underbody, leaf springs • Wind blade, PRSEUS aircraft stringer-and-frame complex preform
Spray Deposition/ Directed Fiber Preforming	• Aplicator System AB • Carbon Conversions 3-DEP • CFP Composites (ROCCS process) • Sotira Composites • Tenax Part via Preform (PvP) • Schmidt & Heinzemann • Bentley Motors/Univ. of Nottingham (Directed Carbon Fibre Preforming)	• Auto and marine parts • Auto door bolster support, upper plenum • Fire-resistance, ships, oil & gas, industrial • Auto decklid, door & tailgate surrounds, sills • Auto assembly carrier — • Auto wheel well, floor pan
Tailored Fiber Placement	• Filacon • Keim Kunststofftechnik • Leibniz Institut for Polymer Research Dresden (IPF) • ShapeTex • TFP Technology • Tajima (stitching equipment) • LayStitch Technologies • ZSK (stitching equipment) • Hightex	• Heated preforms, net-shape with holes • Panels, brackets, stitched foam inserts • Robot arm with integrated heating, pump rotor, recurve bow riser, race car rear wing • Auto parts • Discs, rotors, auto parts, sporting goods — — — • Aircraft omega frame, A350 window frame, NH-90 helicopter longerons

❭ PREFORMING METHODS AND TERMINOLOGY

Automated tape/fiber placement—Dry bindered tow or tape is cut to tailored lengths and applied, typically to a flat layup table in a high-speed process. These automated systems are able to lay fibers in any direction by using a layup table that rotates, a multi-axis robotic placement head and/or a rotating placement head on a gantry. The resulting 2D tailored blanks are then shaped in a preforming tool and finally transferred into a liquid molding press. (Figure 8-3)

Binder—Liquid or powder, typically thermoplastic, which is applied to dry fiber and tape in small percentages (e.g., 1-5% by weight), providing tack so that materials can adhere to layup tables, tools or other preform layers during automated placement processes. The binder melts with heat and pressure in subsequent preforming, producing a consolidated, shaped preform for resin infiltration during liquid molding. Epoxy-based binders are also available that react with and/or dissolve into epoxy liquid molding resins.

Continuous forming—Bindered dry woven or multiaxial noncrimp fabrics and/or unidirectional tapes are heated and fed into a series of rollers (roll-forming). These apply pressure for compaction to set the shape of long profiles such as stringers or beams. Hybrid layups with metal plies can also be processed. (Figure 8-4)

Draping and drapability—The process of bending a reinforcement into a 3D shape. Depending on the bending properties of the reinforcement and the complexity of the preform shape, defects may occur, including wrinkles, gaps between fibers and skewing of fibers, causing deviations from the desired fiber orientations. Draping simulations are commonly performed

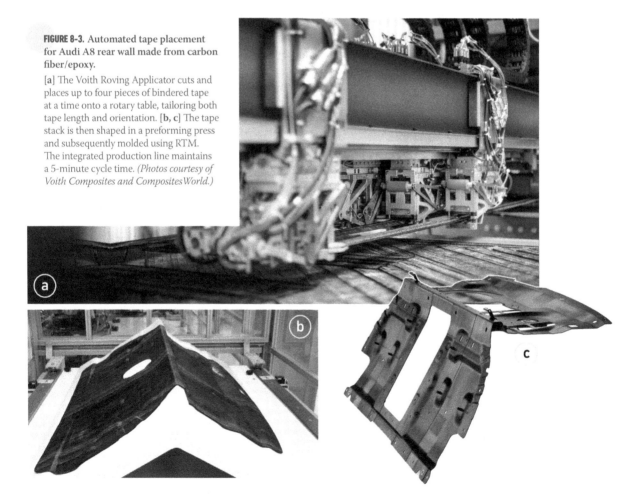

FIGURE 8-3. Automated tape placement for Audi A8 rear wall made from carbon fiber/epoxy.

[a] The Voith Roving Applicator cuts and places up to four pieces of bindered tape at a time onto a rotary table, tailoring both tape length and orientation. [b, c] The tape stack is then shaped in a preforming press and subsequently molded using RTM. The integrated production line maintains a 5-minute cycle time. *(Photos courtesy of Voith Composites and CompositesWorld.)*

as part of the design process and standard drapability tests can be done to determine a fabric's conformability.

Grippers—These are used to pick up 2D tailored blanks, 3D preforms and finished parts. Vacuum grippers are typically used for thin fabrics, films, prepregs and cured laminates, while needle grippers are often preferred for more porous, flexible fabrics that are difficult to pick and place with vacuum.

Spray deposition/directed fiber preforming—A robot sprays chopped fiber with binder (e.g., 3-5% by weight) onto a preforming surface, such as:

- Part-shaped screen through which vacuum is applied
- Magnetized mold (trace amounts of cobalt or ferrous elements are added to binder) to achieve fiber alignment and ≈80% of continuous fiber properties.

See Figure 8-5.

FIGURE 8-4. Continuous roll forming of composite performs.
COPRO Technology's process uses rollers to produce straight and curved profiles with variable curvature, cross-section, width and thickness, as well as localized ply build-ups/patch reinforcements and localized shaping (e.g., bump-outs like joggles in stringers). *(Photos courtesy of DLR (CC by 3.0))*

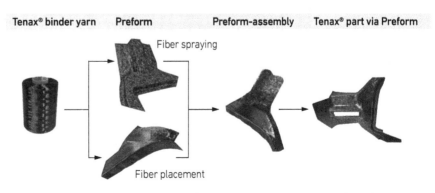

FIGURE 8-5. Fiber spray deposition used in the Teijin Part via Preform (PvP) process. *(Photo courtesy of Teijin)*

Stacked woven or noncrimp fabric—Individual plies are cut and stacked manually or via automated equipment and then shaped in a heated preform press and tool prior to liquid molding. (Figure 8-6)

Tailored fiber placement—Continuous tows or rovings are placed precisely and stitched onto a substrate, typically a veil, woven fabric or dissolvable foil. TFP allows orienting fiber in shapes other than just a straight line. Selective stitching can be used to make the 2D preform more stretchable and easily formed into the final 3D part. Industrialized production is possible by using multiple stitch-heads to produce many identical preforms simultaneously. (Figure 8-7)

FIGURE 8-6. Automated stacking of noncrimp fabric for automotive performs.

Non-crimp fabric [a] being cut, [b] being picked up with grippers, and [c] in the preforming mold after draping. *(Photos courtesy of Dieffenbacher)*

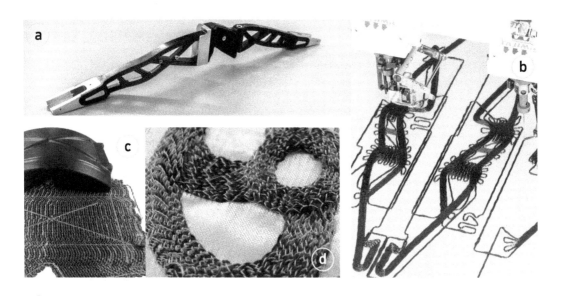

FIGURE 8-7. Tailored fiber placement (TFP) for composite performs.

[a] The complex-shaped riser for a high-performance recurve bow is made from complex TFP preforms [b] and resin infusion. [c, d] ShapeTex can stitch up to 12 preforms simultaneously and use a wide range of fibers and substrates to produce complex parts with almost zero waste. *(Photos courtesy of [a] IPF/Emanuel Richter; [b] IPF/J.Läsel; [c,d] ShapeTex and Composites World)*

Vacuum/Resin Infusion

Vacuum infusion processing (VIP) is a cleaner and more precise process than wet layup because all reinforcements are placed into the mold dry and vacuum bagged, after which resin is pulled into the laminate stack through tubing (resin feedlines) due to the pressure differential created by pulling vacuum under the bag. Cure takes place after the laminate is fully infused with resin. This process is also commonly called resin infusion, vacuum infusion processing, or vacuum assisted resin transfer molding (VARTM). (Figure 8-8)

It should be noted here that industry uses a great number of descriptions and acronyms to identify these (and other) processes and that there is no one standard that governs the use of these terms. It is recommended that the reader understand each process and avoid being overly concerned with the use of the various acronyms.

Resin infusion is used to produce, for example, skateboard decks and other sporting goods, bus bodies, boat hulls and decks, wind turbine blades, bridge decks, 6m x 13m gates for locks in Dutch canals, 5m x 18m decks for scrubbers in power plants, re-entry heat shields on spacecraft, and aircraft structures such as the Boeing 787 movable trailing edge, rear pressure bulkheads for the 787 and Airbus A350, empennage for the Mitsubishi Regional Jet and wings for the single-aisle MS-21 commercial airliner.

FIGURE 8-8. Resin infusion of composite upper body for Proterra Catalyst electric bus. *(Photo courtesy of TPI Composites and CompositesWorld)*

TABLE 8.3 Benefits of Infusion Processing vs. Hand Layup and Potential Issues

Benefits	Issues
Higher fiber-to-resin ratio.	▶ Molds must be vacuum tight – leaks will introduce porosity and voids in the laminate.
More compacted, higher-quality laminates.	▶ Laminates are thinner, so design must be reviewed for stiffness and deflection.
Less wasted resin.	▶ Large amounts of resin are mixed and infused in short time periods; therefore, training, preparation and detailed process control are mandatory to avoid ruining the part.
Cleaner, closed-mold process attractive to workers.	▶ Process is more complex and requires significant amounts of consumable materials (resin feedlines, flow media/mesh, vacuum bag film).
Layup time not limited by resin open time/working time (layup is done dry, before impregnation).	▶ Layup should "soak" under vacuum to remove all ambient moisture (prevent resin volatization).
Resin-infused laminates can achieve aerospace industry quality with appropriate materials and processes.	▶ High-quality surface finish is harder to achieve.
Volatile organic compounds (VOCs) released during cure are contained within vacuum bag.	▶ Resin should be degassed to reduce risk of porosity.
Good for producing low volumes of large structures cost-effectively without an autoclave.	▶ The "one-shot" process can move swiftly once started, affording little time to fix errors.

❯ PROCESS STEPS

For resin infusion processing, workers do not have to apply or roll the resin into the reinforcements but there are many more steps and process materials involved. (Figure 8-9)

Mold Preparation

The mold must be vacuum tight (*see* "Vacuum Leak Tests" below) and should have at least a 6-inch-wide flange to accommodate placement of plumbing—spiral tubing for pulling vacuum along the perimeter of the infusion setup for a resin-break zone, and sealant tape (or a seal system) for sealing the vacuum-bagged layup to the mold.

The mold should be treated with a release agent; special formulations may be required for infusion. (Figure 8-10)

Preforming

Dry fabrics and/or kitted preforms are applied to the mold. Spray adhesive is typically used on vertical and complex-shaped surfaces. *Caution:* the wrong type and/or too much spray adhesive can degrade laminate properties and interfere with resin curing. Preforms should be designed to allow fiber movement in the radii to alleviate bridging in the laminate.

Processing Materials

See Figure 8-11 for illustration of some of the following:

Tacky tape (sealant tape) creates a tight seal for the vacuum bag film around the infusion layup.

Peel ply is the first layer of the "bagging stack" and provides a protective layer that is peeled off the finished part.

Perforated release film is used between the peel ply and the flow media/mesh, enabling the mesh (saturated with resin after infusion) to be removed separately yet leaving the peel ply on the part until later. The film's perforations are essential to maintaining resin flow.

Flow media is typically a plastic mesh made from LDPE or HDPE that helps resin flow from the resin feed line across the laminate.

Resin feed lines are positioned on top of the flow media. These are non-collapsible tubes or omega profiles that supply resin under vacuum pressure to infuse the laminate. These lines

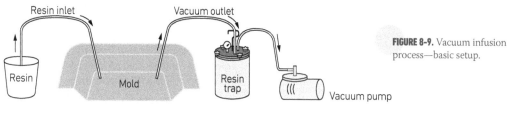

FIGURE 8-9. Vacuum infusion process—basic setup.

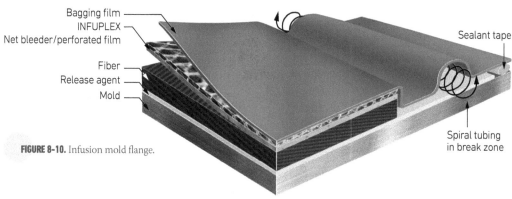

FIGURE 8-10. Infusion mold flange.

under the bag film are connected to the resin supply via sealed ports or resin feed connectors (similar to vacuum ports).

Spiral tubing can be used as resin feed lines that quickly move resin across the laminate. These must be in contact with the flow media. Spiral tubing is also used for vacuum plumbing.

Larger or more complicated infusions require well-planned positioning of multiple resin feed lines and spiral tubing.

Arrangement of Vacuum vs. Resin Feed

The basic concept of resin infusion is to pull vacuum on a bagged preform thereby pulling the resin into the low-pressure areas created by the differential. Thus, resin will flow from the resin feed lines, filling between the fibers in the preform and moving air and gas along the flow-front toward the vacuum lines.

One common configuration is to place the resin feed lines along the center of the part being infused and spiral tubing vacuum channels along the edges. The vacuum tubing is typically sheathed in a semi-permeable membrane or peel ply set-back from the edge of the laminate, creating a break-zone which allows air and gas, but not resin, to move into the vacuum lines. This minimizes the possibility of resin accumulation in the trap (catch pot) prior to the vacuum pump. (Figure 8-12)[1]

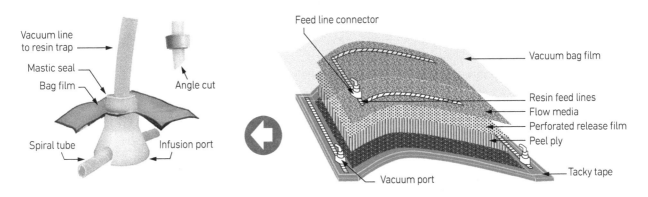

FIGURE 8-11. Infusion processing materials. *(Source: Vacmobiles)*

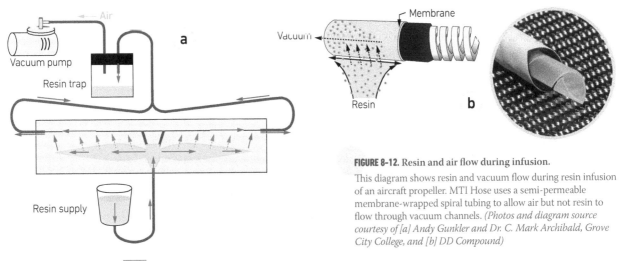

FIGURE 8-12. Resin and air flow during infusion.

This diagram shows resin and vacuum flow during resin infusion of an aircraft propeller. MTI Hose uses a semi-permeable membrane-wrapped spiral tubing to allow air but not resin to flow through vacuum channels. *(Photos and diagram source courtesy of [a] Andy Gunkler and Dr. C. Mark Archibald, Grove City College, and [b] DD Compound)*

1   Andy Gunkler and Dr. C. Mark Archibald, "Composite Propeller Construction"; Proceedings of the 2011 ASEE NC & IL/IN Section Conference, © 2011 *American Society for Engineering Education*, Grove City College, Grove City, PA

Both vacuum and resin feed lines are controlled by at least one clamp or valve, allowing quick on/off control during the infusion.

When vacuum is pulled on the bag, the preform will be compacted in place before resin is introduced into the laminate. Based on standard fluid dynamics, the flow of the resin is dictated by the low-pressure differential created under the bag and will continue until a pressure equilibrium is reached. The distance that the resin will flow is determined by the amount of pressure available (amount of vacuum), the viscosity of the resin, and the permeability of the fiber preform.

Multiple resin inlets are usually installed at select locations to allow for infusion of larger sized parts. As a rule of thumb, a span of no more than 24 inches/61 cm should be attempted between inlets—this may need to be lessened if vacuum or resin flow is compromised.

Channels for Resin Distribution

The use of channels to distribute resin is common practice in infusion molding. Channels are found in materials with an open area that creates a low-pressure cavity when placed under vacuum. From fabrics that intermittently skip a warp yarn, to core materials that are scored, many materials are available with channel attributes that promote resin movement.

While the use of channels can be beneficial to the process, care must be taken to prevent the channels from "race-tracking" the resin ahead of the desired flow-front. This causes dry areas to develop in the laminate (*see* Figure 8-13). Building test sections will aid the operator in determining the best use of these materials.

FIGURE 8-13. Race tracking.
Race tracking occurs when there is a channel or low-pressure area around a feature that allows the resin to flow ahead of the flow-front, resulting in a pressure equilibrium and a dry area left in the preform. *(Photo courtesy of Abaris)*

Interleaf Channel Layers

Materials such as continuous polyamide or nylon monofilament can be used internally as an **interleaf channel** between layers in a laminate to promote resin flow. This is especially useful when infusing layers of tape or spreadtow that do not naturally have much open space within the form. The big advantage to this material is that interleaf layers of monofilament increase the fracture toughness of the laminate by providing a small amount of damping between layers.

Vacuum Leak Check and Moisture Removal

Once the preform materials, bagging stack, and resin-feed and vacuum lines have all been prepared, the bag film is applied, pleated, and sealed. The vacuum and resin feed line connections are carefully inserted through the bag and sealed to prevent leaks from developing. An in-line flow meter is useful in finding leaks in the seal while bagging (*see* Figure 8-14). Resin lines are cut at an angle to be sure the feed line will not seal itself off at the bottom or sides of the container. The feed line is secured down in the resin bucket to ensure that it will not move causing air to be sucked into the line, and thus, into the laminate. Vacuum lines are connected to the resin pot and the pot is connected to the vacuum pump.

Before infusion, a vacuum leak check should be performed. An ultrasonic leak detector along with a flow meter can identify leaks in the bagged layup. A vacuum leak test should also be performed by installing a vacuum gauge inline on the vacuum/resin pot. Pull full vacuum on the bagged infusion layup, then shut the vacuum off using a valve to isolate the bagged part from the vacuum pump. Take readings via the vacuum gauge over a 5-minute period. As a rule, the vacuum leak rate should not exceed 0.5 inch Hg over 5 minutes (1 inch over 10 minutes).

Note: A vacuum leak test should be performed on the mold prior to use to ensure it is vacuum-tight and is not a source of leaks that cannot be mitigated.

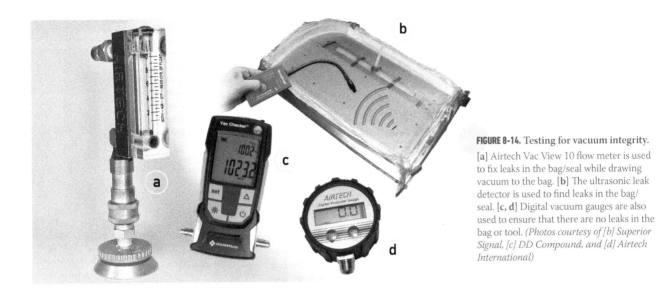

FIGURE 8-14. Testing for vacuum integrity. [a] Airtech Vac View 10 flow meter is used to fix leaks in the bag/seal while drawing vacuum to the bag. [b] The ultrasonic leak detector is used to find leaks in the bag/seal. [c, d] Digital vacuum gauges are also used to ensure that there are no leaks in the bag or tool. *(Photos courtesy of [b] Superior Signal, [c] DD Compound, and [d] Airtech International)*

Once the vacuum-bagged layup has been tested for vacuum integrity, full vacuum is applied and held for a sufficient period of time to compact the laminate and remove any residual moisture in the preform. One trick is to place a loop in the vacuum tubing that runs between the bag and vacuum/resin pot. If moistures is present, small droplets of water will form at the bottom of the loop and when they disappear, the preform is ready to infuse.

Note: it can take up to 24 hours at room temperature to purge moisture.

Mix and Supply Resin

The resin is mixed and degassed (if applicable), and the feed line clamp/valve is opened to begin resin infusion into the laminate. Resin supply should be monitored. If the supply container empties, air bubbles will be sucked into the laminate. How much resin will be needed to complete the infusion should be calculated as part of pre-process planning.

The resin should be visually moving at a constant rate. Most infusions, even of very large structures like bridge decks, can be completed in one hour. Though this depends on many factors, including resin viscosity and overall preform permeability, a common strategy is that shorter infusion time reduces risk. This will be discussed further in the next section.

Once the preform is completely impregnated with resin, the resin feed line clamp/valve is closed. Vacuum is maintained until the resin has gelled.

❱ PROCESS CONTROL AND QUALITY

Vacuum infusion processing provides consistent laminate properties with fiber volumes up to 70%. Because the resin is not mixed until after reinforcements are placed and vacuum bagged, time for preform assembly and layup into the mold is not limited as it is with open mold wet layup. This allows for more precise control of material placement and fiber orientation. The amount of waste is also minimized and volatile organic compounds (VOCs) can be better contained.

The key to successful resin infusion is a very standardized, repeatable process. Most infusions will impregnate the entire part in one flow, at one temperature, simplifying the process and making it less variable.

Trial Panels

A key part of infusion process control is determining the preform permeability, and verifying the calculated volume of resin required to fill the preform at the target temperature. Trial panels, at least one square yard in size, should be made using the resins and reinforcements desired for infusion. For best results, one panel should be made for each variation in material composition of the laminate across the part, as permeability will change with differing stacks. The influence of core materials, channels, and other features of shape should also be evaluated in trials before committing to full scale.

For each panel, a marker is used to trace the resin flow front at one-minute intervals, measuring the flow distance per minute. This can help determine resin feed line spacing. During testing, once resin flow slows to 0.6 inches (15 mm) per minute, it is recommended to add another feed line. (Figure 8-15)

Temperature Control

As with wet layup, there are significant issues regarding the control of temperature, both in the shop and at the mold surface. A cold mold can greatly affect the viscosity of the resin being infused, lessening the distance that the resin will flow. On the other hand, too much heat may cause the resin to gel before sufficiently filling the preform.

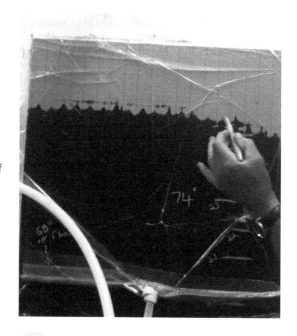

FIGURE 8-15. Trial panel for measuring flow distance per minute.

The flow front location is marked in silver for each minute (e.g., 21, 22, 23) from bottom to top as the test progresses. *(Photo courtesy of Bravolab)*

There is also the significant risk of out-of-control exotherm with the resin contained in the supply bucket or barrel, as it is usually a sizeable volume. These problems must be closely monitored during and after an infusion process.

Process Plan

For large infusions especially, it is essential to have a well-trained team and a detailed plan of clearly sequenced steps (*see* Figure 8-16). Fully understood assignments are tasked to each person, for example:

- Measure, mix and transport resin to the supply/feed line containers.
- Monitor vacuum gauges (installed at multiple locations on the part plus the pump for large infusions).
- Monitor temperature of the resin, mold, part, and the workshop room temperature.
- Monitor flow fronts and have a plan for in-process repairs.
- Turn on/off resin supply.
- Turn on/off vacuum supply.

Reinforcement and Laminate Design Issues

Another issue in vacuum infusion processing has to do with the precompaction of the fibers. At the faying surfaces, the fiber-to-fiber interfaces are so tightly compacted that it is difficult to get resin wetting between fibers. As a result, there may be a reduction in interlaminar shear strength and edgewise compressive properties of the laminate. The use of interleaf channel layers can improve this condition.

Yet another consideration with infusion processing is achieving symmetry with the various forms chosen for this process. It is not unusual to select a multiaxial stitched form that does not produce a symmetric laminate, even when it is flipped over at the mid-plane of the laminate.

INFUSION PROCESS ORGANIZATION

1. Infusion Director controls all operations based on input from the team.

2. Vacuum Monitor manages vacuum level and integrity, records flow distance and time.

3. Resin Team manages measure, batch, mix and delivery to resin station.

4. Resin Monitors track resin level, elapsed time and temperature.

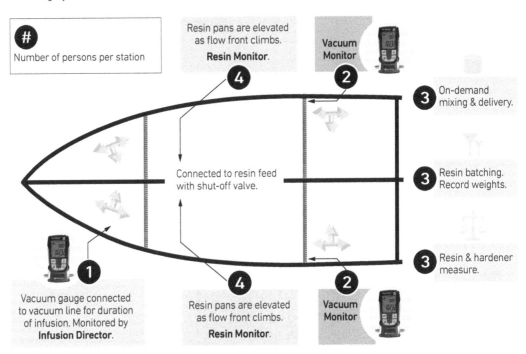

FIGURE 8-16. Infusion process management diagram.

This infusion process management diagram shows stations and tasks for each of the 17 team members with the black circles denoting number of persons per station. The infusion was designed to flow from each side towards the center with a high flow distance per minute, demanding vigilant monitoring. *(Photo courtesy of Bravolab)*

Selection of materials that allow for laminate symmetry is paramount to the dimensional stability of the infused structure.

Resin Transfer Molding

Resin transfer molding (RTM) is a closed mold process where mixed resin is injected into a dry fiber preform. It uses heated matched molds to produce parts with good surface finish on both sides and high fiber volume. Molds are designed to withstand vacuum and pressure, and can be made from composites, nickel, aluminum or steel. RTM is well-suited to complex shapes and large surface areas, and encompasses a wide range of process variations (*see* Figure 8-17 on the next page).

❭ RTM MATERIALS

A wide range of resin systems can be used with RTM including polyester, vinyl ester, epoxy, phenolic, benzoxazine, polyimides, and BMI. Reinforcements may be continuous strand mat, unidirectional (UD), woven and/or noncrimp fabrics (NCF), braided forms, or three-dimensional textile preforms.

Smooth-On

- Composite + countermold or reusable silicone bag
- Static mixing head
- Polyester, vinylester, epoxy
- Noncrimp fabric, core
- < 15 bar/218 psi pressure
- 100s – 1000s parts/yr
- Rail car floors, boats

Jasper Plastics

- Composite, metal matched molds
- Static mixing head
- Polyester, vinylester resin
- Glass, carbon fiber mat, woven fabrics
- < 15 bar pressure
- Up to 20,000 parts/yr
- Truck hoods, roofs, bus bodies, boats, car parts, agriculture equipment

Matrix Composites (Rockledge, FL)

- Steel, aluminum matched molds
- Static or dynamic mixing head
- RTM6 epoxy, BMI, PMR-15
- Carbon fiber woven fabrics and 3D woven textile preforms
- Up to 20 bar (300 psi)
- 100s – 1000s parts/yr, e.g., 2-6 hr cycle time
  - LEAP engine blades, 7m-long A320 multispar flap, A350 passenger doors
- Carbon fiber woven fabrics
- Automotive epoxy resins, e.g., 30-60 min cycle time
  - Spoilers, trim parts

SGL Carbon, Volvo

- Steel matched molds
- Impingement mixing head
- Snap-cure epoxy, polyurethane
- Carbon, glass UD and NCF
- Up to 150 bar in mix head
- 30-120 bar (1740 psi) in mold
- Up to 500,000 parts/yr, 25-sec cycle time
  - Volvo rear axle leaf springs, BMW i3 and i8 body structure

Audi, Voith Composites

- Steel matched molds
- Impingement mixing head
- Snap-cure epoxy, polyurethane
- Carbon fiber bindered UD tape
- < 15 bar pressure
- 10,000 – 300,000 parts/yr, e.g., 2-5 min cycle time
  - Audi A8 rear wall

FIGURE 8-17. RTM procedure.
The procedure for RTM is to place the preform in the mold, close the mold and inject resin into the preform through one or more ports, depending on the size and complexity of the part. Vacuum may also be applied, and the mold may be heated to shorten cure-time. When the resin has cured, a molded part is removed for trimming and final finishing, and the process repeated.

❱ PROCESS PARAMETERS

RTM is usually characterized as a low-pressure process using pressures under 100 psi. Higher pressures may be used, most commonly with matched metal tooling. RTM may use preheated resin and/or a preheated mold—an example would be an epoxy resin preheated to 80°F (29°C) and a composite or aluminum tool preheated to 120°F (49°C). However, high temperature variations have been successfully used to manufacture parts at temperatures close to 707°F (375°C) with long-term service temperatures in that same range. Process parameters are determined by the resin system used in conjunction with cycle time (viscosity and wet-out), desired T_g (service temperature) and post-cure requirements.

〉LIGHT RTM

This process uses a rigid A-side mold (bottom) and a semi-rigid B-side mold (upper)—typically a conformable counter-mold or a selectively reinforced, reusable silicone rubber vacuum bag. Flexibility in the B-side mold enables a tight fit to the compacted preform in the A-side mold, critical for achieving the necessary laminate compaction, resulting in high-quality parts. This process is also called Flex Molding, RTM Light, RTM Lite and LRTM. (Figure 8-18)

Typically, resin is injected using a low-pressure (<5 psi) pumping unit and no additional pressure is exerted on the molds. Cosmetic surfaces can be achieved on both the A- and B-sides of the part. Parts made using LRTM should be designed for this process. Fiber volumes are not typically as high as in RTM or aerospace-grade resin infusion, but theoretically could be increased with appropriate materials, equipment and process control. Start-up costs are moderately higher compared to resin infusion, but LRTM is better suited for producing hundreds to low thousands of smaller parts per year with repeatable, high-quality results thanks to much-reduced bagging and faster resin injection. (Figure 8-19)

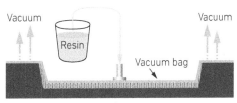

Vacuum infusion processing

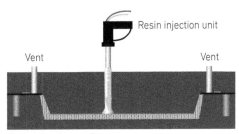

Resin transfer molding

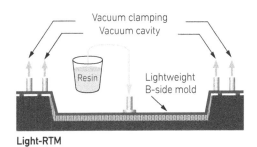

Light-RTM

FIGURE 8-18. Light RTM vs. VIP and RTM. *(Diagram source: Molded Fiber Glass Companies)*

FIGURE 8-19. Light RTM for high-speed train flooring.
SMT is using SAERTEX LEO SYSTEM and an RTM Light process to produce 25,000 m2 of composite railcar floors to replace plywood in Deutsche Bahn's ICE Version 3 high-speed trains. The blue reusable silicon vacuum bags from Alan Harper Composites can be seen applied over layups in the foreground and rolled up in the background. *(Photos courtesy of SAERTEX, Alan Harper Composites and CompositesWorld)*

❱ RTM USING FLOATING MOLDS

The method of low-pressure RTM using glass fiber mat-reinforced polyester and vinyl ester resin for truck/bus, agricultural, industrial and recreational products, traditionally has produced 5,000 to 15,000 parts per year. However, newer methods, such as the **floating mold** technology developed by VEC Technology, can enable production volumes as high as 45,000 to 50,000 parts per year. This process also uses thin-shell composite tooling, which is much more economical than the aluminum, steel and nickel-shell tooling usually used for high-volume RTM production.

In the VEC method, composite upper and lower mold skins are placed into a water bath and fastened to the outer walls of the universal vessel. The water bath fully supports both mold skins, eliminating the need for machined metal tooling. (Figure 8-20)

❱ RTM FOR AEROSPACE APPLICATIONS

RTM was patented by a U.K. aircraft company in 1956. Though it has been used with styrene-based resins since the 1960s and 70s, it was not really adopted for aerospace applications until the 1990s. Due to the need for higher fiber content (e.g., 70% by weight) and < 2% void content, pressures of 100–300 psi are typically used, but this is still considered low-pressure compared to high-pressure RTM (HP-RTM) where in-mold pressures can reach 1,740 psi.

One of the first applications of RTM in aircraft structures was the sine-wave spars for the Lockheed Martin F-22 Raptor's wings. More than 400 parts are made for the F-22 using RTM and BMI or epoxy resin. Table 8-4 gives details for an array of aerospace structures made using RTM.

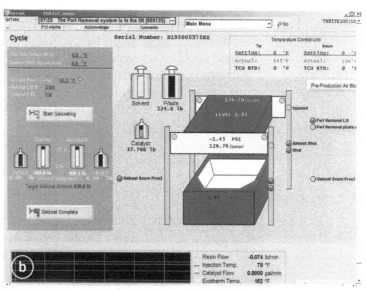

FIGURE 8-20. Resin transfer molding.
VEC Technology uses RTM to manufacture up to 45,000 per year. [a] After kitted reinforcements are placed into the lower mold, [b] VEC's computer-controlled system [c] closes the mold and begins resin injection, which takes approximately 20 minutes. [d] After cure, the gantry-mounted upper mold halves are lifted and parts are demolded. *(Photos courtesy VEC Technology and Composites World)*

TABLE 8.4 Aerospace RTM Applications

COMPANY		PRODUCT/Description	
Lockheed Martin		F-35 vertical tail	
12-ft tall along leading edge, with skins made using 100 plies of carbon fiber. Reduced part count from 13 to 1, eliminating 1,000 fasteners with a cost savings of 60%.			
Matrix Composites Rockledge, FL	Production	Spacers for AEDC wind tunnel compressor	a
610-mm-long by 356-mm-wide by 560-mm-high spacer body integrates monolithic egg-crate into single-piece unit with laminate thickness increasing from 2.5 mm to 16.3 mm. Used IM7 carbon fiber, 4-harness satin fabric and 5250-4 BMI resin prepreg, 60% fiber content. 9-hr RTM cycle with 9-hr post-cure for 150°C service temp. *(Photo courtesy of Matrix Composites, Rockledge, FL and CompositesWorld)*			
Elbit-Cyclone	Development	Passenger door for commercial aircraft	b
Unitized, all-composite passenger door for commercial aircraft: RTM redesign of this primary structure eliminates metal fasteners, cuts weight by 30% and cost by 25%–35% compared to aluminum and "black metal" composite designs. *(Photo courtesy of Elbit Cyclone and CompositesWorld)*			
Israel Aerospace Industries	Development	Helicopter seat	c
Carbon and glass fiber satin weave fabrics reinforced Prism EP-2400 one-component toughened epoxy, heated to 80°C. Mold is preheated to 120°C and increased to 180°C with a pressure of 6 bar (90 psi) during a 2-hr molding cycle. RTM seat cuts cost 30% and weight 7% vs. prepreg. *(Photos courtesy of IAI and Composites World)*			

(continued)

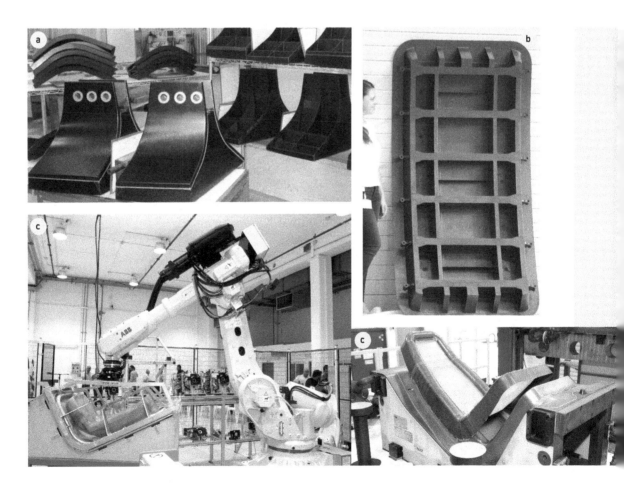

TABLE 8.4 Aerospace RTM Applications *(continued)*

COMPANY		PRODUCT/Description
Aernnova Aerospace	Production	A350 horizontal tail plane leading edge and its load-bearing rib
Airbus Bremen	Production	A320 outboard flap **d**
Replaces 26 carbon fiber prepreg components with a single component, cutting cost by 20%. 7-m-long, varies in width from 80 mm to 200 mm, and 2-5 mm in thickness skin with I-beam spars. 5-harness satin carbon fabric, HexFlow RTM6 epoxy resin. Aluminum tool, preforms and resin preheated to 100°C, ramped to 180°C for 2-hr cure, demolded at 100°C. Radius Engineering RTM press. *(Photos courtesy of Airbus Bremen and CompositesWorld)*		
FACC	Production	A350 spoiler main hinge fitting **e**
PRIFORM dry preform with toughener patented by Cytec (now Solvay) using 6K carbon fiber and thermoplastic fibers. Cycom 977-2 epoxy resin preheated to 60-90°C and increased to 180°C in a steel mold. Cuts weight 25% and saves months of lead-time vs. metal. *(Photo courtesy of FACC and CompositesWorld)*		
Albany Engineered Composites	Production	LEAP engine fan case and 18 fan blades per engine
Both made using carbon fiber woven into a complex 3D textile preform and epoxy resin. Blades molded in a Radius Engineering RTM press. *(Photos courtesy of Albany Engineered Composites and CompositesWorld)* **f**		
Airbus Helicopters Donauworth, Germany	Production	A350 passenger door inner structure
Wickert 2500-kN press used to cure preconsolidated preforms made from 3C-Carbon Composite Co. carbon fiber fabrics injected with HexFlow RTM6 epoxy for six-hour cure cycle per door.		

EVOLUTION TO HP-RTM FOR AUTOMOTIVE APPLICATIONS

RTM gained usage in the automotive industry in the late 1990s to early 2000s. BMW began ramping RTM in 2001, acquiring a Cannon Afros preformer for a development program to shape carbon fiber noncrimp fabrics (NCF) before molding them in an RTM press. It announced this more industrialized system in 2003.

By 2005, a less automated version of RTM had been used heavily in the Dodge *Viper* and also by Sotira Composites for multiple automotive OEMs, including a joint development effort with Ford Motor Co. and Aston Martin which used sprayed fiber preforms. A turning point came in 2007, when Roctool touted a "high-speed RTM" process using its induction-heated molds for a 50% cut in cycle time. By 2010, BMW was using its automated NCF-based RTM process to mass produce CFRP roofs for its *M3* and *M6* models, as well as *M6* bumper supports. That same year, BMW unveiled plans to produce a new electric vehicle—the *i3*—beginning in 2013, which would feature a CFRP Life Module passenger cell made using high pressure RTM (HP-RTM). (Figure 8-21)

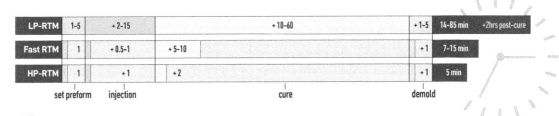

FIGURE 8-21. Evolution of standard LP-RTM to fast RTM, to HP-RTM.

What Defines HP-RTM?

Both low pressure RTM (LP-RTM) and HP-RTM processes use a fiber preform, a closed mold, a press and a resin injection system. HP-RTM, however, swaps a static or dynamic mixing head for an *impingement mixing head*, like that developed for polyurethane (PU) foam applications in the 1960s. In fact, metering/mixing/injection suppliers for the PU and reaction injection molding (RIM) industries were among the early developers of HP-RTM, including Cannon Afros, FRIMO, Hennecke and KraussMaffei. RIM did evolve into an RTM-like process called Structural RIM (SRIM), where typically PU resin was injected into a fiber preform (*see* "Injection Molding," page 255). Fiber content in SRIM is lower than RTM—typically up to 30% by weight, whereas fiber weight up to 75% has been claimed for HP-RTM parts.

TABLE 8.5 Definition of HP-RTM vs. LP-RTM

	Low-pressure RTM	High-pressure RTM
Mixing head	Static or dynamic	Impingement
Injection pressure	10-20 bar (145-290 psi)	Up to 150 bar (2,180 psi)
Pressure in mold	<15 bar (218 psi)	30-120 bar (435-1,740 psi)
Cycle time	30 min to hours	3-15 min
Automation	Spray or stacked fabric preforming Mix, meter, dispense (MMD) system for resin	Automated preforming, robotic handling, automated injection and press, in-mold sensors
Tooling	Composite, aluminum, steel	Steel invar
Resin	Polyester, vinylester, epoxy	Snap-cure epoxy, polyurethane

Snap-cure Resins

New snap-cure resins commercially available by the early 2000s were the final key that unlocked the 5-minute cycle times common for HP-RTM. Development of more reactive, faster-curing resins has been ongoing since the 1990s. BMW chose a fast-cure EPIKOTE epoxy from Hexion in its initial CFRP roofs and a Huntsman Araldite 5-minute cure epoxy for the *i3* Life Module.

By 2012, Hexion had developed a 2-minute cure system (at 120°C) with a 1-minute injection window, while Dow Automotive improved its VORAFORCE 5300 epoxy from a mold cycle < 90 seconds in 2014 to < 60 seconds in 2015.

HP-RTM Variations

The HP-RTM process has been modified to achieve faster impregnation by forcing preform impregnation in the z-direction via compression RTM (C-RTM). Surface RTM (S-RTM) achieves high-quality exterior parts by flow-coating the molded part surface with a thin layer of polyure-thane which is then cured in the mold. (Figures 8-22, 8-23)

❱ LP-RTM AND ULTRA RTM

LP-RTM continues to be used in automotive applications. In 2016, Tier 1 supplier Mubea Carbo Tech had six LP-RTM presses which could process preforms up to 10 mm thick. These presses use mix, meter and dispense (MMD) units for mixing and injecting the resin. Visual carbon parts

COMPRESSION RTM (C-RTM)

- Also called gap impregnation
- Preform inserted into mold
- Epoxy or PU is injected into a slightly opened mold
- Mold is closed (compressed) to impregnate in the z-direction
- Part is demolded

SURFACE RTM (S-RTM)

- HP-RTM process is followed
- Mold is opened a second time and PU surface layer is flooded into the mold
- Mold is closed for second time and PU lacquer distributed/cured

THERMOPLASTIC RTM (T-RTM)

- Thermoplastic matrix (e.g., PA)
- Caprolactam monomer and activator/catalyst are injected
- Preform is impregnated and matrix polymerization to PA completed

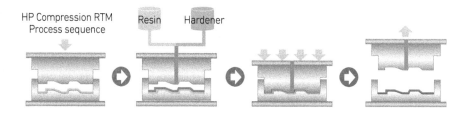

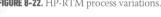

FIGURE 8-22. HP-RTM process variations.

FIGURE 8-23. Surface RTM.
(Photo courtesy KraussMaffei and CompositesWorld)

made in this way typically comprised 2–4 plies with unidirectional reinforcements forming the bottom pair and two woven (commonly twill) fabrics for the aesthetic surface. The standard pressure used was 10–20 bar and cycle time ranged from 30 minutes for small parts to 60 minutes; most automotive parts do not exceed 2m x 2m in size. Table 8-6 on the next page shows where HP-RTM and LP-RTM fit into Mubea Carbo Tech's portfolio of composite production processes.

Ultra RTM is a process developed by automaker Audi that retains the automation and fast cycle time of HP-RTM, but reduces pressure, enabling the use of foam cores and lower-tonnage presses. Audi wanted to use 100–150 kg/m^3 density polymethacrylimide (PMI)-based foam in the CFRP rear wall for its R8 sports car, but the high injection pressure of HP-RTM could damage the foam. Audi found that it could achieve its goal by balancing injection pressure mold gap and press force: injection pressure < 40 bar, mold gap up to 0.6 mm, and maximum press force of 500 metric tonnes. This combination maintained the foam's integrity yet reduced injection time to 15 seconds with an overall cycle time of 5 minutes. The smaller tonnage press costs less and uses less energy.

Even though the CFRP rear wall for Audi's A8 luxury sedan does not use foam, it is still made with the Ultra-RTM process. Pressure was reduced further to less than 15 bar, enabling a reduction in press force from 2,500 kN (for HP-RTM) to 350 kN. The resin used is Dow VORAFORCE 5300 epoxy, curing at 120°C in 120 seconds for an overall part cycle time of 5 minutes. (Figures 8-24, 8-25)

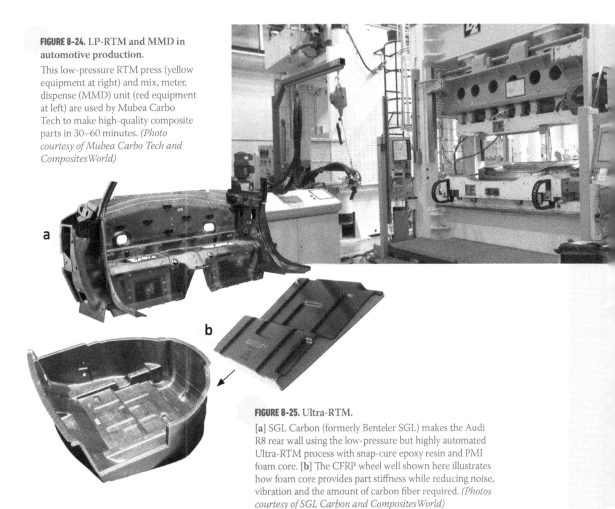

FIGURE 8-24. LP-RTM and MMD in automotive production.

This low-pressure RTM press (yellow equipment at right) and mix, meter, dispense (MMD) unit (red equipment at left) are used by Mubea Carbo Tech to make high-quality composite parts in 30–60 minutes. *(Photo courtesy of Mubea Carbo Tech and CompositesWorld)*

a

b

FIGURE 8-25. Ultra-RTM.

[a] SGL Carbon (formerly Benteler SGL) makes the Audi R8 rear wall using the low-pressure but highly automated Ultra-RTM process with snap-cure epoxy resin and PMI foam core. [b] The CFRP wheel well shown here illustrates how foam core provides part stiffness while reducing noise, vibration and the amount of carbon fiber required. *(Photos courtesy of SGL Carbon and CompositesWorld)*

TABLE 8.6 Mubea Carbo Tech Composite Processes

		Parts/yr	Tooling	Pressure (bar)	Variations
PREPREG	Autoclave	1–100	Al, CFRP	6	-
	Mini-autoclave	500–2,000	Steel, Al	4–12	-
	Prepreg Press	1,000–5,000	Steel	up to 30	-
NON-PREPREG	LP-RTM	1,000–5,000	Steel	10–40	Solid Cored Hollow
	HP-RTM	10,000–50,000	Steel	100	Solid
	Wet Pressing	10,000–100,000	Steel	up to 100	-

Wet Compression Molding

Also called liquid compression molding (LCM), wet pressing or wet molding, this is an old process that has recently been updated. Molded Fiberglass Companies claims it has used the process for volume composites production since 1948. In this early form, the process entailed wetting out a glass fiber mat with polyester resin, then curing and molding it in a compression molding press. This process is still used today, including variations in which chopped fiber and resin are applied robotically onto a heated aluminum tool and then placed into a compression molding press. This process achieves medium to high volumes of transportation and recreational structures with low defect rates. (Figure 8-26)

Wet pressing was reinvented during BMW's industrialization of CFRP parts, used to produce 17 parts for the 2013-launched *i3*, 21 parts for the *i8* and 9 parts for the 2016-launched *7 Series*. BMW's highly automated version of this process begins with ultrasonically welded stacks of carbon fiber noncrimp fabric, which are transported by a robot arm equipped with a needle gripping system. Stacking accuracy and weight are checked and then the preforms are placed onto a table with double or quadruple cavities. Two robots apply liquid epoxy resin to the preforms, a process that is weight-controlled to ensure reproducibility and quality of the final parts. During this open process, resin fumes and volatiles are collected by a ventilation system. (Figure 8-27)

The wet-out preforms are then robotically transferred into the compression molding press, which closes and cures the parts. Parts may then be transferred (using a robot with vacuum grippers) to a cooling press, after which they are trimmed and machined for vehicle assembly.

FIGURE 8-26. Wet pressing with glass fiber.

Glass fiber mat is wet out with resin (foreground) before being placed into a compression molding press (background) for simultaneous shaping and curing. *(Photo courtesy of Protectolite Composites)*

❭ MATERIALS AND MOTIVATION

This modern iteration of wet compression molding uses mostly unidirectional, woven and/ or noncrimp fabrics. Resins are mostly epoxy but may also be thermoplastic. In 2014, Dow Automotive Systems touted a wet compression molding process that applied its VORAFORCE epoxy in 15–20 seconds and cured in 30 seconds for a total cure cycle < 60 seconds. Hexion offered a total cycle time of 75–135 seconds with its fast-cure epoxies for wet compression molding, suggesting further reduction by replacing part of the in-mold cure with a post-cure.

Arkema's liquid ELIUM acrylic-based thermoplastic resin may also be used in wet compression molding at temperatures ranging from 20°C to 80°C. Other options for thermoplastic composites include molding with low-viscosity caprolactam monomer for PA6 or laurolactam monomer for PA12. These are polymerized during compression molding (in-situ polymerization) to form fiber-reinforced PA6 and PA12 thermoplastic composite parts. (Figure 8-28)

FIGURE 8-27. BMW wet compression molding using carbon fiber.

[a] Stacks of carbon fiber noncrimp fabric are wet out with resin, [b] transported into a twin-cavity mold, and then [c] molded in a compression molding press into side sills for the 2016 BMW 7 Series body-in-white called Carbon Core. *(Photos courtesy of BMW Group and Composites World)*

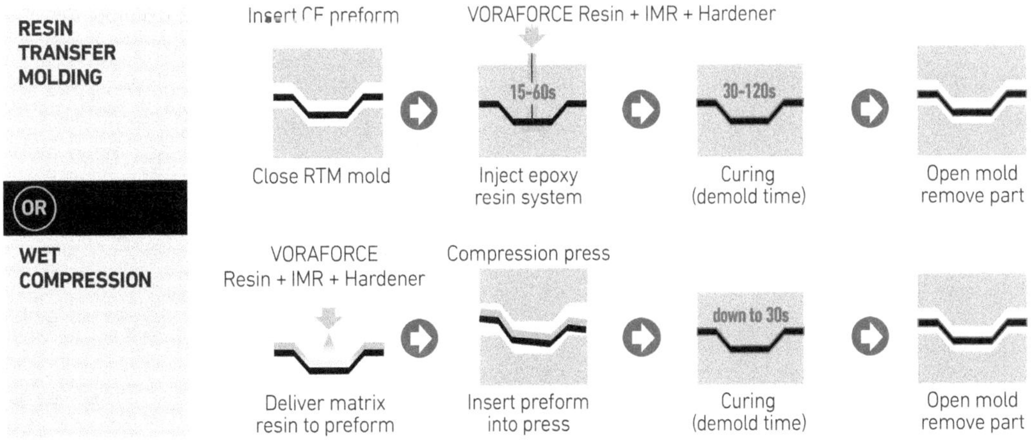

RESIN TRANSFER MOLDING

Insert CF preform → Close RTM mold

VORAFORCE Resin + IMR + Hardener → 15–60s → Inject epoxy resin system

30–120s → Curing (demold time)

→ Open mold remove part

(OR)

WET COMPRESSION

VORAFORCE Resin + IMR + Hardener → Deliver matrix resin to preform

Compression press → Insert preform into press

down to 30s → Curing (demold time)

→ Open mold remove part

FIGURE 8-28. Wet compression molding as faster alternative to HP-RTM (Note: "IMR" = internal mold release). *(Diagram source: Dow Automotive Systems)*

Wet compression molding offers a faster cycle time than HP-RTM by eliminating preforming and resin injection. Instead of preforming, stacks of reinforcement are used, and where resin was injected and cured inside the closed mold, it is now applied outside of the press. This allows parallel processing—application of resin while other parts are curing in the press. The resin can also be more reactive (faster cure) because no latency is required to allow filling a heated RTM mold before cure. Also, the process is well-suited for pressing multiple parts simultaneously. In order to achieve this high-rate production, however, parts must be relatively simple in geometry.

❭ DYNAMIC FLUID COMPRESSION MOLDING (DFCM)

This is an iteration in wet compression molding developed by Huntsman, which uses vacuum to enable greater geometrical complexity as well as higher fiber volume. It claims cycle times of <90 seconds for parts with medium draw or truly 3D (as opposed to 2.5D) geometry.

DFCM is a low-pressure process (typically 30 bar) offering highly structural parts made with fast-cure Araldite epoxy resins that Huntsman claims can achieve fiber volumes of up to 65% with low-waste and low void content. The low pressure means lower equipment investment compared to HP-RTM, and elimination of fiber wash issues (the disruption of the fiber preform due to high injection pressure). (Figure 8-29)

Filament Winding

Filament winding involves feeding continuous fiber reinforcements through a resin bath and winding them onto a rotating mandrel. The fibers are drawn from a creel and through a collector that combines them into a tape. The mandrel rotates as the collector traverses its length. The combination of transverse movement and rotation creates a fiber angle on the structure. This angle can be changed by varying the speed of rotation and the rate of collector movement. The end result creates angles ranging from a hoop wrap, which is good for crush strength, to a helical wrap, which provides greater tensile strength. (Figure 8-30)

Successive layers may be oriented differently from the previous layer. Tension may also be precisely controlled, with higher tension producing a more rigid and higher strength end-

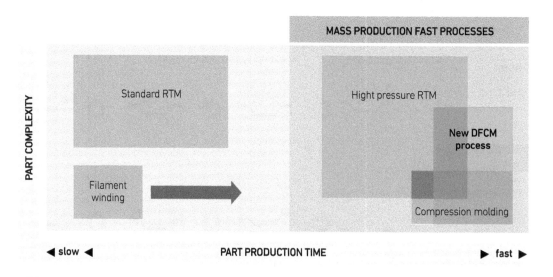

FIGURE 8-29. Comparison of HP-RTM, wet compression molding and DFCM. *(Source for diagram: Huntsman Advanced Materials)*

product, and lower tension leading to one with more flexibility. Filament winding machines can now be controlled using computer numerical control (CNC) technology, managing from 2 to 6 different axes of motion.

Once filament winding layup over the mandrel is complete, the composite structure is cured at room temperature or in an oven for higher temperature cure. After cure, the mandrel is typically removed for re-use, leaving a hollow final product.

Filament winding excels at creating structures with very high fiber volume fractions (60%–80%) and with specific, controlled fiber orientation. These factors, combined with the continuous fibers applied without interruption through successive layers, produce some of the greatest strength-to-weight ratios of all composite manufacturing processes.

Filament winding machines are typically customized to meet specific part and project requirements, resulting in a wide range of equipment sizes and configurations. These range from highly automated systems with multiple spindles, producing pressure vessels for compressed natural gas (CNG) and hydrogen storage, to large machines for space launch vehicles. (Figure 8-31)

FIGURE 8-30. Filament winding.
The photo at right shows fibers being drawn through *(from right to left)* a collector and nip rollers before being applied to a cylinder as a tape. *(Photo courtesy of Connova)*

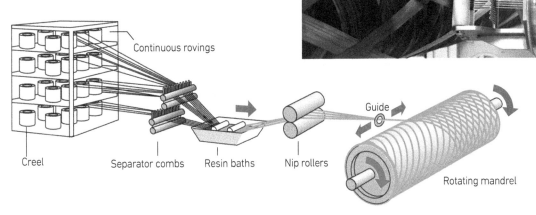

FIGURE 8-31. Variety of filament winding machines.
[a] The 12-foot-diameter composite-wound rocket motor case shown here is produced by Northrop Grumman in Promontory, Utah for its OmegA rocket—each stage comprising one, two or four of these filament-wound segments. [b] This MIKROSAM machine can produce five CNG or LPG tanks simultaneously, as part of an automated production line that can output up to 500,000 tanks/year. *(Photos courtesy of Northrop Grumman and MIKROSAM)*

❭ FILAMENT WINDING MATERIALS

Traditionally, glass fiber has been the most common reinforcement used in filament winding, but carbon and aramid fibers have also been used. Resins employed include epoxy, polyester, vinyl ester, polyurethane and phenolic. Thermoplastic polymers in the form of commingled fiber products and semipreg tapes, as well as carbon fiber/epoxy prepreg are also materials, as are dry fiber tapes, which are vacuum bagged once winding is complete and processed using resin infusion.

❭ EVOLUTION TO ROBOTIC, 3D AND CORELESS WINDING

Robotic Filament Winding

Robots have played an increasing role in filament winding. Though conventional winding equipment has relied on an accumulator traversing the length of the mandrel on a linear rail, newer systems use a robot. MF Tech, a filament winding specialist launched in 2004 in France, has based all of its systems on robots. This includes simply mounting the fiber feed onto a robotic arm, which provides up to eight axes of motion. But there is also a more unconventional setup where the rotating mandrel is carried by a robot, which moves it past a stationary fiber feeding unit (as shown in Figure 8-32). This system can pick up a plastic liner, wind a pressure vessel onto it and then place the wound tank into a curing oven.

More recently, filament winding systems have been developed where *both* the mandrel and fiber feed may be moved and rotated by robots. These robots cooperate to increase the range and accessibility for fiber winding, enabling larger structures as well as more complex fiber layups and shapes.

Cygnet Texkimp has developed a high-speed, 3D winding machine, designed to produce complex layup parts that vary in both cross-section and shape. The system combines rotation and fiber feed into a single mechanism that is robotically moved along the liner, mandrel or core, winding as it goes. The Cygnet 3D winder uses two counter-rotating fiber application rings mounted together on the end of a robotic arm (*see* Figure 8-32).

As the arm maneuvers the rings so that a complex-shaped mandrel is passed through their center, fiber is fed from eight bobbins on each ring—that number is scalable—and is wound onto the mandrel. This 3D winder can apply 24K or 48K dry carbon fiber at a rate in excess of 1 kg/min, varying the angle of each ply and 0° plies via installation of an additional "hoop ring."

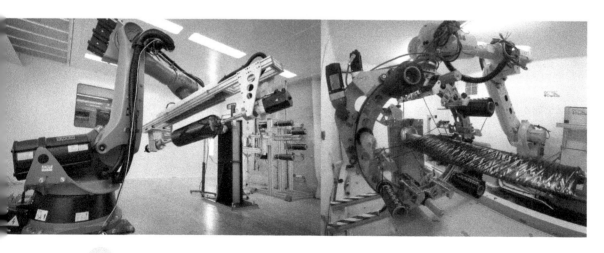

FIGURE 8-32. Robotic 3D winding. *(Photos courtesy of MF Tech and Cygnet Texkimp)*

Hybridized Filament Winding

Robotics are allowing MF Tech to develop very complex wound carbon fiber structures with ribs, beams and machined features for automotive applications. The company is also combining filament winding (FW) with automated fiber placement (AFP) and other processes, as is turnkey supplier MIKROSAM which introduced its own hybrid AFP/FW work cell in 2017.

Winding is also being hybridized with 3D printing. German composites engineering firm CIKONI uses a robotic process to wet-wind carbon fiber/epoxy onto 3D printed plastic, composite or metal cores. Generative 3D winding—with a robot but without a core—is a technology being developed by Daimler (Stuttgart, Germany). Dubbed FibreTEC3D, and moving away from rotational structures altogether, the process wet winds carbon fiber/epoxy around aluminum pegs inserted into a tooling plate, producing load-tailored, lightweight structures without generating waste. (Figure 8-33)

Multi-Filament Winding

Japanese firm Murata Machinery Ltd. has developed a high-speed multi-filament winding (MFW) system that simultaneously applies 48 to 180 tows/fiber inputs. Installed at the Institut für Textiltechnik (ITA) der RWTH Aachen University in 2017, its MFW-48-1200 machine comprises four main components: a creel, a ring-like adaptive nozzle, a rotating mandrel or liner unit and an annular hoop unit. (Figure 8-34)

FIGURE 8-33. Robotic 3D winding hybridized with 3D printing.
CIKONI creates lightweight, cost-effective robotic grippers, fixtures and other structures by filament winding onto cores made from injection molded or 3D printed plastic, as well as metal cores made with conventional machining or additive manufacturing. The core handles compressive loads while the carbon fiber handles tensile loads.
(Photos courtesy of CIKONI)

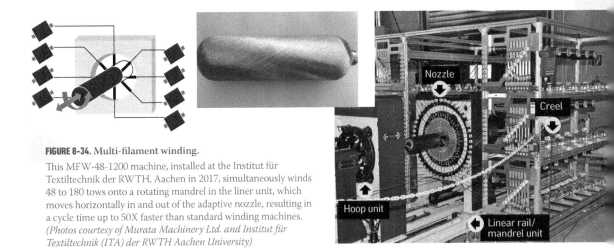

FIGURE 8-34. Multi-filament winding.
This MFW-48-1200 machine, installed at the Institut für Textiltechnik der RWTH, Aachen in 2017, simultaneously winds 48 to 180 tows onto a rotating mandrel in the liner unit, which moves horizontally in and out of the adaptive nozzle, resulting in a cycle time up to 50X faster than standard winding machines.
(Photos courtesy of Murata Machinery Ltd. and Institut für Textiltechnik (ITA) der RWTH Aachen University)

The MFW-48 creel contains 48 bobbins of fiber. The circular adaptive nozzle (also called an *iris*) feeds and applies the 48 fiber inputs from the creel onto a rotating mandrel or liner that moves horizontally in and out of the nozzle. The result is similar to braiding in that the fibers/ tows cross over each other, but without inducing crimp. The winding angle is determined by the relative speed between the rotational and horizontal movement of the liner/mandrel and its diameter. Four additional fiber feeds may be applied at 90° via the hoop unit.

Murata and ITA claim a complete layer can be applied in one pass, resulting in a cycle time 50 times faster than standard winding machines with an increase in torsional stiffness compared to braided laminates. The machine is being developed for applications such as hydrogen storage cylinders.

3D Coreless Winding

German automaker Daimler has developed a 3D winding technology called FibreTEC3D that does not use a core/mandrel, but instead a robot winds carbon fiber wet with epoxy resin around aluminum pins protruding from an aluminum tooling plate (*see* Figure 8-35). The FibreTEC3D manufacturing process and the parts it produces are described as *generative*. This refers to **generative design**, in which a part's geometry is not predefined but instead computer-generated, by applying advanced algorithms to data derived from the application's requirements and constraints—size envelope, loads, strength, stiffness, impact, deflection, cost, etc. This design approach enables the economical production of highly optimized, small-series products.

In FibreTEC3D, the part's digital design directs placement of the pins and generation of the robotic winder code—the combination, in turn, actualizes the part. The entire design process is managed by software. There is no cutting and no waste, so that the cost driver is no longer carbon fiber but instead the aluminum tooling. Materials used so far include 24K carbon fiber (common in automotive) and room-temperature cure epoxy. Wound parts could use a higher-temp cure epoxy and be placed in an oven for post cure.

Although B-staged **towpreg** is typically faster to filament wind, wet winding is preferred for the FibreTEC3D process because the towpreg is too sticky. The wet winding provides a slippery roving that settles into the neutral axis as it is wound, providing excellent resin-to-fiber distribution. Towpreg also requires compression, post-winding to remove entrapped air during layup,

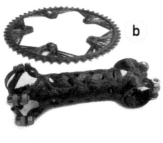
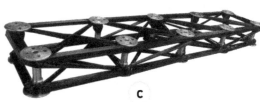

FIGURE 8-35. 3D coreless winding.

[a] Daimler's FibreTEC3D process winds carbon fiber tow impregnated with epoxy around aluminum pins with zero waste, [c] producing parts like this robotic gripper for automotive assembly weighing 50% of a standard aluminum version. [b] The xFK in 3D process commercialized by Automotive Management Consulting is similar, producing a zero-waste, lightweight racing bicycle chain ring, handlebar stem and seat clamp. The Carbone chain ring weighs less than 46 grams, produced by LBK Fertigung using SIGRAFIL® continuous tow from SGL Carbon. *(Photos courtesy of Daimler and Automotive Management Consulting)*

whereas the tension in wet winding compacts the fiber as it is wound. In addition, the fiber is pre-tensioned which improves final part properties.

The xFK in 3D process developed by the German technology firm Automotive Management Consulting is similar, robotically winding an epoxy-saturated fiber around a simple metal positioning fixture. It creates ultra-light, bionic composites with fibers arranged specifically to match each part's loads and desired functions. By winding around metal and/ or plastic inserts for fasteners, the process also enables cost-effective connections, historically a challenge for composites.

PULTRUSION

Composite **pultrusion** is a high-rate, continuous production process that is analogous to aluminum extrusion, but as the name suggests, the fiber reinforcements are pulled (versus pushed with extrusion) through a heated die to form a cured composite profile. Pultrusion can create almost any shape, including a wide variety of bars, beams, stringers, stiffeners, tubes, pipes and angles. These shapes can be solid, cored, or hollow, depending on the type of pultrusion process being implemented. (Figure 8-36)

The pultrusion process typically starts with racks or creels of fiber reinforcement and other materials such as core. As the materials are pulled off the racks/creels, they are guided through a resin bath, or resin is injected directly into the die. In the non-injected thermoset form of the process, reinforcements exit the impregnation area fully wet-out and are then guided through a custom series of tooling called a "pre-former." The preformer helps arrange and organize the reinforcements into the final shape while squeezing out excess resin.

The impregnated and shaped composite is then pulled into a heated die, which may have several different temperature zones to cure the matrix resin. Pulling may be achieved by using friction wheels or friction belts on either side of the profile die to pull on the shape as it exits the die. In some variants the die itself is used to pull the materials along through the process. Another method uses reciprocating clamps that grip the shape and pull it in a continuous fashion in a "hand-over-hand" motion. Once the cured composite profile exits the die, it is then cut into lengths by a cutoff saw. (Figure 8-37, next page)

PULFORMING

In the 1980s, pultrusion pioneer W. Brant Goldsworthy put forward a new process that combined pultrusion technology with step-compression molding processes to pultrude a profile through a cross sectional die, then mold the wet blank in a die to achieve a curved shape in one process. The goal was to make GFRP leaf springs for the automotive industry, but at the time the concept never made it past the research and development phase. As a result, the process known as pulforming emerged as a variant of the pultrusion process that allowed variation of section and shape.

FIGURE 8-36. Typical pultruded composite products.

Small to medium sized pultrusions are shown here, illustrating the variety of cross-sections and shapes possible. This manufacturer can readily fabricate pultruded profiles up to 60 inches in width. *(Photo courtesy of Sumerak Pultrusion Resource International)*

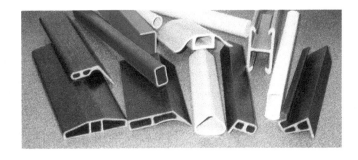

FIGURE 8-37. Typical pultrusion process. *(Photos courtesy of Sumerak Pultrusion Resource International)*
[a] Continuous roving, mat and/or fabric materials travel from creels and through guides toward the resin impregnation and forming station. [b] The material infeed forms the fibers into the final shape before entering the pultrusion die. [c] Material exits the pultrusion die and heating system fully cured to final shape. [d] Dual reciprocating grippers pull the part at continuous speed. Next, a flying cutoff saw (not shown here) cuts the continuous profiles to final length.

❱ RADIUS-PULTRUSION

Radius-pultrusion is a pulforming type of process trademarked by Thomas Technik & Innovation (Bremervörde, Germany) in which the product is shaped in a moving, heated die. (This is different than conventional pultrusion that employs a reciprocating pulling system downstream of the die.) In this case, the moving die is used to receive and shape the incoming material at the beginning of the line, cure the profile, and move it downstream toward a fully automated (stationary) gripper and cutoff saw.

As the die moves, the gripper remains open and the cured shape is moved through it to the automated cutoff saw where it is clamped and cut to length. When the die stops the gripper closes, holding the cured profile while the die opens and moves back upstream to once again shape, cure, pull, and cut the next section of material. This process repeats to make additional lengths in a continuous production manner. (Figures 8-38, 8-39)

❱ PULLWINDING

The pullwinding variant of pultrusion is uniquely applicable to manufacturing round tubes or other tube profiles with very thin wall thicknesses. The process simply integrates a fiber winding system along with conventional longitudinal fibers upstream of a pultrusion die where selected reinforcements are pulled along and wrapped around a feeder mandrel, then fed into a pultrusion die where it is compacted and cured to a final shape. This allows for precise control of the structural properties of the final product by adjusting the amount of lengthwise and crosswise fibers in the profile. (Figure 8-40)

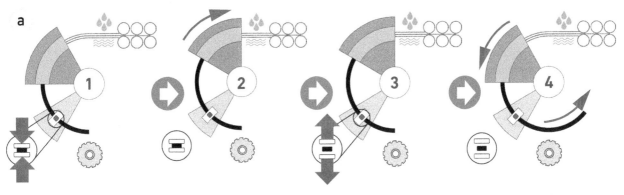

1. Gripper clamps the profile

2. Die moves back to accept fresh material

3. Gripper releases the profile

4. Die advances profile to the saw

FIGURE 8-38. Radius-pultrusion concept.

[a] Moving die, stationary gripper, and cut-off saw: [1] The profile is clamped so that the moving die can open. [2] The die returns to receive fresh material, closes and cures profile. [3] The profile is unclamped. [4] The die moves to advance the profile to be cut at the saw. The process repeats in a continuous production process. [b] Pipe coil prototype made using radius-pultrusion process. *(Source info. for diagram: Thomas Technik & Innovation, Lower Saxony, Germany)*

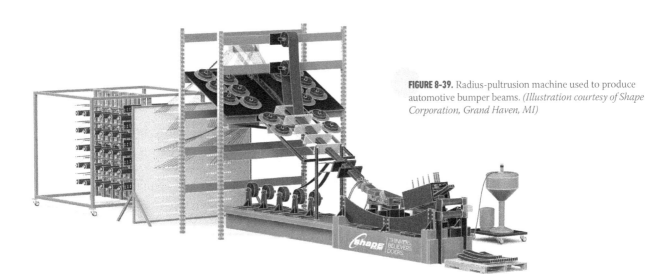

FIGURE 8-39. Radius-pultrusion machine used to produce automotive bumper beams. *(Illustration courtesy of Shape Corporation, Grand Haven, MI)*

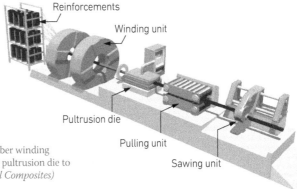

Reinforcements

Winding unit

Pultrusion die

Pulling unit

Sawing unit

FIGURE 8-40. Pullwinding process showing fiber winding around a mandrel prior to moving into the pultrusion die to make a tube. *(Source info. for diagram: Exel Composites)*

) PULPRESS AND PULCORE

The PulPress™ and PulCore™ variants of pultrusion were developed by Evonik Industries (Darmstadt, Germany) in partnership with Secar Technologie, GmbH (Mürzzuschlag, Austria), to produce complex profiles that contain machined high-temperature Rohacell® (PMI) foam core blanks fed upstream into the process. The blanks are shuttled through a winder or braider (PulWinding™ or PulBraiding™) and a pultrusion die, after which the cured product is cut from continuous stock. The process allows for either a constant or changing cross-section and can produce profiles with complex 3-dimensional geometries. (Figure 8-41)

) PULTRUSION MATERIALS

Thermoset resins are most common materials for this process, including polyester, vinyl ester, epoxy, phenolic and polyurethane resins. Thermoplastics may also be used, either deployed in the reinforcement as a powder or prepreg, or it can be injected as a liquid at the die. The most common reinforcements used in pultrusion include fiberglass roving, continuous strand mat, and woven or stitched fabrics. Most any fiber type or combination of fibers, fiber forms, and (high temperature) core materials can be employed in the pultrusion process, depending upon the properties desired in the product.

CirComp has developed a thermoplastic pultrusion process which can produce 10 mm x 10 mm bars with 50K carbon fiber tow, which is spread inline (*see* Chapter 3, "Spread Tow") to enable good impregnation with a PA6 matrix.

) PROCESS PARAMETERS

Typically, pultrusion speeds are faster for thinner profiles. Profiles with a thickness of about 0.25–0.375 inch made with relatively fast-curing resins can be pultruded at roughly 8 to 9 ft/min with smooth surfaces. Thicker profiles ranging from 0.375 to 0.675-inch thickness can be made with a slightly slower-cure resin at speeds up to 6 ft/min with minimally abraded surfaces. Profiles over 0.67 inches in thickness made with medium reactivity resins can be produced at speeds of 0.5 to 5 ft/min and tend to have significantly more surface roughness.

Derakane quotes speeds as fast as 10 m/min (~33 ft/min) for profiles of 1 to 3 mm (0.04 to 0.1 inch) in diameter, and 1.5 to 1.9 m/min (~5 to 6 ft/min) for profiles between 4 and 9 mm (0.2 to 0.35 inch) in thickness.

Pultrusion is considered to be a low-pressure process, however, research has shown that pressures in the range of 3.5 to 14 bar (50 to 200 psi) can be developed with high filler and fiber contents.

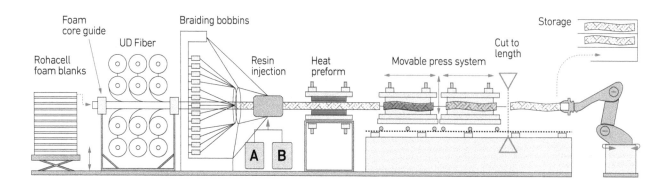

FIGURE 8-41. PulPress process showing fiber braiding around core blanks prior to moving into the pultrusion die. *(Source info. for diagram: Evonik)*

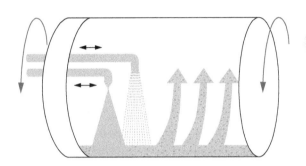

Centrifugal Casting

Also called spin casting, this process is used to make hollow structures such as pipes, tanks and poles. Reinforcements are placed onto the inner walls of a cylindrical mold. Resin is then injected into the rotating mold, where centrifugal force pushes it through the layers of fibers. Ideal for pipes and tanks, the surface against the mold becomes a smooth exterior while excess resin injected into the inside forms a resin-rich, corrosion- and abrasion-resistant interior liner. The mold can be placed into a heated chamber to accelerate cure. (Figure 8-42)

FIGURE 8-42. Centrifugal casting.
(Source info. for diagram: CompositesLab.com)

❱ CENTRIFUGAL CASTING MATERIALS

Reinforcements may include chopped strand mat, continuous strand mat, chopped and sprayed rovings, woven roving and other fabrics. Resins that may be used include polyester, vinylester and epoxy.

❱ ADVANTAGES AND DISADVANTAGES

This is an economical process for producing large and small hollow structures and any fumes/volatiles are contained in the closed mold. However, the process mandates low-viscosity resins and is limited in the shape complexity and properties it can produce.

Injection Molding

Injection molding involves inserting a liquid material into a mold, heating and/or cooling to solidify the molded part and then ejecting it from the mold. Though injection molding can be done with molten metal or rubber, the focus here will be on thermoplastic and thermoset composite polymers.

Injection-molded plastics are used for everything from toys to kitchen utensils to automotive parts. These applications use unreinforced and unfilled thermoplastic polymers. To produce composites, chopped fibers are added to the polymer, increasing strength and stiffness, which enables reduced wall thickness and reduced weight. Fibers are most commonly glass or carbon, but basalt and natural plant-based fibers have also been used. Metal or ceramic fibers or particles may also be added to increase temperature resistance and wear resistance. Such fibers and additives, including nanomaterials, may also help to provide thermal and electrical conductivity.

Typically, fiber content is 30–45% by weight, though higher strength and modulus compounds with content as high as 65% by weight are available. The maximum glass fiber loading that is practical via standard compounding techniques is roughly 50% by volume or 70% by weight. Fiber length is typically 0.25 inch but can span from milled fiber up to 0.5 inch. Compounds with fibers longer than this are usually classified as **long fiber thermoplastics (LFT)**, discussed below.

TABLE 8.7 Advantages and Disadvantages of Injection Molding

Advantages	Disadvantages
• Ability to produce millions of parts quickly	• Significant expense and lead time to develop tooling (though this is improving by using additive manufacturing)
• Low scrap rates vs. machining from blocks of material (but not as low as 3D printing)	
• Very repeatable, high part reliability throughout production volume	• Can be difficult to make changes
• Very good at producing small parts with complex geometries	• Significant investment in equipment (though this is also changing with ability to lease machines)
• Offers the ability to integrate many parts into a single, molded piece	• Not as good at producing large parts with high wall thickness and undercuts

Injection molding compounds are typically supplied as pre-compounded pellets, which are melted and then injected within the molding machine. Because some polymers absorb water from the atmosphere, proper material storage, handling, drying and machine feed conditions must be maintained in order to achieve consistent, high-quality injection molded parts.

Pellets are loaded into the hopper, which feeds them into the heated barrel of the machine where they are melted. A reciprocating screw continues to feed and mix the proper amount of material. The screw ram-injects a shot of the material into the mold cavity where it cools and conforms to the mold shape. After cooling, the finished part is ejected from the mold and the cycle is completed. Injection molding cycles are normally less than 60 seconds long, with cooling comprising more than 50% of the total cycle time. (Figure 8-43)

Examples of applications include injection-molded wood plastic composite cases, gears, car interiors, toys and consumer goods (*see* Figure 3-15); chopped glass fiber reinforced nylon and polypropylene parts such as washing machine tubs and under-hood car parts (*see* Figure 3-19); and nylon reinforced with chopped glass, carbon or natural fibers for automotive parts used in doors, instrument panels, front end carriers and liftgate assemblies.

❭ LONG FIBER THERMOPLASTIC (LFT) AND DLFT

LFT materials typically use fibers that are 0.5–1.0 inch (13–25 mm) in length. However, for both short fiber and LFT compounds, the fibers will be sheared during mixing and injection, reducing length to roughly 1–3 mm for LFT compared to approximately 0.3 mm for short-fiber compounds.

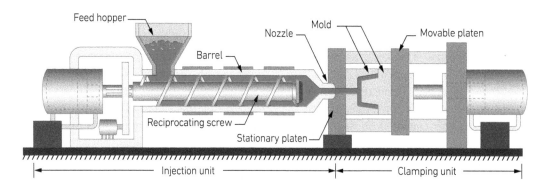

FIGURE 8-43. Injection molding machine.

LFT offers higher mechanical properties than standard short-fiber compounds, including tensile and impact strength, dimensional stability and thermal stability. LFT pellets are made by pultrusion and most commonly use glass or carbon fiber. Hybrid pellets combining glass and carbon fiber have been made as well. Matrix polymers include PA, PA6, PA66, PBT, PC, PET, PP, PPS and TPU.

The process for *direct* long fiber thermoplastic (D-LFT or LFT-D) directly chops fiber and mixes it with resin at the injection machine—i.e., the reinforced thermoplastic is directly compounded and then molded in one operation. D-LFT is described as a cross between injection molding and compression molding. Initially, the process evolved out of reinforced injection molding, and thus comprised feeding the mixed material through a transition tube into an injection molding press screw, like that shown in Figure 8-43. However, increasing fiber length began to reach the limits of injection molding, clogging the equipment. Therefore at the higher end of fiber length, D-LFT consists of compounding the material inline and then placing it as a charge into a compression molding tool and press. (Figure 8-44)

With D-LFT, chopped fiber lengths are extended to as high as 50 mm. PP, PET and PBT are listed as the most common thermoplastic matrix materials and though glass fiber has historically been the most common reinforcement, applications using carbon fiber have been increasing.

One advantage of D-LFT is that molders have more control over materials versus using pre-compounded pellets, including the ability to choose precise polymer and fiber types, as well as fiber length and fiber content, with the latter controlled to a variation of less than ±2% by weight, according to some D-LFT equipment suppliers. Cost savings are also claimed as a key benefit, while properties are equal to or greater than LFT pellets.

❯ DLF AND REGRINDING WASTE

DLF® is an acronym used by Greene, Tweed for its "discontinuous long fiber" technology. However, it does not use injection molding but instead compression molding (*see* Chapter 7) of Xycomp® carbon fiber-reinforced composites that may use PEI, PPS, PEEK or PEKK thermoplastic polymer as a matrix. DLF materials may be formed by chopping prepreg tape. In this way, they are more similar to a long fiber compression molding materials, for example SEREEBO™ (20-mm-long carbon fiber in PA6 matrix) developed by Teijin. While Sereebo™ was developed for automotive applications, DLF is targeted for aerospace and replaces metal in brackets and also in oil and gas tubulars and fittings, cutting weight up to 60% and integrating multiple parts into a single component.

One part of the Sereebo™ process is common to D-LFT long fiber injection molding processing described above: regrinding of part production scrap and mixing with virgin material in a screw extruder. This can be seen in the use of Sereebo™ to mold the 2019 GM

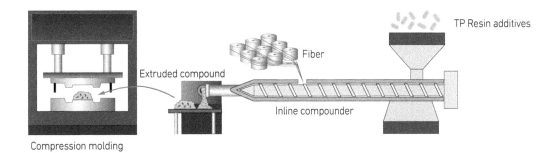

Compression molding

FIGURE 8-44. D-LFT process using compression molding. In this diagram, thermoplastic pellets are melted and mixed inline with chopped fiber and then extruded as a charge, which is placed into a compression molding tool and press.

Sierra Denali pickup bed. The Sereebo™ material in this part comprises 1-inch long randomly oriented fibers in a mat with PA6 polymer already mixed in and 35% fiber content by volume. For this application, 25% of the material is virgin PA6 pellets and 75% is reground scrap from trimming blanks prior to compression molding. Both are fed to an extruder that mixes and dispenses the material charge used to mold each pickup bed part. Both the regrind and the PA6 pellets must be dried and dosed accurately before feeding into the extruder, as noted above for injection molding of D-LFT.

❯ RESIN INJECTION MOLDING/REACTION INJECTION MOLDING (RIM)

Compared to injection molding, which injects viscous thermoplastic polymers melted from pellets, RIM is a thermoset process. It mixes low-viscosity liquid part A and part B monomers to react and form crosslinked resins including polyurethane, acrylic, polyester, vinylester and epoxy. Once correctly dosed and mixed, these polymers expand, thicken and harden after they are injected into a heated mold. (Note the molds in thermoplastic injection molding are cooled, not heated.)

Historically, RIM has mostly used polyurethane (PU or PUR) without any reinforcement to manufacture larger and lighter-weight parts compared to injection molding. There are RIM process variants, including reinforced RIM (RRIM) and structural RIM (SRIM), that are used to make composite parts. (Figure 8-45)

Reinforced RIM (RRIM)

In reinforced RIM (RRIM), part A and part B resin monomers are filled with chopped fibers and/or particulates before impingement mixing and injection into the mold. Parts manufacturers using RRIM claim it is possible to mix the fiber or particle-filled A and B components using a combination of low pressure and high velocity (*see* Figure 8-46). Glass fiber is the most common reinforcement, but carbon fiber is starting to be used. These fibers are usually milled or chopped to very short lengths—between 0.009" and 0.02" (0.2 and 0.5 mm).

Parts made using RRIM feature light weight, flexibility, durability and toughness compared to unreinforced parts, including superior strength and impact resistance. As with injection molding, the addition of fibers means wall thickness can be reduced for less material used and lighter weight. The process is well-suited for large parts such as body and floor panels, bumpers, mudguards and fenders. It may also feature short cycle times, as low as 90 seconds.

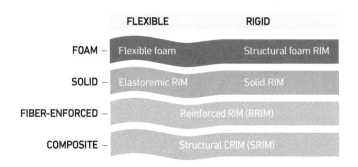

FIGURE 8-45. RIM and process variants for composites.
RIM is used to make a wide variety of parts, ranging from flexible to rigid, unreinforced foams and solids to fiber-reinforced composites. *(Source: RIM MFG www.reactioninjectionmolding.com)*

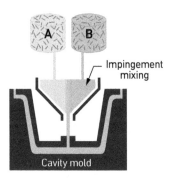

FIGURE 8-46. RRIM uses chopped fiber in A and B monomers.

Another advantage is the possibility to use in-mold coating for both glass and carbon fiber parts, enabling Class A surface finish directly out of the mold and eliminating the need for secondary painting operations. RRIM is used to produce fenders, spoilers and bumpers for small-series production cars, trucks buses and agricultural equipment.

Structural RIM (SRIM)

Just as D-LFT transformed from an injection molding to a compression molding process with increased fiber length, structural RIM (SRIM) moves from RIM to more of an RTM process, using very long or continuous fiber preforms placed into the SRIM mold prior to injection. Preforms are typically continuous strand mats, nonwovens, directed spray preforms or textile fabrics. In order to wet out these preforms it is important to use sufficiently low-viscosity resins. Cycle times are usually 3-5 minutes and the process is used to make large parts similar to RRIM, but characterized by higher stiffness. In-mold painting, however, is not possible. Thus, SRIM parts tend to be used where Class A finish is not required, such as automotive dashboards and shelves.

❭ LONG FIBER INJECTION (LFI)

Not really an injection process where resin is injected into a cavity, chopped glass fiber and PU resin are sprayed onto an open mold coated with in-mold paint (IMP). The two-part matched metal mold is then closed and the resin and fiber mixture is compression molded. Resulting parts with Class A finish can be up to 70% lighter than steel, 20% lighter than aluminum, 40% lighter than sheet molding compound (SMC) and 50% lighter than hand layup/sprayup fiber-reinforced plastic (FRP).

Cycle time is typically 5 to 10 minutes, with the higher end of this range for parts as large as 3.5 m by 3.5 m. Fiber lengths range from 12.5–100 mm (0.5–4 inches) with fiber content up to 50% by volume. LFI uses a relatively low pressure, about 5 bar (75 psi) compared to typical thermoplastic injection molding processes (1,500–2,500 bar/20,000–30,000 psi) and compression molding of sheet molding compound (25–172 bar/500–2,500 psi). This allows use of less costly, smaller 150–500 ton presses.

LFI is used to produce side panels, roofs and hoods for trucks, farm equipment and construction vehicles as well as roofs and rear shelf panels for cars.

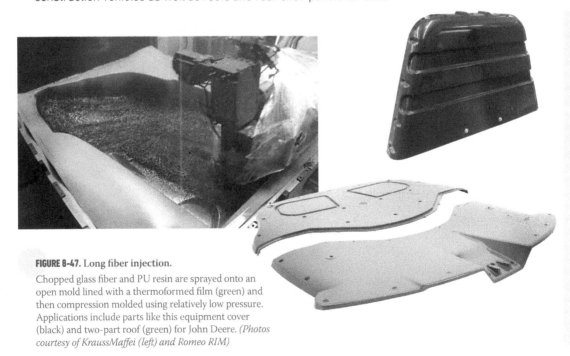

FIGURE 8-47. Long fiber injection.
Chopped glass fiber and PU resin are sprayed onto an open mold lined with a thermoformed film (green) and then compression molded using relatively low pressure. Applications include parts like this equipment cover (black) and two-part roof (green) for John Deere. *(Photos courtesy of KraussMaffei (left) and Romeo RIM)*

Overmolding

Overmolding combines compression molding and injection molding. The process involves forming a composite part from a fiber-reinforced thermoplastic blank in a mold, then immediately injection-molding thermoplastic stiffening (or cosmetic) features to one or both sides of the part. This approach allows high-rate production of lightweight, rigid, and complex structural parts in a one-step process.

Typically, the same thermoplastic matrix is used in both the fiber-reinforced blank and the overmolding compound. The overmolding thermoplastic may be unfilled or filled with chopped fiber. Thermoplastics used in overmolded composites range from polypropylene (PP), polyethylene (PE), polyamide (PA, nylon), PA 6 and polycarbonate (PC) for automotive and consumer goods to polyetheretherketone (PEEK), polyetherketoneketone (PEKK) and low-melt polyaryletherketone (LM-PAEK) for aerospace structures.

Tool designers must consider both molding the preformed blank and integrating the injection-molding requirements into one mold set. The molds are matched metal, and are typically preheated. (Figure 8-48)

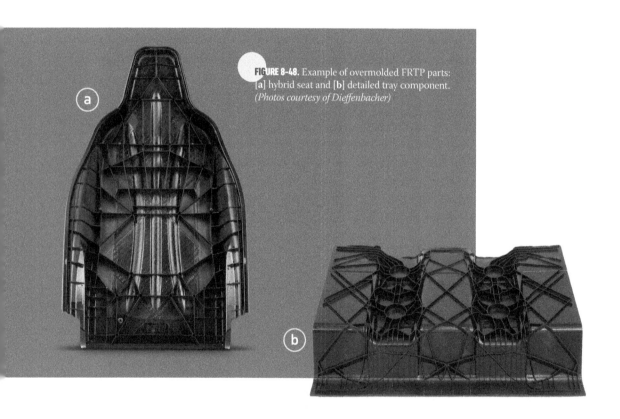

FIGURE 8-48. Example of overmolded FRTP parts: [a] hybrid seat and [b] detailed tray component. *(Photos courtesy of Dieffenbacher)*

Introduction to Tooling

CONTENTS

Key Factors

Unlike metals that require bending, forming, casting, or machining to achieve a desired shape, composites are typically molded. A mold, mandrel, or die-set is designated for this purpose. These tools must produce a dimensionally accurate part, often through an elevated temperature and pressure process.

In addition to the primary molding process, **tooling** may be required to facilitate many post-molding operations such as trimming, drilling, bonding, fastening, and assembly of composite structures. Good quality, precision tooling is the key to reliable and repeatable production. Without tooling, every part is a one-of-a-kind piece.

While we examine some of the more common types of tools and materials in this chapter, it should be noted that this only scratches the surface of what is possible. Part, process, and tool designs for composites are only limited by the imagination.

Tool and Part Design

As composite engineers design larger and more complex monocoque structures, the need for more innovative tooling and manufacturing methods will emerge. Building multi-piece molds for cocuring entire assemblies requires traditional "tool designers" to become "mold-designers," thinking in terms of eliminating as much of the secondary assembly as possible. This must be coordinated carefully with the structural design and manufacturing engineering groups to best assure manufacturability.

This shift in approach requires the industry to recognize that the extra costs associated with designing, building and using these complex molds are cost effective in the long run. These molding costs can easily be offset by the fact that there will be far less assembly and assembly tooling required to make larger molded structures. As a result, the cost of inspection and quality assurance is also shifted in this direction toward the **layup – assembly** approach and away from the "layup, machine, and then assemble" parts approach.

❯ DESIGN CONSIDERATIONS

Some factors to consider in tooling design are as follows (*see* Figure 09-1):

- Production rate requirements—tool life expectations—number of units needed.
- Process selected for manufacturing part or structure.
- Operator friendly: simple, uncomplicated design/fabrication/usage.
- Cost, both initial tooling cost and the amortized cost per part in production.
- Design of part removal features and methods in the molds.
- Tool compensation for thermal expansion, shrinkage, and spring-in effects.
- Weight, transportation, and handling
  - Protection from operator induced damage (OID).
  - Protection from forklift induced damage (FLID).

❯ PRODUCTION RATE REQUIREMENTS

Tools are designed to meet production quantities that are outlined for the project. A tool designed to make 20 parts may be very different from one designed to make 20,000 parts. Very-high-rate tooling such as that used in the automotive industry may be designed to make

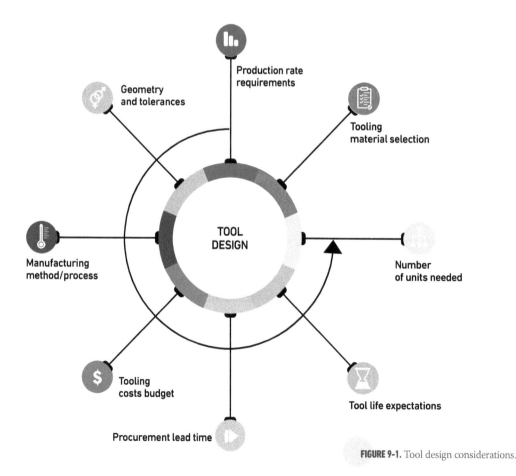

FIGURE 9-1. Tool design considerations.

hundreds of thousands of parts per year. The production *rate* requirements often drive the choice of which processes will be used to make the parts and ultimately the tool design, tool materials, and the resulting tooling costs.

❯ TOOL TYPES AND FUNCTION

For any given composite part or assembly, a series of tools are required to control critical operations in the manufacturing process. Each tool is designed around how it will function. For example, a layup mold or mandrel that might be used for hand layup of prepreg materials may be required to control only one surface of the part and accommodate a vacuum bag, allowing the operator to layup, bag, and cure a dimensionally accurate part.

Compare this to a mold or die-set used for **resin transfer molding (RTM)**, where the tool must control both surfaces of the part. It must allow for the positioning and clamping of a fiber-preform, have ports for resin injection, a flash zone, and vents or a vacuum cavity.

The tool function is often described in the tool description or name. Many companies use an abbreviated description or a tool code to designate each tool and its function. Common types of tools used in conventional fiber-reinforced thermoset matrix composite manufacturing are listed below. (Note that tool codes are not standard to every manufacturer.)

- Layup mold or mandrel (LM)
- Trim fixture or jig (TF, TJ)
- Hand router fixture or jig (HRF or HRJ)
- Drill fixture or jig (DF or DJ)
- Bond assembly fixture or jig (BAF or BAJ)
- Combination fixture or jig (CF or CJ)
- Semi-rigid tools and cauls (Caul)
- Elastomeric vacuum tool (EVT)

❯ TOOLING MATERIAL PROPERTIES

The material used to construct tooling is critical to the performance of the tool, especially when the tool is subjected to a change in temperatures. For example, aluminum has a high **coefficient of thermal expansion (CTE)** compared to that of a carbon fiber-reinforced epoxy laminate. Because of this, care must be taken when selecting aluminum for a mold or mandrel that will be used to make composite parts at elevated temperatures.

Aluminum does have very good heat transfer properties that may be advantageous in an RTM or press-molding process. On the other hand, CFRP composites and Invar both have a low CTE and would be a more favorable choice for dimensionally accurate parts cured in an oven or autoclave. A thorough understanding of material properties is required to properly design functional tooling.

Tooling Materials

The materials used in tooling are shown in Table 9.1, and their thermal properties in Table 9.2.

TABLE 9.1 Tooling Materials

Metals • Steel • Aluminum • Nickel • Invar 36 or 42	**Monolithic graphite**
	Ceramics
Composites – made with EP, CE, BZ, or BMI resins • Glass fiber reinforced plastic (GFRP) • Carbon fiber reinforced plastic (CFRP)	**Carbon foam**
	Polyurethane foam
	Epoxy tooling board
Fiber reinforced elastomer (FRE) • Silicone or fluoroelastomer matrices	**Thermoplastics (3D printed)**

TABLE 9.2 General Comparison of Tooling Material Thermal Properties

Material	Specific Gravity g/cm^3	Specific Heat Btu/lb/°F	Thermal Mass Btu/lb/°F	Thermal Conductivity Coefficient Btu/ft^2/hr/°F/in	Coefficient of Thermal Expansion in/in/°F
Aluminum	2.70	0.23	0.62	1395	12.9 x 10^{-6}
A36 Steel	7.75	0.12	0.93	347	6.5 x 10^{-6}
P20 Steel	7.86	0.11	0.86	288	7.1 x 10^{-6}
Nickel	8.90	0.10	0.89	500	7.4 x 10^{-6}
Invar 36	8.05	0.12	0.97	72.6	0.8–2.0 x 10^{-6}
Invar 42	8.11	0.12	0.97	72.6	2.2–2.6 x 10^{-6}
GFRP	1.80–1.90	0.3	0.54–0.57	21.8–30.0	8.0–9.0 x 10^{-6}
CFRP	1.50–1.60	0.3	0.45–0.48	24.0–42.0	0.1–3.0 x 10^{-6}
Ceramic Washout	0.4–0.6	0.2	0.08–0.12	4.8	4.0–5.0 x 10^{-6}
Ultracal 30® Gypsum Plaster	1.65	0.26	0.43	2.01	8.3 x 10^{-6}
Monolithic Graphite	1.74–2.0	0.3	0.52–0.60	13.3–18.3	0.1–1.0 x 10^{-6}
CFoam	0.32–0.48	0.19	0.06–0.09	4.85	2.7–3.2 x 10^{-6}
Med. Density Fiberboard (MDF)	0.75	0.41	0.30	0.17	6.6 x 10^{-6}
CT850 Ceramic Tooling Block	0.35–0.88	0.23	0.08–0.20	1.87	3.3 x 10^{-6}
Polyurethane Tooling Board	0.29–0.8	0.76	0.22–0.61	0.15–1.88	12.0–35.0 x 10^{-6}
Epoxy Board	0.67–0.79	0.24	0.16–0.19	2.0	18.0–20.0 x 10^{-6}
ULTEM 1010™ Resin (PEI)	1.27	0.25–0.29	0.32–0.36	1.53	24.5–26.1 x 10^{-6}
Dahltram C-250CF	1.22	0.40	0.49	0.18–0.62	6.0–61.0 x 10^{-6}
Dahltram C-350CF	1.15	0.40	0.46	0.16–0.68	4.0–42.0 x 10^{-6}

Calculating the Coefficient of Thermal Expansion (CTE)	
Example	Carbon fiber tool laminate CTE (in/in/°F) = 2 x 10^{-6} = 0.000002 Aluminum tool CTE (in/in/°F) = 12.9 x 10^{-6} = 0.0000129 Cure temperature = 350°F Desired part length = 48.00 in
Calculating the Effect	Temperature differential (cure temp - ambient): DT = 350°F – 70°F = 280°F DT = 177°C – 21°C = 138°C
Relative Expansion	Carbon fiber tool, 0.000002 in/in/°F x 280°F = 0.00056 in/in Aluminum tool, 0.0000129 in/in/°F x 280°F = 0.0036 in/in
Total Expansion	Carbon fiber tool, 0.00056 in/in x 48.00 in = 0.027 in Aluminum tool, 0.0036 in/in x 48.00 in = 0.173 in

TABLE 9.2 General Comparison of Tooling Material Thermal Properties *(continued)*

After-effect of temperature on a carbon fiber reinforced plastic (CRFP) tool/part

After-effect of temperature on an aluminum tool/CRFP part

❱ THERMAL CONDUCTIVITY

Thermal conductivity is the ability of a material to transfer heat. Aluminum and copper are very good thermal conductors. Ceramic is not a good thermal conductor. Carbon fiber reinforced plastic (CFRP) has a higher **thermal conductivity coefficient (TCC)** than glass fiber reinforced plastic (GFRP) tooling. Glass tends to be a good insulator; therefore, CFRP tooling materials will conduct heat more efficiently than GFRP materials. Other metals also offer a higher TCC than composite materials but likely have a much higher thermal mass.

❱ THERMAL MASS

Thermal mass is the ability of a material to absorb and store heat energy. The amount of energy required to heat a given material is calculated by multiplying the density and the specific heat of the material. The thermal mass value of the tooling materials affects the tool's ability to heat and/or cool at the rates required by the process specifications. As a result, tools made with heavy materials (metals) tend to heat more slowly than tools made of CFRP or GFRP materials. This information is important to producing molds and mandrels that will heat at a given rate using the energy available in the press, oven, or autoclave. (Figure 9-2)

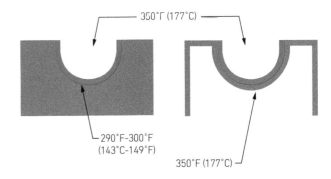

FIGURE 9-2. Effect of mass on the heat-up rate of tooling.
High thermal mass on left, low thermal mass on right.

Metal vs. Composite Tooling

❱ TOOLING BOARD MATERIALS

Tooling board materials are used to manufacture different types of tools, from master models and patterns (used to make composite tools) to prototype molds and various jigs and fixtures. The materials are supplied in planks that are bonded together with adhesive and machined to shape. Types of tooling board materials include polyurethane foam, epoxy block, ceramic block, and carbon foam. Each material (including the adhesive used to bond the boards) has vastly different thermal properties and care should be taken when selecting the proper material for any given application. (Figures 9-3, 9-4, 9-5)

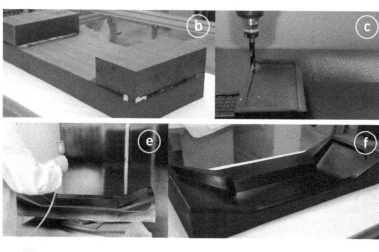

FIGURE 9-3. Typical tooling board fabrication method.

[a–f] Tooling board stock, adhesive bond the planks, machine tool to near-net size (minus sealer), seal coat the surface, resulting finished tool. *(Source: DUNA USA)*

FIGURE 9-4. Carbon foam is used to make both prototype and durable molds and fixtures.

[a] A prototype tool is machined from a bonded block assembly or a billet, finished with a polymeric coating and ready to make a part. [b, c, d] A permanent tool is machined with an offset from the desired net surface, then prepreg carbon materials are applied and cured, and then the final surface is machined and sealed for use in production. CFoam materials can be repurposed after use. *(Photos courtesy of CFoam)*

METAL TOOLS

Metal tools (apart from Invar) can also have a much higher thermal expansion rate than the composite, leading to problems with dimensional accuracy in the cured parts, since the parts cure at the expanded dimensions of the hot tools and then the tools are cooled back to room temperature before demolding. This may lead to physical stress on the part and difficulty removing the part from the mold. Conversely, this characteristic may help when molding parts such as golf club shafts, as the internal metal mandrel will contract on cool-down, pulling away from the part.

Tools designed for high-temperature cure cycles and high production rate requirements are usually made from metal. In aerospace, Invar 36 (or 42) alloy is used for this purpose, as it has a very low CTE. Invar tools are very durable and can be machined accurately to tight tolerances. However, they tend to be very heavy, and the high thermal mass and low TCC can lead to very slow heat-up rates for the materials at the tool face (as opposed to those next to the bag side). This can lead to reduced resin (or adhesive) flow, surface porosity, and/or poor bonding to core materials in the final part. Heated tooling or adjustments to the allowable delta temperature (oven or autoclave to part) can help mitigate this effect. (Figure 9-6)

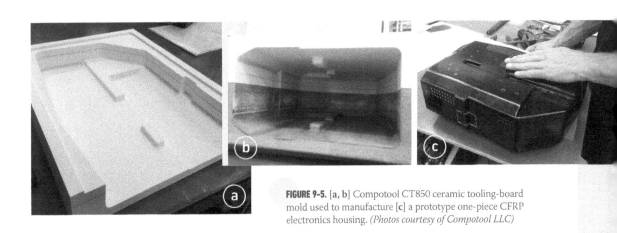

FIGURE 9-5. [a, b] Compotool CT850 ceramic tooling-board mold used to manufacture [c] a prototype one-piece CFRP electronics housing. *(Photos courtesy of Compotool LLC)*

FIGURE 9-6. Large Invar tooling used for production of carbon fiber wing skins for commercial aircraft. *(Photos courtesy of Ascent Airspace – Coast Composites)*

Nickel alloys are the tooling materials of choice for achieving a high-quality surface finish on molded composite parts for automotive or other industries that require aesthetic surfaces. Nickel mold or dies can be polished to a very fine surface finish that is replicated on the molded part. Nickel has a higher CTE than Invar but less than that of aluminum. The thermal conductivity coefficient of nickel is also relatively high compared to steel and Invar, thus making it a good choice for self-heated tooling. (Figure 9-7)

Aluminum is lightweight and easy to machine and is frequently prescribed for prototype tooling but can also be used for production tools where the high CTE or the high TCC of the material is beneficial to the molding process. Aluminum tools can be finished with a black-anodized surface to provide better wear and corrosion resistance. (Figure 9-8)

Steel is primarily used to manufacture molds and die-sets for press molding, compression molding, sheet molding, RTM and HP-RTM processes, but can also be used to make molds and mandrels for conventional oven and autoclave processing. A36 and other low carbon steel alloys are typically chosen for structural applications in which weldability and formability are desired, while P20 (chrome moly) steel is often selected for molds where surface quality and high wear resistance and may be required. (Figure 9-9)

In general, steel is low cost and approximately half the CTE of aluminum yet still considerably higher than CFRP. It is also heavy and takes a great deal of energy to heat to elevated process temperatures.

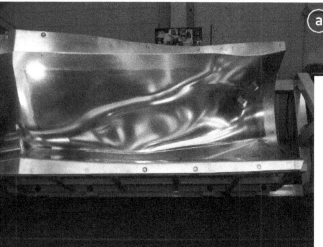

FIGURE 9-7. Polished nickel tooling for [a] aerospace and [b] automotive applications. *(Photos courtesy of Weber Manufacturing, Inc.)*

FIGURE 9-8. Black anodized aluminum mold used to manufacture a CFRP monocoque control surface for a general aviation aircraft. *(Photo courtesy of Radius Engineering, Inc.)*

❱ COMPOSITE TOOLS

Composite tools solve many of the metal tool problems, while introducing different problems of their own. They are much lighter, allowing the tools to better achieve desired heating and cooling rates. The lower weight makes them much easier to move around in the shop and may in some cases eliminate the need for a forklift or crane to move small to medium-sized tools.

Composite tools are also easy to duplicate. Multiple units of the same, exact mold can be made from a single pattern. This is a big advantage over metal tooling, as the duplicates are dimensionally identical to each other and are not affected by machine setup errors or compounding setup tolerances. Often, a permanent pattern will be used for fabricating multiple units of the same mold to support rate production. (Figure 9-10)

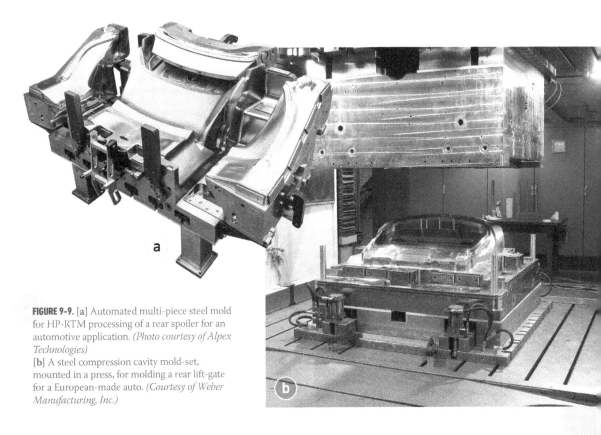

FIGURE 9-9. [a] Automated multi-piece steel mold for HP-RTM processing of a rear spoiler for an automotive application. *(Photo courtesy of Alpex Technologies)*
[b] A steel compression cavity mold-set, mounted in a press, for molding a rear lift-gate for a European-made auto. *(Courtesy of Weber Manufacturing, Inc.)*

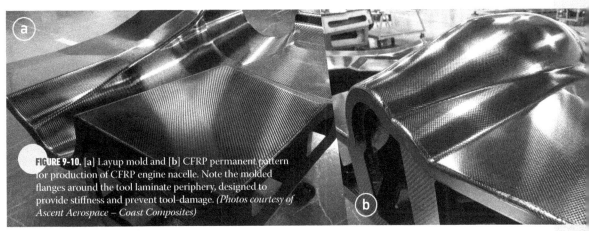

FIGURE 9-10. [a] Layup mold and [b] CFRP permanent pattern for production of CFRP engine nacelle. Note the molded flanges around the tool laminate periphery, designed to provide stiffness and prevent tool-damage. *(Photos courtesy of Ascent Aerospace – Coast Composites)*

On the downside, composite tools can be easier to damage than metal tools, and may ulti-mately develop microscopic cracks (microcracks) in the tool laminate due to differing rates of thermal expansion between the fibers and the matrix. These microcracks in a layup mold will eventually grow together through enough thermal cycles and cause vacuum leaks throughout the tool. This can be detrimental to the process of vacuum-bagged or autoclaved prepreg composite parts.

For this reason, composite tools can have a limited production life compared to metal tools. It is very unusual to get over 300 cycles from an epoxy matrix composite tool before the leak rate becomes noticeable or even unacceptable. Often the number of cycles is much less, down to just a few cycles in particularly bad cases, depending on the tool design, resin type, fiber volume, and the quality and workmanship used in the construction of the facesheet laminate. (Note: many composite tools are damaged when demolded from the pattern. Extreme care is necessary to prevent fracture and microcracking.) It may be possible to achieve more cycles from tools made with BMI or Benzoxazine matrices, or with nanoparticle-fortified tooling prepregs.

Due to this issue, composite tools are not normally prescribed for "high-rate" production such as in the automotive or sporting goods industries, as production quantities are simply too high to make composite tooling practical. However, in the aerospace and marine industries, where production quantities regularly run in the hundreds, composite tools are quite often the best choice. With proper planning, multiple units of composite tools may prove to be less expensive and more productive than multiple units of metal tools.

Molds made from composite materials tend to be more robust when designed with self-supporting (monocoque) features that also move the damage-prone edges away from the part layup/vacuum bag seal area, as shown in Figure 9-10.

❱ HYBRID TOOLS

The marriage of Invar and CFRP for creating durable, lightweight, low CTE production molds has long been considered by industry as a solution to the microcrack/leak problem. Two different approaches have emerged to produce these hybrid tools: (1) an Invar-coated surface on a CFRP composite facesheet and substructure, or (2) a machined CFRP overlay on a rela-tively thin Invar facesheet and substructure.

Ascent Aerospace manufactures HyVarC® hybrid Invar composite layup molds for industry. The molds consist of a <0.2 inch (or 5 mm thick) thick Invar facesheet with a thicker CFRP overlay that can be machined to dimensional surface requirements. The resulting tools are approximately 50% lighter with a 20% shorter lead time than its Invar counterpart and still provide for a vacuum-leak free surface. (Figure 9-11)

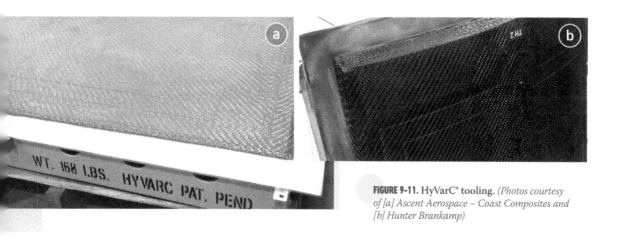

FIGURE 9-11. HyVarC® tooling. *(Photos courtesy of [a] Ascent Aerospace – Coast Composites and [b] Hunter Brankamp)*

❯ 3D PRINTED TOOLING

Models, patterns, molds and mandrels have long been fabricated from thermoplastic resins (ABS, PC, PEI, PPS, PESU, etc.) using 3D printing or **additive manufacturing (AM)** processes. The biggest benefit of AM tooling is the ability to directly fabricate a tool from raw materials using a CAD data file without the need for a physical model or parent tool. Tooling is fabricated in a rapid fashion (usually overnight), limited only by the size of the printing platform and the speed in which the print head can extrude resin. (Figure 9-12)

In recent years, AM equipment, materials, and processes have matured to the point where production of large scale tooling is a reality, with tools made using Big Area Additive Manufacturing (BAAM) or Large Format Additive Manufacturing (LFAM) processes that utilize a combination of additive (3D printing) and subtractive (machining) processes. Short carbon fiber reinforced thermoplastic pellets are used for the raw material and printed to post-net size. The resulting tool is then machined to final size and finished with a surface sealer or film barrier to provide vacuum integrity. (3D printed materials are inherently porous.) Tools of this type have been used for autoclave processing at 350°F (122°C) and 90 psi. (Figure 9-13)

FIGURE 9-12. Vacuum-bagged, 3D-printed ULTEM 1010™ resin mold shown in comparison to a typical CFRP mold for the same part. *(Photo courtesy of Stratasys)*

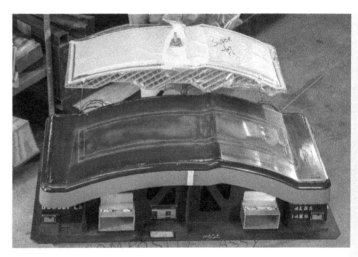

FIGURE 9-13. [a] Large scale and [b] mid-scale CFRTP molds fabricated using a Big Area Additive Manufacturing (BAAM) machine at the Oak Ridge National Laboratory (ORNL). [c] CFRTP leading edge mold using Large Format Additive Manufacturing (LFAM). *(Photos [a, b] courtesy of Oak Ridge National Laboratory, U.S. Dept. of Energy, and [c] Additive Engineering Solutions (AES))*

❯ SELF-HEATED TOOLING

Integrated heating and cooling systems have been long used for tooling used outside of standard oven or autoclave processes. Designs often involve either electric or fluid heating systems that allow for uniform heat transfer to the tool surface of a mold, or to localized regions of a bond fixture.

One method of electric heating is to run low voltage power through an electrically conductive tool surface. Monolithic graphite, carbon foam, and carbon fiber are all conductive materials that have been utilized as electric conductors. A modern example of this idea is shown in Figure 9-14, where an isolated layer of carbon foam (from CFoam, LLC) is used as the conductive layer across the tool surface area. Current is run from a power transformer/controller, through buss attachments on the tool, creating a controllable, heated tool surface. This method requires little power to heat the tool when compared to an oven or autoclave.

A layer of embedded PowerFabric Plus® from LaminaHeat LLC (Greer, SC) can also be utilized to heat a mold, mandrel or fixture uniformly from a near-surface location in the tool laminate construction. The PowerFabric Plus® fabric is comprised of a thin (0.3mm) flexible carbon fiber veil connected to copper foil strips that are sandwiched between layers of E-glass fabrics, which electrically insulate the layer from adjacent materials. (Figure 9-15)

FIGURE 9-14. CFoam self-heated tool.

[a] Isolated carbon foam layer at tool surface, [b] electric buss installed and tested, and [c] final tool assembly. *(Photos courtesy of CFoam, LLC.)*

FIGURE 9-15. [a] PowerFabric Plus® fabric with visible copper strip, [b] wind blade mold built with interlaminate (PowerFabric Plus®) heater layer, and [c] heated wind blade tooling. *(Photos courtesy of LaminaHeat, LLC)*

Surface Generation Ltd. (Oakham, UK) provides fluid-heated (air-heated) tooling that is designed for precise thermal management of the process with custom hardware and control software. The idea is to selectively heat and cool "zones" of a tool to maintain accurate temperatures for curing thermosets or crystalizing thermoplastic polymers. This technology allows such "pixilated" temperature control up to 1,000°C. (Figure 9-16)

Other fluid heating systems use oil or water and require plumbing on the backside of a metal tooling shell. Common designs employ copper tubing for best heat transfer from the fluid to the shell. The plumbing is attached to an independent heat exchange unit with a temperature control system for the fluid exchange. (Figure 9-17)

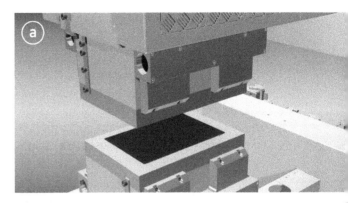
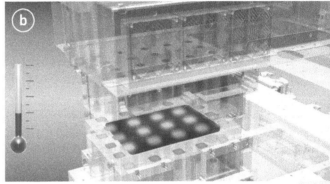
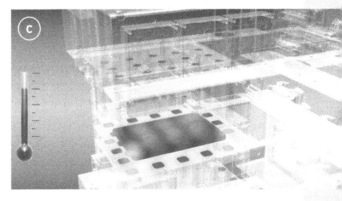

FIGURE 9-16. [a] Surface Generation's production to functional specification (PtFS) concept with [**b, c**] precise digital control of heating and cooling to multiple zones of the tool surface in the molding environment. *(Photos courtesy of Surface Generation Ltd.)*

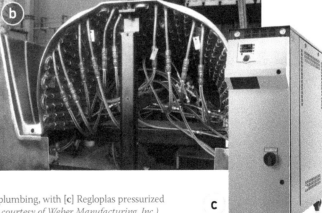

FIGURE 9-17. [**a, b**] Nickel tooling with copper plumbing, with [**c**] Regloplas pressurized water temperature control unit. *([a, b] Photos courtesy of Weber Manufacturing, Inc.)*

❯ RECONFIGURABLE/ADAPTIVE TOOLING

Useful for a host of different applications, automated **reconfigurable, or adaptive tooling** has long been used in the manufacture of high performance sails, architectural products, thermoplastic sheet forming, core-forming and other forming and molding practices.

The greatest benefit of this method is a single tool can be used for multiple part numbers. The data files control the configuration and can be changed or revised as needed without having to build new tools. In this process, a bed of computer-controlled pins position a flexible, wire-frame reinforced face sheet to the desired shape. Surface tolerances of ±0.005 inches are achievable with the assistance of a laser tracker or projector for fine adjustments. (Figure 9-18)

Elastomeric Mandrels, Bladders, and Cauls

Elastomeric mandrels are used to mold hollow-cavity shapes such as hat sections, as well as I-beams, J-sections, and other stiffener configurations in cobonded and cocured composite structures. Most flexible mandrels are made from silicone rubber or other elastomeric formulations. (Figure 9-19)

Care must be taken to compensate for thermal expansion and mandrel shrinkage over time. Mandrel manufacturers such as Rubbercraft (Long Beach, California) have reduced these effects through a hot molding process designed to stabilize the product. Some thermal expansion of the mandrel can aide in laminate compaction when used in a trapped molding application.

Elastomeric bladders are often used for more complex geometries in hollow-cavity or trapped tooling shapes where internal pressure is desired in the process. Bladder molding provides more uniform pressure than trapped solid mandrels and is easier to extract from complex shapes. (Figure 9-20)

Additional tooling is often required for gripping and extracting mandrels and bladders from hollow-cavity shapes. Metallic mechanical fittings may be molded or attached to one or both ends of these mandrels to aid in this purpose. It also should be noted that silicone rubber

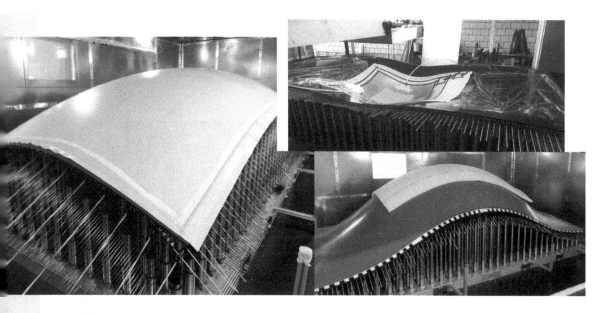

FIGURE 9-18. Examples of adaptive tooling used at CurveWorks to form and/or mold various compound shapes. *(Photos courtesy of CurveWorks BV)*

mandrels have a low tear strength at elevated temperatures and that provisions should always be made to cool to room temperature before extracting.

Semi-rigid caul tooling is often prescribed to control the shape of a laminate on the bag-side of a part in a localized region; e.g., along-hat-section stiffeners to minimize wrinkles and bridging in the corners. (Figure 9-21)

Reusable Vacuum Bags

Also called elastomeric vacuum tools (EVTs), reusable or permanent vacuum bags are typically made from silicone rubber or fluoroelastomer materials and may be attached to a frame for ease of handling. EVTs are either applied or permanently mounted to a mold or mandrel for use as a vacuum bag membrane for infusion or oven or autoclave processing.

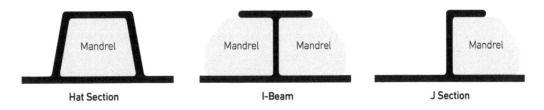

FIGURE 9-19. Elastomeric mandrel varieties.

FIGURE 9-20. Hollow-cavity mandrel.

FIGURE 9-21. Semi-rigid caul tooling.

FIGURE 9-22. [a] EVT for large wing skin tool and [b] for small component tool with complex geometry, and [c] standard EVT used as a vacuum debulk table in production or the lab. *(Photos courtesy of Torr Technologies, Inc.)*

EVT designs vary but most include an extruded rubber vacuum seal that is either bonded around the periphery of the rubber bag or to the mold surface. The bag itself is usually made from calendared silicone rubber sheet, uncured rubber sheet (molded and vulcanized to shape), or liquid rubber rolled or sprayed to a desired thickness. Selected fiber reinforcements embedded in the rubber may be included to prevent tearing. Vacuum fittings, resin ports (for infusion), thermocouple plugs, and other necessary hardware can be installed or co-vulcanized in place during the construction of the EVT.

While reusable vacuum bags can be formed to very complex shapes, it is not recommended, as thermal expansion and shrink of the rubber materials over time may contribute to processing defects such as bridging in corners. Instead, these tools are best used for simpler shapes.

Washout and Breakout Mandrels

Washout materials have evolved over the years from eutectic salts to ceramics to more recently, 3D printed ceramic or polymeric materials. Washout mandrels are particularly useful for producing parts that have complex configurations where the tooling would otherwise be trapped inside the part after processing.

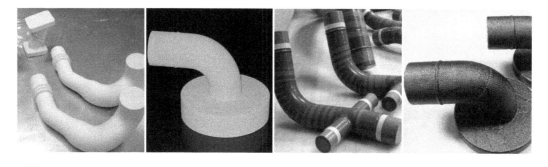

FIGURE 9-23. Ceramic washout mandrels made from AquaPour™ castable ceramic materials from Advanced Ceramics Manufacturing, used to manufacture composite ducting such as the carbon fiber ducts seen on the far right.

FIGURE 9-24. Stratasys' sacrificial tooling uses proprietary, dissolvable (or break-away) thermoplastic materials to simplify the production process for many complicated composite part applications. Additive manufacturing sacrificial tooling eliminates both machining from a bulk form and the need for additional tooling to prepare the mold (e.g., a cavity mold to cast a conventional wash-out material) and permits direct printing of the final shape. *(Photos courtesy of Stratasys)*

Gypsum plaster is used to make sacrificial breakout mandrels for similarly complex parts as those made using washout mandrels but at a higher risk of damaging the parts when breaking away the mandrel from the part after processing. For this reason, the industry is trending more toward washout materials and leaning away from breakout materials. (Figures 9-23, 9-24)

Jigs and Fixtures

Metals, composites, and (3D printed) thermoplastic materials are used to fabricate jigs and fixtures of all types. Post-molded parts are indexed to the jig or fixture using molded index features or tooling pins and bushings that align the part to the tool. Designs often employ either mechanical clamping devices or vacuum-chuck work holding features to securely clamp parts to the tools for machining, drilling, assembly, or other secondary (post-molding) manufacturing operations. (Figure 9-25)

See Table 9.3 on the next page for a tools and materials reference table.

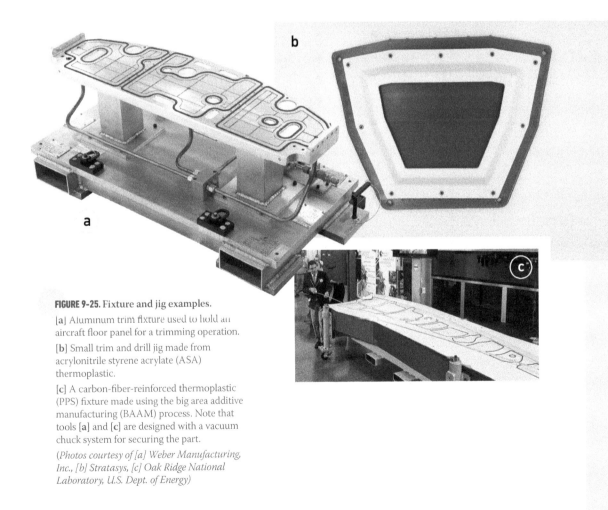

FIGURE 9-25. Fixture and jig examples.

[a] Aluminum trim fixture used to hold an aircraft floor panel for a trimming operation.

[b] Small trim and drill jig made from acrylonitrile styrene acrylate (ASA) thermoplastic.

[c] A carbon-fiber-reinforced thermoplastic (PPS) fixture made using the big area additive manufacturing (BAAM) process. Note that tools [a] and [c] are designed with a vacuum chuck system for securing the part.

(Photos courtesy of [a] Weber Manufacturing, Inc., [b] Stratasys, [c] Oak Ridge National Laboratory, U.S. Dept. of Energy)

TABLE 9.3 Reference Table of Tools and Materials

	Master models and patterns	Prototype molds and mandrels	Production molds and mandrels	Transfer tools and permanent patterns	Press and RTM molds	Washout and breakout mandrels	Cauls, bladders, flexible mandrels and EVTs	Jigs and fixtures
Aluminium	■	■	■		■		■	■
Steel		■	■		■			■
Invar	■	■	■	■	■		■	■
Nickel		■	■		■			
CFRP		■	■	■			■	■
GFRP		■		■			■	■
Gypsum Plaster	■			■		■		
Cast Ceramic	■					■		
Monolithic Graphite		■	■	■				
Epoxy Board	■	■						
Polyurethane Board	■	■						■
Ceramic Block	■	■						■
CFoam and FRCfoam	■	■	■	■				
Elastomer & FRE		■	■				■	
Med. Density Fiberboard (MDF)	■	■						
AM Thermoplastic & FRTP	■	■	■	■		■		■

10

Inspection and Test Methods

CONTENTS

Destructive Coupon Testing

Any claims as to the properties of a given composite material must be validated through destructive coupon testing that (1) is performed according to an accepted standardized method and (2) is supported by statistical data analysis. Tests developed for metals and plastics typically cannot be used for composites. Standard methods specifically used for testing composites have been developed by ASTM International, International Standards Organization (ISO), Suppliers of Advanced Composite Materials Association (SACMA), and a variety of composites and plastics industry associations and manufacturers.

The goal of a **coupon test program** is to generate data that is representative of the actual composite structure being built, throughout its operational life, taking into consideration all possible environmental conditions. Issues such as specimen preparation, environmental conditioning, test setup and instrumentation are all vital to obtaining accurate and repeatable data. The discussion here will present an overview of some of the common tests performed to generate mechanical property data for composites, which includes the following lamina (single-ply) and laminate (two or more plies) testing categories:

- Tensile
- Compressive
- Shear
- Flexure
- Fracture toughness
- Fatigue

Table 10.1 shows the mechanical properties and test methods from which the *Composite Materials Handbook-17* (CMH-17) accepts data on composites.

TABLE 10.1 Standard Test Methods Recommended by CMH-17

Test Category	Standard Test Method
Material Screening / Process Control	
In-Plane Shear (±45° laminates only)	ASTM D3518
Short Beam Strength	ASTM D2344
4-point Flexure	ASTM D7264
Lamina/Laminate Mechanical Tests	
In-Plane Tension	ASTM D3039
In-Plane Compression (combined loading)	ASTM D6641
0° property data reduction from 90/0 tension / compression tests	(No recommendation)
In-Plane Shear	ASTM D7078
Out-of-Plane Shear	ASTM D5379
Out-of-Plane Tension	ASTM D7291
Open-Hole Tension	ASTM D5766
Filled-Hole Tension	ASTM D6742 combined with ASTM D5766
Open-Hole Compression	ASTM D6484
Filled-Hole Compression	ASTM D6742 combined with ASTM D6484
Single-Shear Bearing (single or double fastener)	ASTM D5961 Procedure B or C
Double-Shear Bearing ("pin" bearing)	ASTM D5961 Procedure A
Bearing/Bypass Interaction	ASTM D7248
Fastener Pull-Thru Strength	ASTM D7332
Compression after Impact	ASTM D7136 (impact)
Compression after Impact	ASTM D7137 (residual strength)
Mode I Fracture Toughness	ASTM D5528
Mode II Fracture Toughness	ASTM D7905
Mode III Fracture Toughness	(No recommendation)
Mixed Mode I/II Fracture Toughness	ASTM D6671
Mode I Fatigue Delamination Onset	ASTM D6115
Mode I Fatigue Delamination Growth	(No recommendation)
Tension/Tension Fatigue	ASTM D3479
Tension/Compression Fatigue	(No recommendation)
Bearing Fatigue	ASTM D6873

Source: CMH-17-1G, Vol. 1, Chapter 2: Guidelines for Property Testing of Composites.

) TENSILE

ASTM D3039 "Standard Test Method for Tensile Properties of Polymer Matrix Composite Materials" testing is achieved using a rectangular coupon that is held at the ends by wedge or hydraulic grips, and then loaded in uniaxial tension using a universal testing machine. The coupon may be straight-sided for its entire length or width-tapered from the grip-ends into the gage section. (This is often called a *dog-bone* or *bow-tie* coupon, and is not applicable to unidirectional materials.)

The gage section is the part of the specimen over which strain or deformation is measured. In tensile testing, this material response is measured using strain gages or extensometers. End tabs made from glass-reinforced epoxy are typically bonded onto the specimen to better distribute the load and prevent stress concentrations induced by the grips. Common properties derived from tensile testing include tensile strength, ultimate tensile strain, **Poisson's ratio**, and modulus of elasticity (tensile modulus). (Figure 10-1)

ASTM D5766 "Standard Test Method for Open-Hole Tensile Strength of Polymer Matrix Composite Laminates" (a.k.a. the notched tensile strength test) uses a 36-mm (1.47-inch) wide, quasi-isotropic coupon with a 6-mm (0.236-inch) diameter hole in the center, providing a 6:1 width to diameter (w/D) ratio for testing. This test can be useful in determining the knock-down in strength and ultimate strain values of a composite laminate by introduction of holes in the laminate. This standard is used with ASTM D6742 ("Standard Practice for Filled-Hole Tension and Compression Testing of Polymer Matrix Composite Laminates") to assess filled-hole tensile (FHT) test values. (Figure 10-2)

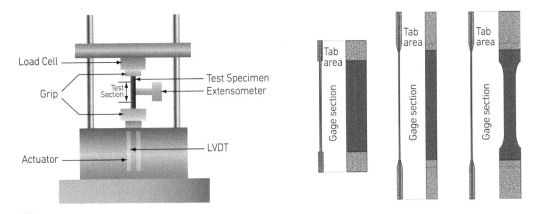

FIGURE 10-1. Tensile test equipment and typical ASTM D3039 coupon configurations.

FIGURE 10-2. Center section of open-hole tension specimen. *(Photo courtesy of Wyoming Test Fixtures)*

❱ COMPRESSIVE

Compressive testing of FRP composites has historically been accomplished by one of two methods: ASTM D3410 "Standard Test Method for Compressive Properties of Polymer Matrix Composite Materials with Unsupported Gage Section by Shear Loading," which introduces shear loads through tabbed ends of a specimen that is gripped in a fixture; or ASTM D695 "Standard Test Method for Compressive Properties of Rigid Plastics," which applies direct end-loading to a tabbed specimen that is indexed and supported in a test fixture. Both methods apply compressive loads to a gage area of the specimen where the strain is measured as the failure occurs. (Figure 10-3)

Today, the ASTM D6641 "Standard Test Method for Compressive Properties of Polymer Matrix Composite Materials Using a Combined Loading Compression (CLC) Test Fixture" method combines both end-loading and shear-loading in a combined loading test. The laminate test specimen is mounted in a CLC fixture that provides control over the end-loading to shear-loading ratio. The fixture is then installed between compression platens which apply the load. Properties obtained by this test method include ultimate compressive strength, ultimate compressive strain, compressive modulus of elasticity (linear or chord), and Poisson's ratio in compression.

ASTM D6484 "Standard Test Method for Open-Hole Compressive Strength of Polymer Matrix Composite Laminates" (a.k.a. notched compression test) uses 3 to 5mm (0.12–0.19 inch) thick by 300 mm (11.81 inch)-long, quasi-isotropic coupon mounted in a special test fixture, in

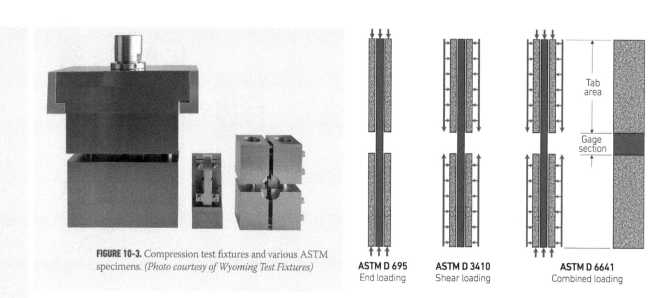

FIGURE 10-3. Compression test fixtures and various ASTM specimens. *(Photo courtesy of Wyoming Test Fixtures)*

ASTM D 695 End loading

ASTM D 3410 Shear loading

ASTM D 6641 Combined loading

Tab area

Gage section

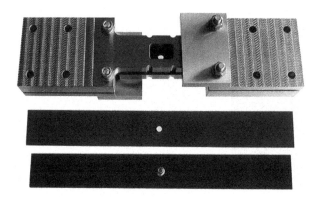

FIGURE 10-4. Open and closed hole compression fixture and specimens. *(Photo courtesy of Wyoming Test Fixtures)*

order to establish engineering data for use in establishing design allowables. This standard is used with ASTM D6742 ("Standard Practice for Filled-Hole Tension and Compression Testing of Polymer Matrix Composite Laminates") to assess filled-hole compression (FHC) test values. (Figure 10-4)

) SHEAR

A simple test, ASTM D3518 "Standard Test Method for In-Plane Shear Response of Polymer Matrix Composite Materials by Tensile Test of a ±45° Laminate" uses a modified ASTM D3039 tensile test coupon laid-up with a fully balanced and symmetric stack of ±45° plies. The test is conducted much like the D3039 method, using a tensile testing machine. Strain gages are oriented in both longitudinal and transverse directions at the center of the section. The resulting measurements provide in-plane shear stress versus strain response, chord modulus of elasticity, and in-plane shear strength of the specimen. (Figure 10-5)

ASTM D5379 "Standard Test Method for Shear Properties of Composite Materials by the V-Notched Beam Method," or the Iosipescu Shear Test (after one of its founders), uses a flat rectangular coupon with symmetrical V-notches on opposing sides of the coupon at its midpoint (*see* Figure 10-6). The coupon is loaded at the edges through a specially designed fixture where the upper head of the fixture is attached to and pushed downward by the cross-head of the testing machine. The fixture thus introduces a shear load into the coupon between

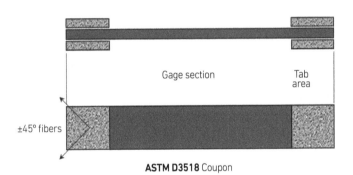

Gage section   Tab area

±45° fibers

ASTM D3518 Coupon

FIGURE 10-5. In-plane shear coupon (ASTM 3518).

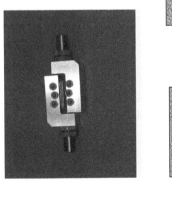
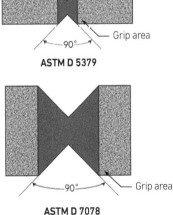

Grip area
90°
ASTM D 5379

90°   Grip area
ASTM D 7078

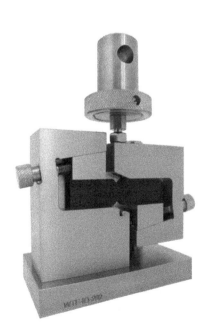

FIGURE 10-6. Iosipescu V-notched beam shear (ASTM D5379) and V-notched rail shear (ASTM D7078) fixtures and specimens. *(Photos courtesy of Wyoming Test Fixtures)*

the two notches, and deformation is measured by strain gages. Tabs may be bonded to the ends of the coupon for increased stability, and in-plane or out-of-plane (interlaminar) shear may be measured, depending on the orientation of the material relative to the loading axis.

ASTM D7078 "Standard Test Method for Shear Properties of Composite Materials by V-Notched Rail Shear Method" uses a similar but larger gage area specimen than the (D5379) Iosipescu test. It loads the specimen through its faces rather than its edges using a tensile test machine. Properties obtained using this method include shear stress versus engineering shear strain response, ultimate shear strength, ultimate engineering shear strain, and shear chord modulus of elasticity.

ASTM D2344 "Standard Test Method for Short-Beam Strength of Polymer Matrix Composite Materials and Their Laminates," commonly known as the short-beam shear (SBS) test, measures the interlaminar shear strength of fiber-reinforced composites using a short and relatively thick specimen (4:1 length to thickness ratio) that has both balanced and symmetric elastic properties. The short, deep "beam" is meant to minimize bending stresses while maximizing interlaminar shear stresses. To perform the test, the coupon is mounted across two dowels and loaded at the midpoint, then contact stresses are introduced at the load points which interfere with the strain distribution. This produces a failure that is rarely pure shear in nature; thus it has been recommended that this test be used for qualitative purposes only. (Figure 10-7)

FIGURE 10-7. Short beam shear test (ASTM D2234). *(Photo courtesy of Wyoming Test Fixtures)*

FIGURE 10-8. Flexure test (ASTM D7264). Procedure A, 3-Point loading *(left)* and Procedure B, 4-Point loading *(right). (Photos courtesy of Wyoming Test Fixtures)*

) FLEXURE

The ASTM D7264 "Standard Test Method for Flexural Properties of Polymer Matrix Composite Materials" is used to evaluate the long beam flexural properties of FRP composites, including strength, stiffness, and load/deflection behavior. Two methods are used: Procedure A, three-point loading with the load point at the center of the specimen mounted on a two-point support span; and Procedure B, four-point loading using two load points that are equally spaced from the adjacent supports. The distance between load points is equal to one-half of the support span (*see* Figure 10-8). This test method uses a standard span-to-thickness ratio of 32:1. (In comparison, the older ASTM D790 flexure standard was set at 16:1.) These procedures are utilized for both coupon and structures testing.

Since the physical properties of materials can vary depending on temperature, a thermal chamber may be used to test materials at temperatures that simulate the in-service environmental conditions. Standard test fixtures are installed inside the chamber, and testing is conducted the same as it would be at ambient temperature. The chamber has electric heaters for elevated temperatures and uses carbon dioxide or other gas as a coolant for reduced temperatures.

❱ FRACTURE TOUGHNESS

Fracture in a composite structure is usually initiated by a crack or flaw that creates a local stress concentration. Fracture toughness is a generic term for the measure of resistance to extension of a crack or initial flaw. As shown in Figure 10-9, fracture toughness testing can be one of three types, depending on the mode in which the crack propagates:

- Mode I, Opening/tensile mode
- Mode II, Sliding/shear mode
- Mode III, Tearing mode

ASTM D5528 "Standard Test Method for Mode I Interlaminar Fracture Toughness of Unidirectional Fiber-Reinforced Polymer Matrix Composites" is the most common standardized fracture toughness test for composites and is often called the double-cantilever beam (DCB) test. This test method is limited to unidirectional carbon fiber and glass fiber tape laminates because use of woven and multiaxial reinforced composites may increase the tendency for the delamination to grow out-of-plane, compromising the accuracy of the test results. The coupon is sized 125 mm long, 20 to 25 mm wide, and 3 to 5 mm thick (5 inches long, 0.8 to 1 inch wide, and 0.12 to 0.20 inches thick). A non-bondable film insert is placed at the mid-plane during manufacture in order to create an initial crack length of roughly 63 mm (2.5 inches). (Figure 10-10)

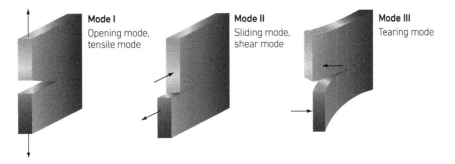

FIGURE 10-9. Fracture toughness modes of crack propagation.

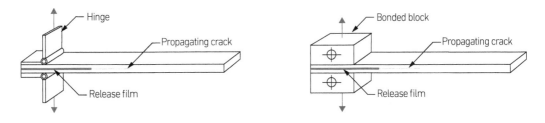

FIGURE 10-10. Schematic of double cantilever beam (DCB) fracture toughness test specimens (ASTM D5528).

ASTM D6671 "Standard Test Method for Mixed Mode I-Mode II Interlaminar Fracture Toughness of Unidirectional Fiber Reinforced Polymer Matrix Composites" is used to determine interlaminar fracture toughness (G_{cf}) of continuous FRP composite (hinged DCB) specimens. This is done at different ratios of Mode I to Mode II loading using the Mixed-Mode Bending (MMB) Test. The test specimens are clamped in the fixture, both to the base and the lever-loading beam. The other end of the specimen rests on a rotating pin support, which can be adjusted to a desired support span length. The center bracket on the loading beam is adjusted to center the loading pin to the support span. Various ratios of Mode I to Mode II can be induced in the test specimen by adjusting the saddle at the left end of the loading beam, thus moving the load point on the lever. (Figure 10-11)

❱ FATIGUE

Most structures do not operate under a constant load. In fact, stresses on a composite structure are constantly changing. A good example is the cyclic loading forces on an airplane wing from air turbulence, changing wind direction, changing air speeds, rain, etc. These factors can make it nearly impossible to determine the stresses acting on an aircraft wing at any given instant in time. Thus, a complete fatigue testing program for such a structure is comprised of statistical analysis where the worst-case states of stress (e.g., landing, take-off and turbulence), as well as the mean and amplitude stresses, are measured and factored in order to determine the long-time fatigue behavior. (Figure 10-12)

To measure the fatigue resistance of materials, a cyclic load is applied to a specimen until it fails. The number of cycles to failure, called the fatigue life (N_f), is recorded. The logarithm of this life is plotted against the stress (or sometimes the log of stress) to develop an S-N curve (stress versus number of cycles to failure) as shown in Figure 10-13. This curve defines the fatigue resistance of a material. As can be seen in the figure, there is a great deal of scatter in the fatigue life. Consequently, a large number of tests must be run in order to determine fatigue resistance.

ASTM D6115 "Standard Test Method for Mode I Fatigue Delamination Growth Onset of Unidirectional Fiber-Reinforced Polymer Matrix Composites" uses a DCB specimen and deter-

FIGURE 10-11. MMB test fixture with specimen.
(Photo courtesy of Wyoming Test Fixtures)

FIGURE 10-12. Flexure fatigue test fixture and specimen.
(Photo courtesy of Wyoming Test Fixtures)

mines the number of cycles *(N)* for the onset of delamination growth based on the opening Mode I cyclic strain energy release rate *(G)* of the test specimen. This test is useful for establishing design allowables used in damage tolerance analyses of composite structures made from these materials.

ASTM D3479 "Standard Test Method for Tension-Tension Fatigue of Polymer Matrix Composite Materials" is used to determine the fatigue behavior of polymer matrix composite materials subjected to tensile cyclic loading. This test method is limited to un-notched test specimens subjected to constant amplitude uniaxial in-plane loading. Note: This test method is less useful for interrogation of matrix fatigue, which is of primary concern to engineering. Within this standard, two different procedures are used:

- **Procedure A:** The servo-hydraulic test machine is controlled so that the test specimen is subjected to repetitive constant amplitude load cycles; i.e., constant load.
- **Procedure B:** The test machine is controlled so that the test specimen is subjected to repetitive constant amplitude strain cycles; i.e., constant strain.

Resin, Fiber and Void Content

The importance of achieving a high fiber volume fraction and low void content in any FRP laminate is paramount to the performance of that structure, both in testing and in service. Fully understanding these properties is the priority of the material and process engineer. There are two basic methods for determining the resin, fiber and void content in a composite:

- Matrix ignition loss (burn-out)
- Matrix digestion (acid digestion)

The following section on matrix ignition loss and digestion methods explains standards and ways used by industry in determination of these properties.

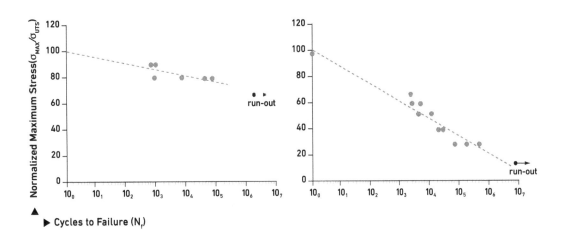

FIGURE 10-13. S-N curves for [02/902]S cross-ply laminates HTA/F922 carbon/epoxy *(left)* and E-glass/F922 epoxy *(right)*.

Carbon/epoxy cross-ply laminates show excellent fatigue resistance. The fatigue life exceeded 4 × 106 cycles at loads approaching 90% and 70% of the ultimate tensile strength (UTS) for [0/90] 4S T300/924, and 70% UTS for [02/902] S, respectively. End tabs are prone to disbond during testing; thus, prevention of end tab disbonding is essential for ensuring reliable fatigue data. *(Source: NPL Report Measurement Note CMMT(MN) 067)*

❭ MATRIX IGNITION LOSS AND MATRIX DIGESTION METHODS

ASTM D2584 "Standard Test Method for Ignition Loss of Cured Reinforced Resins" uses a pre-dried and weighted composite specimen, which is placed in a crucible of known weight and then inserted into a furnace, where the organic resin matrix will burn off. After burning, the crucible is cooled in a desiccator and reweighed. Ignition loss is the difference in weight before and after burning, and represents the resin content of the composite sample. The remaining weight determines the fiber reinforcement and any inorganic fillers in the composite. The fiber volume fraction is calculated as:

$$V_f = \frac{\rho_m W_f}{\rho_f W_m + \rho_m W_f}$$

where:

V_f = volume fraction of fibers
W_f = weight of fibers
W_m = weight of matrix
ρ_f = density of fibers
ρ_m = density of matrix

This method assumes that the fiber reinforcement is unaffected by combustion and is commonly used with fiberglass-reinforced composites, as well as other composite reinforcements that are insensitive to high temperatures, such as glass and quartz (silica).

ASTM D3171 "Standard Test Methods for Constituent Content of Composite Materials" is designed for two-part composite material systems (those without fillers) and determines the constituent content (i.e., resin, fiber, and void) of FRP composite specimens. This standard uses one of two approaches:

- **Method I.** The matrix is removed by digestion or ignition using one of eight procedures, leaving the reinforcement essentially unaffected and enabling calculation of the reinforcement or matrix content (by weight or volume) as well as percent void volume. Procedures A through F are based on chemical removal of the matrix, while Procedure G removes the matrix by igniting it, and Procedure H carbonizes the matrix in a furnace.

- **Method II.** Applicable only to laminate materials of known fiber areal weight, and calculates the reinforcement or matrix content (by weight or volume), as well as the cured ply thickness, based on the measured thickness of the laminate. Void volume is not measured using this method.

To determine the fiber volume fraction using acid digestion, the matrix is dissolved using an appropriate solvent and the remaining fibers are then weighed. Chemical digestion assumes a chemical has been selected that does not attack the fibers. Common choices are hot nitric acid for carbon fiber/epoxy, sulfuric acid and hydrogen peroxide for carbon fiber/polyimides and PEEK, and sodium hydroxide for aramid composites. Others are described in the ASTM D3171 standard.

Alternatively, a computer photomicrographic technique may be used to determine the fiber volume fraction of a laminate. The number of fibers in a given area of a polished cross section are counted and the volume fraction determined as the area fraction of each constituent.

The void volume of the composite is calculated, where possible, by using the known composite density, cured resin density, fiber density, and determined resin and fiber contents by weight. The density properties are usually available from the materials supplier and should be obtained for the specific lot of material being tested. ASTM D2734 "Standard Test Methods for Void Content of Reinforced Plastics" specifies these calculations in detail.

Other standard material test methods are referenced in Table 10.2 below.

TABLE 10.2 Standard Test Methods for Material Properties (as recommended by CMH-17)

Test Category	Standard Test Method
PREPREG TESTS	
Resin Content	ASTM D3529
Fiber Areal Weight	ASTM D3529
Volatiles Content	ASTM D3530
Resin Flow	ASTM D3531
Resin Gel Time	ASTM D3532
Moisture Content	ASTM D4019
Tack	ASTM D8336
Drape	(No recommendation)
Percent of Cure	ASTM D7750
Insoluble Content	ASTM D3529
DSC	SACMA RM 25
Rheology	ASTM D7750
DMA (RDS)	SACMA RM 19
HPLC	SACMA RM 20
IR	ASTM E1252/E168
LAMINA PHYSICAL TESTS	
Moisture Conditioning	ASTM D5229
Glass Transition Temp., dry	ASTM D7028
Glass Transition Temp., wet	ASTM D7028
Density	ASTM D792
Cured Ply Thickness (CPT)	ASTM D3171/Method II
Fiber Volume	ASTM D3171/Methods I or II
Resin Content	ASTM D3171/Methods I or II
Void Content	ASTM D3171/Method I
Equilibrium Moisture Content	ASTM D5229
Moisture Diffusivity	ASTM D5229
CTE , out-of-plane	ASTM E831
CTE, in-plane	ASTM E228
Specific Heat	ASTM E1269
Thermal Conductivity (low to moderate TC materials)	ASTM C177
Thermal Diffusivity (and indirect measurement of high TC materials)	ASTM E1461

Fire, Smoke and Toxicity (FST) Requirements and Heat Release Testing

Composite structures must be tested to meet fire, smoke, toxicity (FST) and heat release requirements, according to their application.[1] The regulations and standards for composites used in commercial aircraft differ greatly from those used in rail and bus, ships, submarines, military aircraft, and the construction/building industry. Fireworthiness and fire safety standards also vary greatly by country.

Fire properties of composite materials measure their ability to ignite, burn and generate toxic substances. The properties commonly tested include ignition time, heat release rate, heat of combustion, flame spread, smoke density and smoke toxicity. The FST and heat release tests used for commercial aircraft and rail applications in the U.S. are discussed in this section.

❭ COMMERCIAL AIRCRAFT

In 1988, the Aviation Safety Research Act was passed, mandating that the Federal Aviation Administration (FAA) conduct long-term investigations into aircraft fire safety, including fire containment and the fire resistance of materials in aircraft interiors. One of the goals of this program is to enable the design of a totally fire-resistant cabin for future commercial aircraft. Thus, FAA regulations are continually changing with technology, trending toward more stringent standards and tests more reflective of actual service. For example, the FAA regulation for non-metallic materials in pressurized aircraft cabins have been reduced from a total and peak heat release of 100 kW/m^2 to 65 kW/m^2. An upper limit on smoke density has also been enforced.

Current FAA FST requirements are described in Title 14 of the Code of Federal Regulations (14 CFR) Part 25, specifically in regulation §25.853 and Appendix F. Part 25's Appendix F (Section I) describes a vertical Bunsen burner test—commonly referred to as vertical burn—to determine the resistance of cabin and cargo compartment materials to flame. A specimen is held in a vertical position by a device inside a cabinet and a Bunsen burner is placed beneath it for 60 seconds (i) or 12 seconds (ii). The burner is then removed and the specimen is observed. The following data are recorded as shown in Tables 10.3 and 10.4.

TABLE 10.3 Data Recorded for 14 CFR Part 25 Vertical Bunsen Burner Test

Ignition Time	Length of time burner flame is applied to specimen
Flame Time	Time that specimen continues to flame after burner is removed
Drip Extinguishing Time	Time that any flaming materials continues to flame after falling from specimen
Burn Length	Distance from original specimen's edge to farthest evidence of damage to specimen

TABLE 10.4 Requirements for Passing 14 CFR Part 25 Vertical Burn Test

Test	Flame Time (sec)	Average Drip Extinguishing Time (sec)	Average Burn Length
(i) 60 seconds	<15	<3	6 in. (152.4 mm)
(ii) 12 seconds	<15	<5	8 in. (203.2 mm)

1 Some text in this section is adapted from "Passenger Safety: Flame, Smoke Toxicity Control," Jared Nelson, *Composites Technology Magazine*, December 2005; reproduced with permission.

In Part 25 Appendix F, Sections IV and V discuss two separate tests: (1) the OSU (Ohio State University) Rate of Heat Release; and (2) Specific Optical Density of Smoke Generated by Solid Materials. The OSU Rate of Heat Release is designed to test whether materials will flashover (the temperature at which the heat in an area is enough to ignite all flammable material simultaneously), or readily contribute to a fire in a crash situation. It provides information on fire size and growth rate. The rate of heat release measures the rate at which a burning material releases heat, using the principle of oxygen consumption (calorimetry). This small-scale component test, published under ASTM E906, uses the OSU calorimeter and measures a specimen's heat release when exposed to a constant, external radiant heat source producing a heat flux (heat transfer per unit of cross-sectional area) of 35 kW/m^2. Passing criteria can be seen below in Table 10.5.

The smoke density test, previously known as the ASTM F814 (now retired), closely follows ASTM E662, which is often referred to as the "NBS smoke density chamber" (NBS: National Bureau of Standards). This test measures the smoke generated by a burning material expressed in terms of specific optical density (D_s); it is designed to help improve the ability for passengers to exit an aircraft cabin during a fire. D_s readings are taken at 1.5 minutes into the test and 4 minutes. (*See* Table 10.5.)

TABLE 10.5 Requirements for Passing 14 CFR Part 25 OSU Heat Release Test and Smoke Density Test

Total Heat Release within first 2 minutes	≤ 65 kW·min/m^2
Peak Heat Release	≤ 65 kW/m^2
4-minute Smoke Density (D_s)	≤ 200

ASTM E1678 outlines testing for toxic gas emission and measures the lethal toxic potency of smoke produced from a material that is ignited while exposed to a radiant heat flux of 50 kW/m^2 for 15 minutes. Test specimens are no larger than 3 by 5 inches (76 by 127 mm), with a thickness no greater than 2 inches (51 mm), and are intended to represent actual finished materials or products used, including composite and combination systems. Concentrations of the major toxic gases are measured over a 30-minute period. These measurements are then used in predictive equations to determine the analytical toxic potency of a material, based on data for carbon monoxide, carbon dioxide, oxygen, hydrogen cyanide, hydrogen chloride, and hydrogen bromide. Thus, the test's predictive ability is limited to those materials whose smoke toxicity can be attributed to one or more of these toxicants.

❯ RAIL

In the U.S., the Federal Railroad Administration of the Department of Transportation (DOT) regulates the safety of passenger trains, buses and other "people movers." Fire safety requirements are found in Title 49 of the Code of Federal Regulations (49 CFR) Part 238, wherein Appendix B dictates flame spread and smoke requirements in areas where composites are used based on two standards developed by ASTM.

ASTM E162 "Standard Test Method for Surface Flammability of Materials Using a Radiant Heat Energy Source" measures flame spread. This test is performed using a 15.2 cm by 45.6 cm (6 inch by 18 inch) panel tilted at a 30° angle. The panel's top edge is exposed to a 670°C/1,238°F heat source placed at a distance of 11.9 cm/4.7 inches. The test is run either until the flame reaches the bottom edge, or (if it does not) for 15 minutes. The flame spread index (I_s) of the material is calculated based on the distance the flame traveled and the amount of heat generated, with a passing criteria of ≤ 35.

Smoke density is measured using the same standard as for commercial aircraft, ASTM E662. Smoke toxicity is not currently regulated in 49 CFR Part 238, but can be measured using either the Boeing Specification Support Standard or the Bombardier SMP 800-C test. Both tests use a smoke chamber. The Boeing test method employs a flamed heat source, to gauge toxic fume concentration. The Bombardier method, however, uses colorimetric tubes or absorptive sampling. In the first instance, colorimetric tubes are placed into the smoke chamber. Each tube contains a specific fluid that reacts with a specific gas, resulting in a change of fluid color. Gas concentration is determined using a color band scale. Alternatively, absorptive sampling uses light spectroscopy. Each gas has a unique spectroscopic signature that permits technicians to identify the gas and determine its level of concentration.

Both tests gauge the concentrations of six gases—carbon monoxide (CO), hydrogen fluoride (HF), hydrogen chloride (HCl), hydrogen cyanide (HCN), nitrogen oxides (NOx) and sulfur dioxide (SO_2). The Bombardier test also tests for carbon dioxide (CO_2) and hydrogen bromide (HBr). As in the smoke density test, concentration levels are recorded at specified time intervals. For the Boeing test, the maximum allowable levels for the six gases are 3,500, 200, 500, 100, 100 and 100 parts per million (ppm), respectively. For the Bombardier test, carbon dioxide and hydrogen bromide are specified at 90,000 and 100 ppm, respectively. Otherwise, permitted maximums are identical to the Boeing test, except for hydrogen fluoride, which is reduced to 100 ppm. Ultimately, end-users and local municipalities must determine acceptable toxicity levels.

ASTM E162 and ASTM E662 criteria were originally developed using unsaturated polyester resin as a baseline. Since then, some local municipalities have developed more stringent criteria. After a 1979 passenger railcar fire in San Francisco's Bay Area Rapid Transit (BART) system, it was determined that many interior components (of composite and other materials)—the seats, in particular—did not meet applicable ASTM requirements. Since that time, BART has replaced seating with nonflammable, nontoxic materials, and its cars not only meet but beat ASTM criteria. Its suppliers now must conform to standards that exceed federal requirements. For example, while ASTM E662 specifies the four-minute smoke density at less than 200, BART dictates that the level must remain at 100 or below.

Outside of the U.S., most countries have developed FST requirements similar to U.S. limits. However, following a fire in the Kings Cross station of the London Underground in 1987, the British government implemented much more stringent flame spread and smoke density requirements, which are outlined in British Standard (BS) 6853 and BS 476, respectively. Unlike the U.S. requirements, British requirements specify different criteria depending on the transportation environment. For instance, trains that travel underground have more stringent smoke density requirements than trains that travel above ground.

Since the establishment of ASTM standards, fire-retardant-filled resin systems have been developed that are far more flame/smoke resistant than the polyesters used to establish ASTM flame/smoke criteria. A common solution is the addition of fire retarding materials to neat resins to improve FST performance. Intumescent coatings may also be used (intumescence describes the phenomenon of swelling). Intumescent systems with flame-retarding capacity usually contain three components: an acid, carbon (not carbon fiber), and a catalyst, which interact to form a char layer. This then effectively eliminates an at-risk coated material (e.g., a balsa-cored sandwich panel) as a fuel source for the flame. Recently, intumescent materials have been incorporated into the composites themselves, to take advantage of a char barrier without resorting to secondary coating operations.

❱ BUILDING MATERIALS

Fiber reinforced plastic (FRP) composite building materials are used for stand-alone architecture, interior and exterior walls, and facades worldwide. When it comes to fire and smoke requirements, there is as much attention given to building materials as those used in the aircraft, rail, and transportation industries. As a result, there are many international standards and building codes that govern the use of composite materials in construction, far too many to list here.

ASTM E84 "Standard Test Method for Surface Burning Characteristics of Building Materials" (a.k.a. "tunnel test") is a fire test response standard used to provide surface flame spread and smoke density measurements in comparison to select grade red oak and fiber cement board surfaces under specific fire exposure conditions. This is applicable to exposed building surfaces such as walls and ceilings and is tested in the ceiling position, face down to the flame source.

For exterior or interior architectural surfaces, FRP composites systems must achieve a minimum performance level consistent with a Class A rating per the International Building Code (IBC) Section 803.1 and tested (per ASTM E84) to meet a flame spread index of < 25 and a smoke developed index of < 450. If the FRP-composites system meets these levels, it will satisfy many of the IBC requirements for an FRP component. When FRP composites are used on more than 20% of the flat side of the exterior of the of a building, and the building is over 40 feet (12.2 m) tall, a full-scale test of the exterior FRP assembly must be tested.

Numerous techniques are available to improve the flammability properties of FRP composites. A popular technique is to combine a halogen and a synergist into the resin. Through combustion, halogenated resins disrupt the combustion process and reduce the spread of flame. While halogenated resins perform well and receive a Class A rating in the ASTM E84 test, they receive a Class D rating in single burn item (SBI) testing used in the EU and China. Another method involves adding fillers such as aluminum trihydrate (ATH) into the resin. When heat is applied, these fillers give up their water and act as a heat sink to cool the composite, thus reducing thermal decomposition. The primary issue with fillers is that they do not disperse well in some processes such as vacuum infusion or light RTM. The intumescent-char method described in the rail section above is also deployed in building materials as a method of reducing the transfer of heat within the structure.

Non-Destructive Testing

A variety of non-destructive inspection (NDI) techniques are available for defect or flaw detection in composites in both manufacturing and damage assessment for repair. Each method has its own strengths and weaknesses, and more than one method may be needed to properly define location, size, and depth of flaws or anomalies in a composite structure. Regardless of method, a good knowledge of the underlying structure is necessary for proper evaluation. The following is a list of current methods and techniques used by industry.

❭ VISUAL INSPECTION

Visual inspection can be a powerful and often under-rated technique for detecting surface voids (porosity), inclusions, disbonds, delamination, or other flaws or damage in a composite structure. Many anomalies are easy to detect with the naked eye and can often be enhanced with a magnifying glass and/or a light source. For example, a slight wave or ripple on the surface may indicate an underlying delamination or disbond or a light spot or "whitish" area on a fiberglass part may indicate trapped air, a resin-lean area, or a delamination. If a defect is suspected, additional inspection using more advanced methods are in order. (Figure 10-14 on the next page)

❭ TAP TESTING

Tap testing is probably the most common inspection technique (other than visual) for assessment of suspected defects or blatantly damaged structures. By tapping gently on the surface

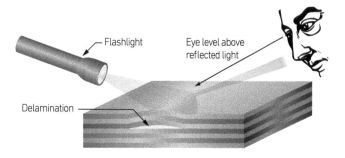

Flashlight

Eye level above
reflected light

Delamination

FIGURE 10-14. Visual inspection.
The "flat angle" or "grazing light"
method of visual inspection involves
examining parts at a very flat angle,
towards a light source, and looking
for the subtle shadows caused by
minor waves from damage or defects.

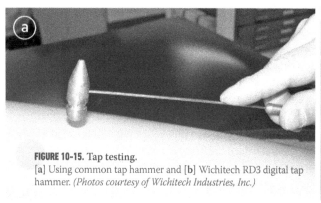

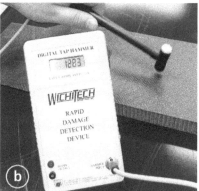

FIGURE 10-15. Tap testing.
[a] Using common tap hammer and [b] Wichitech RD3 digital tap
hammer. *(Photos courtesy of Wichitech Industries, Inc.)*

of a composite laminate, one can hear a change in sound from a clear sharp tone to a dull thud.
By tapping back and forth over the area in question, and making a small mark at the point
where the tone just begins to change, it is possible to outline large, irregularly-shaped areas of
delamination or disbond. However, there are many limitations to tap testing, including:

- Limited to near surface defects.
- Cannot easily locate very small defects.
- Requires knowledge of the underlying structural detail of the part.
- Manual method less effective for quantifying the degree of damage.

The Wichitech RD3 digital electronic tap hammer allows for a quantitative and record-
able indication of the local stiffness of the part being tested. The display reads the millisecond
response output of the hammer as it is lightly impacted on the part surface. (Figure 10-15)

❱ ULTRASONIC TESTING

Ultrasonic testing (UT) or inspection is used to detect flaws in a wide variety of materials,
including metals and composites. It can be performed using portable battery-operated equipment, enabling parts to be inspected anywhere in the manufacturing facility or out in the field
on a part that is in service.

An ultrasonic testing instrument typically includes a pulser/receiver unit and a display
device. The pulser/receiver unit includes a transducer or probe that converts an electrical
signal into a high frequency sound wave (the Piezoelectric effect) and then sends that wave into
the structure being tested. A defect in the structure, such as a delamination, causes a density
change in the material and will reflect sound waves back to the transducer. The transducer
then converts the received sound waves (vibrations) into an electrical signal, which is shown on

the UT display device. Inspection data may be displayed as an A-scan (time-based waveform display), B-scan (distance vs. time graph) or C-scan (plan view grayscale or color mapping). (Figure 10-16)

The A-scan presentation is the basis for all pulse-echo ultrasonic instruments because this presentation provides the time of flight (depth) and amplitude (signal strength of the echo). As phased array instruments are becoming the instrument of choice for both manual scanning and for inspection systems, the cross-sectional view, or B-scan is becoming far more common.

The B-scan is a series of A-scans that are either timed or encoded to produce the cross-sectional B-scan view. The standard ultrasonic array image is a B-scan built from the array of ultrasonic transducer elements (often 64 elements, but 128 is becoming more common). The main advantage with the ultrasonic array instruments is that the transducer and its multiple elements provide a wider inspection swath with very high resolution. (*See* "Phased Array UT Technology," on the next page.)

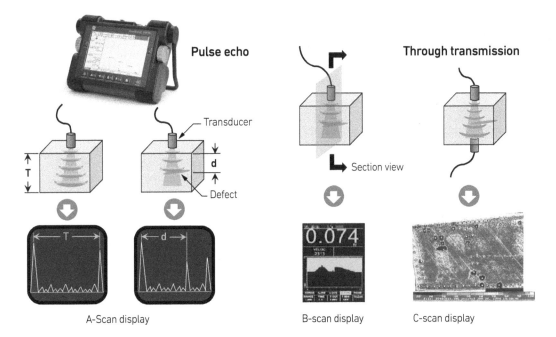

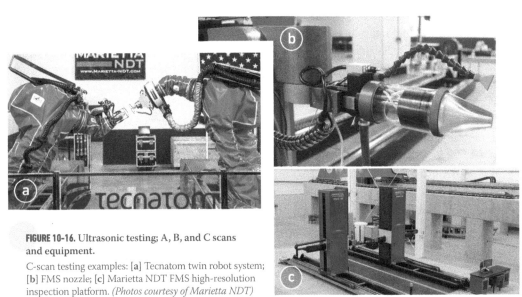

FIGURE 10-16. Ultrasonic testing; A, B, and C scans and equipment.

C-scan testing examples: [a] Tecnatom twin robot system; [b] FMS nozzle; [c] Marietta NDT FMS high-resolution inspection platform. *(Photos courtesy of Marietta NDT)*

The C-scan image is the product of an ultrasonic scan system. The most common application is for newly manufactured components. Interpretation of these displays requires extensive training, and a skilled, certified operator is mandatory. Under most circumstances, UT equipment measures not only the location but also the depth of the damage, which cannot be done visually or by tap testing. However, there are limitations, including:

- Requires a **couplant** between the transducer probe and the part being inspected, as sound does not transmit high frequency energy through air to solid materials. Couplant may be water or a viscous liquid or gel and must be harmless to the component being tested.
- Calibration standards are required for each type of material and thickness to assure the operator that the setup of the ultrasonic instrument can detect the required conditions with the chosen transducer.
- Interpretation of the returned signal demands careful analysis that requires extensive skill, training and experience.

There are two basic techniques used in ultrasonic inspection: *pulse-echo* and *through-transmission*. Table 10.6 outlines the differences between these two methods.

TABLE 10.6 Differences Between Two Main UT Inspection Methods

Pulse-Echo	Through-Transmission
• One transducer generates and receives the sound wave.	• Uses two transducers: one to generate the sound wave and another to receive it.
• Usually limited to detecting first occurring defect—sound wave echoes back preventing detection of anything beyond.	• Defects located in the sound path between the two transducers will interrupt the sound transmission.
• Better at detecting thin film inclusions in composites.	• Interruptions are analyzed to determine size and location of defects but not depth.
• Good calibrations standards provide the material velocity of the component being tested. By knowing speed of the sound wave through the composite, the depth of the defect can be determined.	• Less sensitive to small defects than pulse-echo.
	• Better at detecting defects in multilayered structures and in quantifying porosity.
	• Requires access to both sides of the part.

❱ PHASED ARRAY UT TECHNOLOGY

Phased array UT dramatically reduces the time required for inspection while providing excellent detection of small defects, including the location and depth. The phased array probe uses an array of multiple transducer elements aligned in a single housing. By firing the elements at slightly different time intervals, the sound waves can be focused (depth) or steered (left, right, at an angle, etc.) toward a specific location, allowing the operator to see a live comparison between individual elements or element groups. The transducer probe is also much larger than traditional UT probes, 4 inches wide (versus 0.4 inches), capable of a wide swath, and because of the fine resolution of the multiple transducer elements, a higher probability of detection (POD) of defects. (Figure 10-17)

❱ RADIOGRAPHIC TESTING AND INSPECTION (X-RAY, CT)

Adapted from "X-rays for NDT of Composites," CompositesWorld Magazine, Jan 2017.

Radiographic testing (RT) is a noncontact NDT method with a long history in composites. It uses short wavelength electromagnetic radiation—X-rays or gamma rays—or beams of atomic particles (e.g., neutrons). Compared to the wavelength of visible light (6,000 angstroms), the very

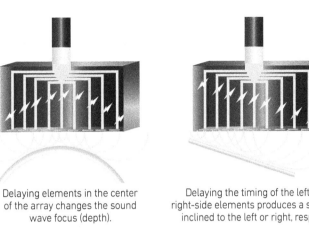

FIGURE 10-17. Phased Array UT Inspection. (*Courtesy of GE Inspection Technologies*)

Delaying elements in the center of the array changes the sound wave focus (depth).

Delaying the timing of the left-side or right-side elements produces a sound wave inclined to the left or right, respectively.

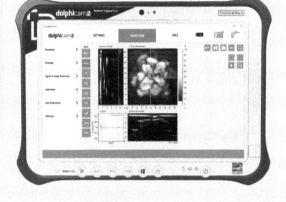

a

FIGURE 10-18. The Dolphicam2 (DolphiTech AS, Raufoss, Norway).

The Dolphicam2 uses a 2-dimensional transducer array that does not require an encoder to define position of each element. [a] The Dolphicam probe is connected to [b] a Windows-based tablet running software that provides the display. [c] In this blow-up of the display are the A-scan, B-scan, C-scan, and B-scan images shown in real time to the operator.

b

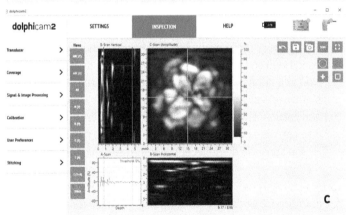

c

short wavelengths of X-rays (1.0 angstrom) and gamma rays (0.0001 angstrom) enable them to penetrate materials that light cannot. Their value in NDT is based on differential absorption of the radiation; that is, different parts of a test piece will absorb different percentages of penetrating radiation due to variations in material density, thickness or composition. Unabsorbed radiation passes through the part and is captured on film or photosensor, and then shown on a fluorescent screen or monitored by various electronic radiation detectors.

Conventional (X-ray) RT is most useful when the parts are neither too thick nor too thin. Gamma rays are better for thick parts, because they have shorter wavelengths, while low-voltage radiography may be used for thin parts (1-5 mm thickness). RT can detect voids and porosity, inclusions, trans-laminar cracks, resin-to-fiber ratio, non-uniform fiber distribution and fiber misorientation, such as fiber folds, wrinkles or weld lines. It is advantageous for larger 3D woven composites, which may be too thick for meaningful UT results. In typical RT results, one can see warp, weft and tow features and quantify their directions and positions in a part.

Although delaminations typically can be seen in conventional RT *only* if their orientation is not perpendicular to the X-ray beam, new techniques of computed/digital laminography are overcoming this limitation, using X-rays to scan and visualize the interiors of large composite components. (*See* Figures 10-18 on previous page, 10-19 below.)

There are a variety of RT methods for different applications:

- **Film radiography.** An X-ray source placed on one side of the object is used to expose a film on the other side of the object for a fixed period. The film is then developed using chemicals.
- **Computed radiography (CR).** Basically, the digital replacement of traditional film. Instead of using chemicals to develop a film, a laser scanner digitizes the image.
- **Digital radioscopy (DR).** This form of X-ray imaging uses digital imaging detectors connected to a computer, which enables the user to look at images in real time.
- **Computed tomography (CT).** This method requires the object (or the source and detector) to rotate up to 360°. Images are taken during the rotation, and then software reconstructs virtual slices of the object. It is an excellent technique for 3D visualization of a part's interior features.
- **Digital laminography (DL).** For large composite parts, this method typically does not require the part to be moved. The relative movement of the detector and the source can be on a circular trajectory or a vertical and horizontal path. The result after reconstruction is a set of images in different planes across the part.

There are also new methods being adapted to composites. Back-scatter X-ray, typically used in full-body airport security scans, detects the radiation reflected from the target instead of what passes through, enabling inspection where access is limited to a single side of the component. Neutron radiography has been used to detect moisture ingress, honeycomb cell corrosion and adhesive/composite hydration in aircraft flight-control surfaces. It provides images similar to conventional RT but offers higher resolution, detecting elements such as hydrogen and carbon. Thus, organic materials and water are clearly visible in neutron radiographs, which often provide complementary information to normal X-ray inspection.

Perhaps the most beneficial improvement in RT in the past decade has been the accelerated development in X-ray detectors and computer processing of data and images. Data now can be captured at very high resolutions. A resolution of one millionth of a meter is possible in micro-focused CT scanning for example. Data also can be processed into 3D images as well as numerically analyzed. Further, a variety of quantitative analyses are possible:

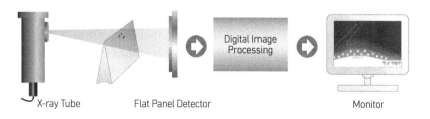

X-ray Tube   Flat Panel Detector   Monitor

Digital Image Processing

FIGURE 10-19. Conventional RT (X-ray) inspection.

- **Defect analysis.** CT scanning can detect and quantify percentage porosity in a composite part, measuring void diameter and volume, as well as distance from the edge of the part. It also can be used to walk through the layers of apart from any direction, enabling visualization of how porosity, weave structures and fiber orientation change from top-to-bottom and side-to-side.

- **Resin/fiber analysis.** CT can detect different densities of material within a part, enabling the removal of resin digitally to analyze fiber distribution in the as-made part, including calculation of resin/fiber content and measurement against set thresholds to flag where the part is not hitting spec. Fiber orientation and fiber length distribution also can be analyzed and flagged against spec.

- **Wall thickness.** The thickness of every wall throughout an entire part can be measured, regardless of the number of cavities, stringers, complex ribs/ stiffeners, etc., returning a visual plot similar to finite element analysis, with colored indicators showing where wall thickness is compromised by porosity.

- **Dimensional check.** CT part scan is compared to its CAD file (CATIA, Solidworks, NEX, etc.) to check actual against design dimensions. While only 6–12 points are typically recorded with a coordinate measuring machine (CMM), CT measures thousands of points, including contours and interiors. When the measurement template is created, every part afterward is analyzed in seconds, not hours. Output includes exact deviation at every location and comparison of dimensions from part to part to check for manufacturing consistency, how parts have changed over production time or due to design or process iterations, etc.

- **Data for process simulations.** CT analysis can be used to create polygon files, STLS or point clouds, similar to measurements from vision-based technologies, except that information is provided for the part's interior as well as its exterior. For example, files including porosity, void and fiber orientation can be imported into Moldflow injection and compression molding software to run simulations using actual part information versus theoretical parts.[2]

- **Damage mechanism and failure analysis.** CT also is a powerful tool in structural and failure analysis, enabling the identification of failure modes in even the most complex structures, without destructive testing that might compromise data.

Figure 10-20 shows a small composite block sample in which the x-ray and imaging settings are optimized to enhance the contrast between the carbon fibers and epoxy resin. This enables virtual segmentation and removal of the resin material to expose the fiber structure, data which is extremely valuable in evaluating structural properties of materials and different manufacturing processes. Software tools are now commercially available that can numerically evaluate fiber consistency and orientation over an entire structure.

FIGURE 10-20. Composite block sample showing enhanced contrast between the carbon fibers and epoxy resin. *(Photos courtesy of NTS)*

2 Moldflow® plastic injection and compression mold simulation software from Autodesk, Inc. (San Rafael, CA)

The images in Figure 10-21 show a high-resolution CT scan of a not-yet fully cured prepreg composite. In the single cross-section, the brighter areas are the uncured resin material, and small openings and voids can be seen inside. Numerical analysis can provide far more data, including fiber volume fraction, both locally, and over a larger area.

❯ THERMOGRAPHIC NONDESTRUCTIVE TESTING

Thermographic Nondestructive Testing (TNDT) uses a heat source (typically a heat gun or flash lamps) to heat the composite part being inspected. As the part cools down, it is monitored with an infrared (IR) camera and digital processing equipment. Irregularities in the temperature distribution across the surface indicate the presence of defects.

TNDT offers a quick and safe method for inspecting large areas and can provide a subsurface image of the entire structure. It requires fewer safety precautions than x-ray inspection and works well in detecting disbonds, delamination, inclusions, and variations in thickness and density. It is not as sensitive as ultrasonic inspection in detecting disbonds and delamination, yet it does not require contact or couplant and can be performed with single-side access (i.e., part-installed). This is now achievable with portable equipment that is easy to set up and break down. (Figures 10-22, 10-23)

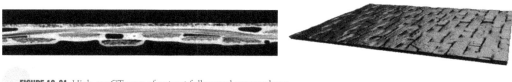

FIGURE 10-21. High-res CT scan of not-yet fully cured prepreg layer. *(Photos courtesy of NTS)*

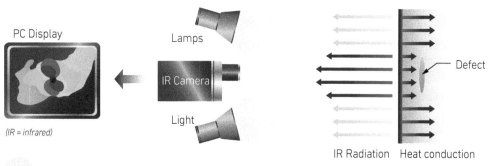

FIGURE 10-22. Basic principles of thermal imaging.

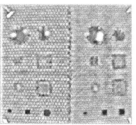
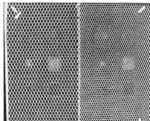
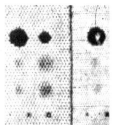

Flash thermography with thermographic signal reconstruction (TSR) processing

X-ray

Through-transmittion (UT)

FIGURE 10-23. Thermal imaging vs. X-ray and UT inspection.
Three different inspection displays of the same panel having carbon fiber face skins and Teflon inserts embedded in its aluminum honeycomb core. *(Photos courtesy of Thermal Wave Imaging, Inc.)*

Active Thermography[3]

TNDT is a widely-used inspection method that creates a subsurface image of a part by monitoring its surface temperature as it responds to thermal excitation. Although it can be applied to metals, ceramics and polymers, it is particularly well-suited to the surface and thermal properties of fiber reinforced polymer (FRP) composite materials.

In a typical TNDT inspection, internal changes in the thermal properties of the part (e.g., due to delamination, impact damage, trapped water or porosity) act to obstruct or accelerate the flow of heat from the surface into the part, causing transient anomalies in the temperature distribution to appear as hot or cold spots in the IR image (*see* Figure 10-24). Detection limits vary with flaw size and type, the thermal properties of the matrix and the type of equipment used, but in general detection is limited by the aspect ratio (diameter to depth ratio) of the flaw, and the requirement that the flaw diameter is greater than its depth.

A basic TNDT system for inspection of in-service composite parts can use a hot air gun or heat lamp and a handheld, uncooled IR camera. This type of system is typically limited to detection of large and/or severe defects during maintenance, where portability and simplicity are of primary importance. Results are interpreted by an inspector viewing the IR image of the cooling part in real time, calling out hot or cold spot indications.

Advanced TNDT systems used in composite manufacturing employ more precise heat sources (e.g. flash lamps or electromagnetic induction), sensitive high-speed IR cameras and dedicated signal processing to perform material characterization and quality control, and to detect smaller, deeper defects compared to the basic configuration. In advanced systems, the time-history of each pixel on the IR camera (typically > 300,000 pixels) is processed independently using a method called Thermographic Signal Reconstruction (TSR®), and then reassembled into an image of the subsurface structure or map of material properties. The TSR® processed signal is sensitive to defect conditions that are undetectable in the original image sequence, and consumes less storage and computational space, so that signals from many shots of a large structure may be combined to form a single data set and subsurface image. (Figures 10-25, 10-26, 10-27)

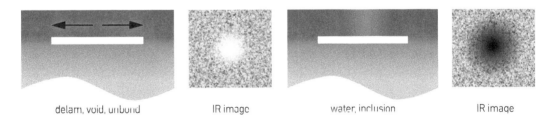

delam, void, unbond   IR image   water, inclusion   IR image

FIGURE 10-24. Heat applied to the surface of a composite sample propagates into the bulk.

An insulating flaw (e.g., a void or delamination) obstructs the flow of heat into the part, causing a transient hot spot to appear in the IR image (*left*). A thermally absorbent flaw (e.g., water, or a metal inclusion) will act as a heat sink, resulting in a transient cold spot (*right*).

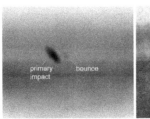 

primary impact   bounce

FIGURE 10-25. Impact damage in a CFRP–aluminum core rudder.

Heat gun excitation *(left)*: Impact damage and a faint bounce are detectable. Flash excitation *(right)*: In addition to the impact and bounce, details of the honeycomb structure, ply orientation and layup are evident in the TSR®-processed, flash-excited image. *(Photos courtesy of Thermal Wave Imaging, Inc.)*

3 Text and photos in this section are provided courtesy of Dr. Steven Shephard, Thermal Wave Imaging, Inc., Ferndale, MI, and are used with permission.

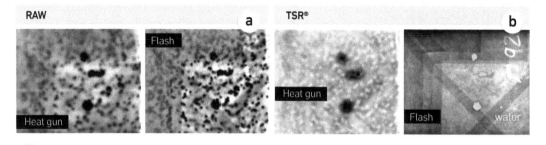

FIGURE 10-26. Trapped water in a CFRP-Nomex® core rudder.

[a] Water filled cells appear similar to excess resin in unprocessed images using heat gun (*left*) or flash (*right*) excitation. [b] Water and resin indications are distinguished after TSR® processing. Water is automatically highlighted in the TSR® flash image (*right*). (*Photos courtesy of Thermal Wave Imaging, Inc.*)

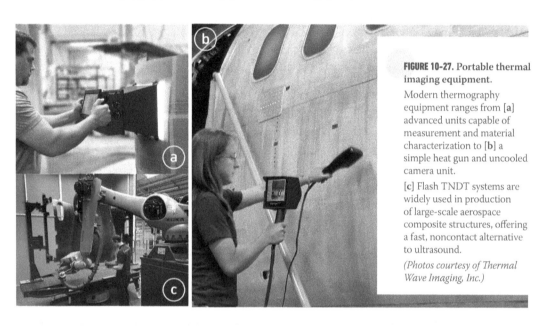

FIGURE 10-27. Portable thermal imaging equipment.

Modern thermography equipment ranges from [a] advanced units capable of measurement and material characterization to [b] a simple heat gun and uncooled camera unit.

[c] Flash TNDT systems are widely used in production of large-scale aerospace composite structures, offering a fast, noncontact alternative to ultrasound.

(*Photos courtesy of Thermal Wave Imaging, Inc.*)

❯ LASER SHEAROGRAPHY

NOTE: In this section, photography and text is provided courtesy of Mr. John Newman, President of Laser Technology, Inc., and used with permission.

Shearography nondestructive testing (NDT) is based on the phenomenon that when an object is subjected to a change in load, subsurface anomalies can produce slight local deformations on the surface. These out-of-plane deformations may be as small as several nanometers but are easily imaged by a shearography camera. For example, when a carbon fiber laminate panel is exposed to infrared radiation, the change in temperature causes primarily in-plane thermal expansion. In a panel with a delamination, the laminate on the heated side of the panel expands more and sooner than the laminate on the cooler side of the panel away from the heat. This differential expansion coupled with the defect boundary causes the material over the defect to deform out of plane.

A shearography camera is used to capture a reference image before the application of the thermal stress to the panel, and then a second image as the heat is flowing through the laminate plies. These images are processed to produce an image of the delamination that can be measured and located on the part. In addition to thermal stress, the range of shearography techniques include pressure changes, vacuum stress, vibration stress using both sonic and ultrasonic signals, mechanical load changes, varying magnetic field loads and microwave stress changes.

Shearography uses a common path, laser-based imaging interferometer to detect, measure and analyze surface and subsurface anomalies in structures by imaging submicroscopic changes to a test part surface when an appropriate stress is applied. Laser shearography is non-contact (except for portable vacuum shearography), non-contaminating, and near real-time. And while laser light is a non-penetrating radiation, shearography systems are capable of inspecting composite structures for both surface and subsurface defects such as impact damage, disbonds, delaminations, near surface porosity, wrinkled fibers, fiber bridging, foreign objects (FO), heat damage and cracks.

Shearography nondestructive testing methods are mature and can be highly effective solutions for a wide range of composite NDT applications. Common successful applications include metal and composite honeycomb or foam-cored panels with metal or composite face

FIGURE 10-28. Shearograph showing impact damage to a solid laminate panel. A tool drop on the composite skin aircraft control surface.

(All photos in Figures 10-28 through 10-34 are courtesy of Laser Technology, Inc.)

FIGURE 10-29. Laser shearograph of sandwich structure. Shearography of titanium honeycomb and face sheet shows individual cells and disbond at center.

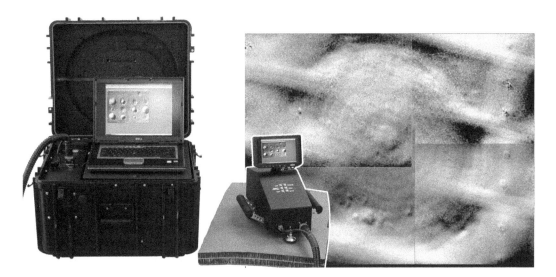

FIGURE 10-30. LTI-6200 portable thermal shearography system.

FIGURE 10-31. Shearograph of engineered scarf joint repair. Completed on an aircraft solid laminate with bonded stringers on the far side shows porosity as small pillowing bubbles at the bottom center and left center.

sheets, bonded elastomers or cork, solid composite laminates and fiber wound structures such as composite over-wrap pressure vessels (COPVs). Shearography NDT offers the inspector a convenient, non-contact optical inspection technique for composites in aerospace, automotive, wind energy, naval and civil engineering applications both for new production and in-service inspections.

A shearography NDT system consists of a laser light source, a shearing image interferometer, an image processing computer, display monitor and a means to provide a controlled and repeatable stress to the test object. Shearography cameras are relatively resistant to environmental vibration and motion.

Large production systems with 30 feet (9.14 m) gantry scanners have been built as well as portable systems for on-vehicle inspection. Shearography is relatively insensitive to test part bending or deformation due to the applied stress but still highly sensitive to local deformation caused by a defect.

Shearography cameras are sensitive to changes in the distance from the object surface to the camera and in practice these z-axis surface deformations may be as small as 2-20 nanometers depending on environmental noise. Test parts can be inspected with a few images using a large field of view (FOV) camera lens setting or for greater resolution, many images with a smaller FOV can be recorded and automatically stitched together. The optimal FOV for a shearography test depends on the maximum allowable defect size, camera resolution, laser illumination power, the ability to uniformly apply a stress change to the test part over the FOV and background noise.

A wide range of stressing techniques are used in these systems to detect defects in materials and structures, including heating or cooling the test part, vacuum, pressure, mechanical bending, sound, ultrasonic signals and microwaves. The applied load changes are typically very low, compared to normal operating conditions of the test piece. For example, COPVs are typically pressurized to 1/1,000 of the nominal working pressure or less to detect cracks, impact damage and fiber bridging.

❯ HOLOGRAPHIC LASER INTERFEROMETRY (HLI)

NOTE: In this section, illustrations and text are provided courtesy of Dr. Rikard Heslehurst, and used with permission.

Holographic interferometry is a non-contact optical NDI technique that provides visual representation of the out-of-plane deformations of an object. The out-of-plane deformations appear as bright and dark fringes superimposed on a three-dimensional image of the object being

FIGURE 10-33. LTI-5200; testing an FRP face sheet to foam core hull section on a Marine vessel.

FIGURE 10-32. LTI-5200 portable vacuum shearography systems, designed for large honeycomb and foam-cored structures.

FIGURE 10-34. LTI-9000 production shearography system.

interrogated. Object interrogation is performed by loading the object in some form (mechanically, thermally or surface pressure) and exposing a holographic plate to two half-exposures of laser light. The two exposures are done at different load states. *See* Figure 10-35.

Holographic interferometry is very sensitive to surface movement. The differential height between two bright fringe lines is half the laser light wavelength. Using a He-Ne laser, the image on the right shows height relief between two fringes of the order of 0.3 micro-meters (0.0003 mm or 0.0076 inch). Thus, very small defects can be identified with HLI, such as weak bonding behavior. It also shows how the surface is reacting to the presence of the defect that might be hidden within the object. For example, a holographic interferogram will show the behavior of a skin surface in a sandwich structure with damage on the other skin.

The holographic interferogram shown in Figure 10-36 was produced from initially placing a release film between the bondline of a thin sheet of aluminum and an aluminum plate. Note the fringe lines are uniform and further apart around the edges where the aluminum plates are well bonded. The center area depicts a disbonded area whereas the fringes are spaced closer together and in a less uniform manner. The partially bonded area exhibits fringes that are spaced randomly and indicate either a weak bond or a partial disbond in this area. (Figures 10-35, 10-36)[4]

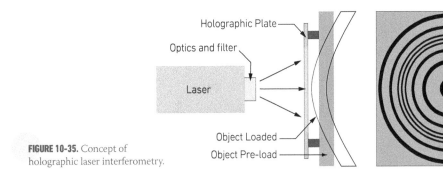

FIGURE 10-35. Concept of holographic laser interferometry.

Holographic Plate

Optics and filter

Laser

Object Loaded

Object Pre-load

Original holographic image

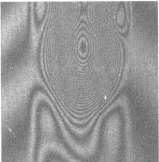

Enhanced holographic image

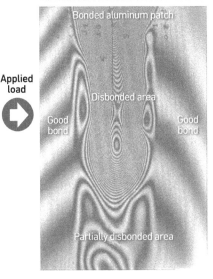

Bonded aluminum patch

Applied load

Disbonded area

Good bond

Good bond

Applied load

Partially disbonded area

FIGURE 10-36. Actual HLI on aluminum plate (*right*).

---

4 (Both illustrations are from *Application and Interpretation of a Portable Holographic Interferometry Testing System in the Detection of Defects in Structural Materials*, Heslehurst, R.B., Ph.D., School of Aerospace and Mechanical Engineering, University College, University of New South Wales, 1998.)

❱ FOURIER TRANSFORM INFRARED SPECTROSCOPY

Fourier transform infrared (FTIR) spectroscopy is a nondestructive testing method used to analyze the chemical functions in a material by sensing vibrations that characterize chemical bonds based upon absorption of infrared radiation by a material. A portable spectrometer is used to target infrared radiation on an FRP sample (or structure) to measure both the wavelengths at which the material absorbs radiation and the intensity of absorption. A mathematical process (Fourier transform) converts raw data into a spectrum for analyses. (Figures 10-37, 10-38)

The wavelengths absorbed by the material are characteristic of the chemical groups present in the material and the absorption intensity at a specific wavelength signifies the concentration of the chemical group responsible for the absorption. This method can be used to determine the following characteristics of a material:

- Composition of the material
- Surface cleanliness/contamination
- Heat damage (oxidation)
- Chemical degradation

FIGURE 10-37. A 4100 ExoScan FTIR being used to measure heat damage on a composite aircraft part, which had been damaged by an engine fire. *(Photo courtesy of Agilent Technologies)*

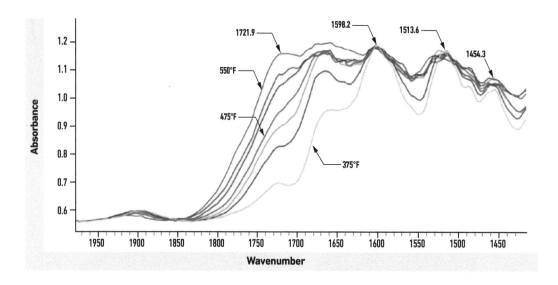

FIGURE 10-38. Measurements of a heat damaged composite aircraft part measured with the 4100 ExoScan FTIR. *(Photo courtesy of Agilent Technologies)*

❱ DYE PENETRATION

This technique is only mentioned because it is quite effective and well-known as a non-destructive method of inspecting metal parts for tiny hairline cracks. However, it is not to be used for composites. Although this technique will also show where cracks have propagated in composites, the dye penetrant liquid contaminates the composite part and any potential bondline one might have in a repair scenario. Therefore, using this method eliminates the possibility of performing a bonded repair on the part. It can be useful on parts that will never be repaired, such as in a crash investigation, but it should not be used to assess the extent of damage in composite parts that will be repaired and returned to service.

❱ COMPARISON OF NDI TECHNIQUES

As shown in Table 10.7, each NDI method has its own strengths and weaknesses, and more than one method may be needed to produce the exact damage assessment required.

TABLE 10.7 Effectiveness of Various Damage Inspection Methods

	Visual	Tap test	A-Scan	C-Scan	X-rays	Thermal imaging	Holographic laser interferometry	Laser shearography
Surface delaminations	Good	Excellent	Good	Excellent	Good	Excellent	Excellent	Excellent
Deep delaminations	N/A	Poor	Excellent	Excellent	Good	Good	Good	Good
Full disbond	Good	Good	Excellent	Excellent	Good	Excellent	Excellent	Excellent
Kissing disbond	N/A	Poor	Poor	Poor	N/A	N/A	Excellent	Good
Core damage	Good	Good	Poor	Excellent	Excellent	Good	Excellent	Good
Inclusions	Good	Good	Excellent	Excellent	Excellent	Excellent	Excellent	Good
Porosity	Good	N/A	Good	Excellent	N/A	N/A	Good	N/A
Voids	Good	Good	Good	Good	Good	Good	Excellent	Good
Backing film	N/A	Good	Good	Good	Good	Good	Good	N/A
Edge damage	Excellent	Good	Good	Excellent	Excellent	Good	Excellent	Excellent
Heat damage	Good	Good	Good	Good	N/A	Good	Excellent	N/A
Severe impact	Excellent	Excellent	Excellent	Excellent	Poor	Excellent	Excellent	Excellent
Medium impact	Excellent	Excellent	Excellent	Excellent	N/A	Poor	Excellent	Excellent
Minor impact	Poor	Poor	Poor	Poor	N/A	Poor	Excellent	Excellent
Uneven bondline	Poor	N/A	Poor	Poor	Poor	Poor	Good	N/A
Weak bond	N/A	N/A	N/A	N/A	N/A	N/A	Good	N/A
Water in core	N/A	Good	Poor	Excellent	Good	Excellent	N/A	N/A

● Excellent ◖ Good ○ Poor

Adhesive Bonding and Joining

CONTENTS

Adhesive Bonding vs. Fastening Composites

In general, an adhesive bonded joint will transfer loads more efficiently than a fastened joint. In a well-bonded joint, the entire bondline area transfers loads between substrates. In a fastened joint, the loads are transferred primarily at each fastener location where the fasteners clamp the substrates together. Typically, a number of fasteners are required along the joint to accommodate these loads, and each through-fastener creates a hole in the composite laminate that weakens the structure at each location.

Long-term durability of an adhesively bonded joint is heavily dependent on many factors such as surface preparation, adhesive selection, timely application of the adhesive, uniform bondline thickness control at a targeted thickness, uniform pressure along the entire joint, and proper cure of the adhesive. If any of these process factors are substandard, the adhesive bond will be compromised.

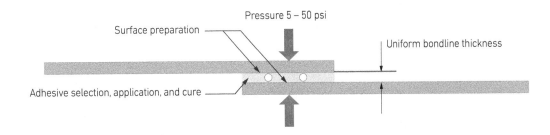

FIGURE 11-1. Basic bonding requirements that require strict process control.

Bonding Methods

❯ COCURING

Cocuring is the act of curing a composite laminate while simultaneously bonding it to some other uncured structure, or to a core material such as honeycomb, foam, or end-grain balsa. All matrix resins and adhesives are cured during a single process. Cocuring creates bonds that are more intimate than secondary bonds and it requires fewer cure cycles to produce a finished part.

Cocured structures generally are lighter than secondary-bonded structures. However, cocuring can be challenging with complex parts that have many details, which often require well-designed, high-performance layup tooling.

❯ COBONDING

Cobonding is the curing together of two or more elements, where at least one has already been fully cured and at least one is uncured. An additional adhesive layer is usually required at the interface between the uncured and pre-cured materials. Cobonding requires meticulous surface preparation on the pre-cured surface and often requires that one structure see more than one cure cycle. Cobonding is commonly prescribed for structural repairs to composites.

❯ SECONDARY BONDING

In secondary bonding, two or more already-cured composite parts are joined together by the process of adhesive bonding, during which the only chemical or thermal reaction is the curing of the adhesive itself. Secondary bonding requires meticulous surface preparation on the mating surfaces. It often requires precise tooling to control the location of the mating parts during the curing process.

Types of Adhesives

❯ LIQUID ADHESIVES

Liquid adhesives are usually lower in viscosity—less than 6,000 cP—and used in a thinner bondline, typically .002–.010 inch thick. They are often used with scrim cloth, non-woven mat, or **microbeads** to achieve thickness control. (*See* "Bondline Thickness Control Media" starting on page 312.) Liquid adhesives include room temperature and elevated temperature curing materials.

❯ PASTE ADHESIVES

These adhesives have a thixotropic paste consistency, where viscosity is usually greater than 10,000 cP, and are used in slightly thicker bondlines, typically .005–.020 inch thick. They are also common in liquid shim/gap filling applications that are greater than .020 inches thick (this usually requires fasteners). Some paste adhesives may not wet-out well on a cured composite substrate, as wetting of the adjacent surface may be inhibited by fillers in the adhesive formula. This is due to a strong chemical attraction of the adhesive resin to the filler that reduces chemical activity with the intended surface. (*See* "Surface Preparation", page 312.)

Typically, scrim cloth, non-woven mat, or microbeads are used for thickness control. However, wetting of a fibrous material may be difficult with pastes. Microbeads may be a better alternative with thicker paste adhesives. Both room temperature and elevated temperature curing paste adhesives are available.

❱ FILM ADHESIVES

Film adhesives require frozen storage and out-time tracking like prepregs. The film thickness is precisely controlled by using a knit or non-woven mat as a carrier for the adhesive, with a range of .004–.012 inch (0.10–0.30 mm) thicknesses usually available. Note that these films are commonly specified not by thickness, but by aerial weight in pounds per square foot (psf) or kilograms per square meter (Kg/m²). Film adhesives are used for many bonding applications and require elevated temperature cure.

Supported Film Adhesives

These film adhesives have a "carrier," usually of a very thin **non-woven mat** or knit material to support the adhesive and aid in handling (*see* Figure 11-2). The carrier is also useful for maintaining bondline thickness control.

Unsupported Film Adhesives

These film adhesives do not have a carrier, and therefore will not control bondline thickness. One use for an unsupported film adhesive would be in a honeycomb sandwich structure where very low weight requirements might mandate a minimal areal weight requirement, or where the use of a carrier would otherwise not be desired within the laminate design. Unsupported film adhesives are traditionally hard to handle as they elongate and tear easily.

Reticulating Adhesives

These are typically unsupported film adhesives—designed to be heated, usually with a hot air gun, so that the hot adhesive runs to the edges and down into the cell walls of a honeycomb core before the face sheets are laid down. The cell walls become coated with adhesive, giving the desired fillets. Little extra adhesive is left over in the middle of the open cells after reticulation.

Another application is for acoustic panel bonding where the reticulated adhesive leaves open holes on the perforated skin, allowing for the attenuation of sound. Because there is no bond strength over the open spaces, a lighter adhesive may be used in some of these applications, saving weight while maintaining bond strength.

❱ CORE SPLICE ADHESIVES

Core splice adhesives are not considered a structural bonding adhesive but rather a core assembly adhesive. They are typically an expanding/foaming type of adhesive, increasing in volume by more than 200%. Core splice adhesives are used for bonding core assemblies together and for gap-filling applications. They are typically supplied frozen in sheets, tapes and SemKit® (pre-measured, quick-mix) tubes and require elevated temperature and vacuum for proper processing.

a   **b**   **c**

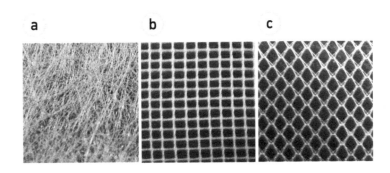

FIGURE 11-2. Adhesive carriers: [a] non-woven mat, [b] scrim, and [c] knit. Note that scrim cloth is usually used with liquid or paste adhesive applications, while non-woven and knit carriers are typical to film adhesives.

❱ BONDLINE THICKNESS CONTROL MEDIA

The ideal thickness of a cured adhesive bond depends on the type of adhesive selected. With a thin, low-viscosity adhesive, a thinner bondline is desired. With a thicker paste-adhesive, a slightly thicker bondline may be required. Most high-strength structural aerospace-quality adhesives perform in a range from about .002 to .020 inches (0.05 to 0.5 mm).

It is important to produce a bondline within the functional range of the adhesive and with a uniform thickness. Bondline thickness control is obtained with carriers in the adhesive (as in Figure 11-2) or with the addition of very small, precisely controlled diameter microbeads in the adhesive mix.

Note that these are not the same as **microballoons**, which are not precisely size-controlled like the microbeads. *See* Figure 11-3.

Microbeads are typically added to liquid or paste adhesives in very small quantities, around 0.5% by weight. Some adhesives are available with the appropriately sized microbeads already mixed in by the adhesive manufacturer.

Surface Preparation

❱ METAL VS. COMPOSITES

Surface preparation of metal surfaces often involve a series of chemical pre-treatments that are designed to first remove the weak oxidized outer layer of the metal substrate, then etch a highly mechanical micro-surface into the freshly exposed metal-oxide layer. This provides the maximum surface area possible on which to bond either an organic corrosion-resistant primer, or an adhesive in a final bonding operation.

Surface preparation of (and ultimately bonding to) a fiber-reinforced polymeric surface depends less upon mechanical attachment, and more upon a strong chemical attachment precipitated by the attraction and eventual sharing of electrons at the interface between the adhesive and the substrate. This is known as a **covalent bond**. Achieving a covalent bond between the adhesive and the substrate surface is dependent upon the ability to raise the *surface free-energy* (SFE) value of the composite surface to comparable or higher value than the *surface tension* (ST) value of the adhesive. (Note: the SFE of a solid is the same property as the surface tension of a liquid.) This electron exchange is crucial to achieving wet-out and a high-performance intermolecular bond to a composite surface. The stronger the attraction, the higher the strength of the bond.

FIGURE 11-3. Microbeads versus microballoons.

[a] Microbeads are uniform in size having diametric tolerances as close as ± .0002 inches, thus allowing for precise bondline control. [b] Microballoons are non-uniform in size, but typically are within a specified range to obtain a desired mixing consistency.

It may be noted that most metals, metal oxides, and ceramics exhibit a high SFE value, usually > 500 mJ/m^2. Most polymers do not and are typically < 50 mJ/m^2. For this reason, even the slightest compromise in the SFE of a polymeric surface can greatly affect the wetting potential of adhesive on that surface.

Typically, surface preparation to a fiber-reinforced polymeric substrate is accomplished using peel ply, plasma treatment, or light abrasion of the surface, and then cleaning (if necessary, without affecting the SFE) so that it is free from dust or debris prior to bonding.

❯ CONTAMINATION CONCERNS WITH GLOVES

The use of rubber gloves to protect the operator from fiber-dust, chemicals, solvents, etc. is highly recommended for all processes involving surface preparation and handling of parts and adhesives. With that said, it should also be noted that gloves can be a source of contamination to a freshly prepared surface or the adhesive itself, and that caution should be taken to avoid glove contact with either. In addition, the same gloves should not be used for multiple operations, as cross-contamination can occur. It is suggested that a new pair of gloves be assigned to each operation to minimize this risk.

❯ MECHANICAL ABRASION VS. PEEL PLY

Achieving a suitable SFE on the composite surface is the goal, not mechanical roughness. The idea is to shear the top layers of molecules on the surface in order to create an energized surface without damaging or breaking the underlying fibers. Deep mechanical scratches from heavy abrasives might not only damage fibers, but also can be a route for moisture (or other fluid) to ingress into the bondline over the life of the structure.

A peel ply is often used along the bonding surfaces of the composite to keep the surface clean until ready for bonding and to provide for enhanced surface energy when removed. Upon removal, the peel ply fractures the resin matrix layer, leaving a fresh, clean, and well-energized surface.

Peel ply-fractured surfaces alone do not necessarily produce the highest SFE nor the best bond-strength in lap-shear testing, but they provide very consistent results. Many manufacturers prefer to use a peel ply-only preparation for bonding to get this consistency.

❯ RELEASE-COATED PEEL PLY FABRICS VS. NON-COATED PEEL PLY FABRICS

Release-coated peel ply fabrics usually leave traces of release agent on the surface of the part and are not recommended for parts that will be bonded or painted. Teflon coated fabrics will easily release from the surface of a part after processing but inherently leave a low SFE. For this reason, non-coated fabrics are preferred for use along the bondline. The issue with non-coated polyester or nylon fabrics is that they tend to stick well to the part laminate, especially after elevated pressure processing, and can be difficult to remove afterward.

❯ PREPREG PEEL PLY

Prepreg peel ply (a.k.a. wet peel ply) has emerged in recent years as an alternative to dry peel ply for parts made with prepreg materials. Made with a specially formulated adhesive matrix resin that is impregnated into a fine polyester fabric, prepreg peel ply materials require less stripping force and yield a more desirable bonding surface than dry counterparts.

In many cases, upon removal of the peel ply, no further processing steps are required prior to bonding. This eliminates the need for additional abrasion and solvent wipe operations and provides for consistency in the process. Bond testing of prepreg peel plies had fracture toughness levels 2–3 times higher than bonds made with either a dry fabric or fabric pre-impregnated with a toughened matrix resin.

❭ SURFACE ABRASION MATERIALS AND METHODS

The risks of damage associated with additional abrasion to the peel ply surfaces must be acknowledged and controlled. The use of common abrasion materials and techniques must be examined to determine what works best with the specific structures to be bonded. In addition, the potential for contamination from the abrasion material itself is a consideration when choosing a material or technique. Operator inconsistency is a risk to repeatability.

Surface abrasion can be accomplished with:

- ScotchBrite® Pads or disks (ultra-fine or fine)
- Fine grit sandpaper (#400–#600 grit)
- Abrasive media blasting (#80–#220 mesh), garnet or alumina grit

Hand-abrading with ScotchBrite® type abrasives are widely recognized as the preferred material of choice for pre-bond abrasion of composite materials and present the least risk of fiber damage, while abrading with sandpaper or performing any type of media blasting presents a greater risk for damage. The risk of contamination with melted nylon from the ScotchBrite® type disks is present when using high-speed grinders or sanders. Contamination from the binders or non-clogging agents on sandpaper is also a risk.

Because polymeric surfaces are generally considered to be low energy surfaces (< 50 mJ/m^2), it is necessary to select a compatible adhesive that has a suitable surface tension value for achieving wet-out on the energized polymeric surface. It is also important to note that the "raised energy" condition, once achieved, is short-lived: within an hour or so, the effect is lost. Thus, surface preparation and bonding should be accomplished in a timely manner to achieve the best results.

❭ SURFACE TREATMENTS

There are a variety of surface treatments used to increase the wettability and polar nature of both thermosets and thermoplastics, including plasma treatment and flame treatment. Although these treatments can be effective, they also have some disadvantages. As noted, treated surfaces will maintain desirable properties for only a short time, and must be protected from contamination until the adhesive is applied and the joint is closed. In addition, the use of high voltage and/or flame also presents significant hazards that might not be practical in certain manufacturing environments or for field repair situations. Plasma nozzle standoff distance and rate of application are critical to most plasma processes.

Plasma Treatment

Plasma treatment takes a gas such as oxygen, argon, helium, or air, and excites it by applying a high voltage between electrodes in a vacuum chamber. The substrate surface is then bombarded with the generated ions, forming reactive groups that attract the adhesive. Plasma treatment successfully changes the mechanical and chemical properties of the surface, at the same time cleaning the surface, promoting the best surface wetting and crosslinking.

Blown-Ion Plasma Treatment

Blown-ion plasma treatment uses air instead of specific gas and directs a concentrated discharge that bombards the surface material with a high-speed discharge of ions. Positive ion bombardment results in a micro-etching, or ablation effect, which removes both organic and inorganic contaminants from the surface. Air plasma surface treatment is inexpensive, effective, easy to use, and works well on thermoset composites.

Flame Plasma Treatment

In flame plasma treatment, an oxygen rich flame is aimed at the substrate surface, oxidizing it and creating desirable polar groups, which create a surface that promotes wetting and adhesion. Flame plasma treatment is typically much hotter than other plasma treatment methods and is usually applied to thermoplastic surfaces.

Cleaning

❱ SOLVENT CLEANING

Many abrasive surface preparation specifications still require solvent wiping as part of the process. If solvent cleaning is specified, then the double-wipe method using a reagent grade solvent and an approved wiping cloth would likely be the method of choice. The double-wipe technique uses one wet cloth followed immediately by one dry cloth to absorb the solvent before it can evaporate. (Figure 11-4)

Note: Solvent storage and deployment must be strictly controlled to avoid the risk of moisture ingress or other contamination.

Solvent Grades

Commercial grade: Inexpensive, but has an undocumented contaminant level. Normally this is suitable for shop use. (Not recommended for cleaning prior to bonding.)

Reagent grade: Expensive, but certified as clean. Typically contaminants are listed in parts per million (PPM) in the certification documents. Strongly recommended if solvent is to be used in a pre-bond cleaning operation.

Common types of solvent, in order of the authors' preference:

- Acetone (Ace)
- Isopropyl Alcohol (ISP)
- Methyl Ethyl Ketone (MEK)
- Methyl Isobutyl Ketone (MIBK)

Non-Solvent Wiping Technique

A non-solvent wipe alternative is recognized as a possible preferred method for cleaning composite surfaces prior to bonding. Careful vacuuming and/or dry wiping with approved wipes to remove surface debris is recommended. Multiple passes may be required to remove the dust and debris to a suitable level.

Since no solvent is used, health, safety and waste disposal problems are minimized. In addition, there is no risk of creating porosity in the adhesive bondline due to residual solvent flashing during high-temperature cures. Some test results indicate bond strengths and longevity using the dry-wipe method are equal to, or better than those using solvent wiping.

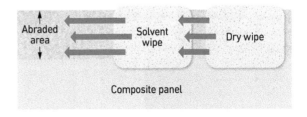

FIGURE 11-4. Double wipe cleaning method.
Wipe in one direction from end to end along the joint, avoid circular wiping, and use clean wipe surfaces for additional passes.

Wipes

Prebond wiping should always be done with an approved, non-contaminating wipe. Industrial woven-fabric rags and cloths might have chemicals or **size** in the material that can be redistributed on the freshly prepared bond-surface. This is especially true when using solvents that break down these chemicals into solution.

Lint-free cloth exists that might still have a "size." Also, size-free cloth exists that may leave lint or traces of fibers on the surface when used. The latter might be of lesser concern because the chemical bond would not necessarily be affected by trace amounts of lint or fiber, whereas it may be drastically affected by the presence of incompatible chemicals.

❱ CONTACT ANGLE MEASUREMENT

In lieu of "old-school" water-break testing of an FRP composite, which has the potential to contaminate the entire bond surface, verification of surface readiness can be accomplished using contact angle measurement verification methods. These methods systematically test multiple micro-locations along the bond surface so as to minimize the risk of hydrating or contaminating the entire surface, or losing SFE due to the passage of time needed for drying.

Contact angle measurement uses a micro-drop of liquid, usually highly purified water, to evaluate how well a surface has been prepared for bonding. A surface with a low SFE tends to be hydrophobic, that is, it will cause the drop of water to bead up, resulting in a contact angle measurement greater than 90°. A contact angle measurement of 0° represents a perfectly hydrophilic surface—one on which the water does not bead up but instead flows out and wets the entire surface (*see* Figure 11-5 below).

FIGURE 11-5. Contact angle measurement.

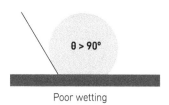

θ > 90°

Poor wetting

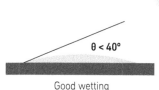

θ < 40°

Good wetting

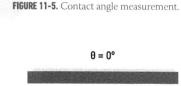

θ = 0°

Maximum wetting

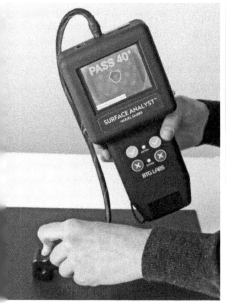

FIGURE 11-6. The Surface Analyst™ uses a ballistic method to deposit a 2μL water drop for accurate, repeatable results in 3 seconds.

An immediate "pass" or "fail" for bonding is enabled by inputting the allowable range of angles, determined by measuring a surface prepared to specification vs. the same unprepared or poorly prepared. *(Photos courtesy of Brighton Science, Cincinnati, Ohio)*

It should be noted that it is not possible to achieve a perfect hydrophilic effect on an FRP composite surface with water at room temperature, as the ST value of water is ≈ 70 mJ/m² and the SFE value of the FRP surface is < 50 mJ/m², thus the water will never achieve a 0° contact angle. To verify the SFE level of a surface, a portable inspection device (*see* Figure 11-6) is used to measure and calculate the contact angle of the liquid. A contact angle of ≤ 40° is desirable for most adhesives.

Bonding to Core Materials

❱ HONEYCOMB CORE

Bonding to honeycomb core is best accomplished with a film adhesive that is specifically designed for such use. Both supported and unsupported film adhesive are used for this purpose. The essential ingredient for a good honeycomb core bond is the formation of uniform fillets of adhesive at the skin and inside the core cells. These fillets give the bond a large surface area, and the adhesive is subsequently loaded in shear, which is the most desirable loading mode. (Figure 11-7)

❱ FOAM CORE

Bonding to foam core is very different than bonding to honeycomb. A good bond to a foam core depends on having a suitable amount of adhesive to wet the porous foam surface and actively (both chemically and mechanically) attach to the material. Often, the adhesive bond to a foam core is much stronger than the foam itself. Frequently, failures to foam core sandwich panels occur in the foam along a line through the mean depth of the surface pores of the material. This "zip-line" failure effect is well recognized by those who work with foam core sandwich structures. (Figure 11-8)

FIGURE 11-7. Bonding honeycomb core.

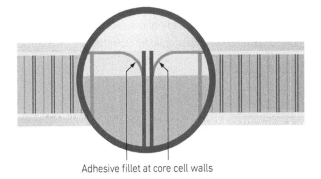

Adhesive fillet at core cell walls

FIGURE 11-8. Adhesive bond to foam core materials.

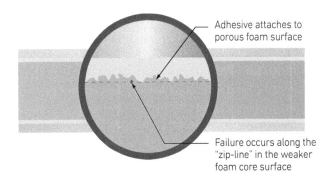

Adhesive attaches to porous foam surface

Failure occurs along the "zip-line" in the weaker foam core surface

Joining Thermoplastic Composites

Most thermoplastic polymers are not easily bonded using epoxies or other common adhesives due to inherent characteristics of non-polar or low SFE surfaces. Many of these issues can be resolved through chemical or plasma surface treatment and the use of select adhesives, or by fusion bonding (welding). Much current activity surrounding thermoplastic composites (TPCs) in both the automotive and aerospace sectors show continued growth in this area and as a result, new and better processes for joining are on the horizon across all industries.

❭ CRYSTALLINE VS. AMORPHOUS THERMOPLASTICS

Case studies from Airbus, Toray, Fokker and other companies have found that amorphous thermoplastics such as PEI can be satisfactorily adhesively bonded using conventional adhesive systems.

Crystalline thermoplastics, such as PPS, PAEK, and PEEK, present more challenges. Studies performed in the late 1980s and early 1990s showed that the use of FM300 film adhesive (a common epoxy system used in composites) combined with a plasma treated surface gave the greatest lap shear strength of 41.6 MPa for AS4 carbon fiber/PEEK laminates. Grit blasting and acid etching produced lap shear strength values of only 20 MPa.

However, Airbus has found that even with sufficient surface treatment, there are still issues with hot/wet performance of adhesive bonds with crystalline thermoplastic composites. Airbus investigated the use of adhesive bonding with PPS structures, using a variety of surface treatments. These included alumina blasting, corona discharge, flame treatment and adhesion promoters, and also a variety of adhesives such as two-part epoxies, two-part urethanes, and epoxy film adhesives. They found that none of the adhesives and/or surface treatments produced results that met their minimum hot/wet lap shear requirements of 1,000 psi. Thus, Airbus has not approved any adhesive bonding techniques for manufacture or repair of crystalline or semi-crystalline thermoplastics such as PPS and PEEK.

❭ FUSION BONDING (WELDING)

Fusion bonding (i.e., welding) involves heating and melting the thermoplastic polymer on the bond surfaces of thermoplastic composite (TPC) components and at the same time pressing these surfaces together for consolidation and solidification. Unlike an adhesive bond, which under a microscope retains a clearly defined joint line formed by the adhesive, materials that have been fusion bonded thoroughly intermix or fuse together, essentially becoming a homogenous part.

Welding has several additional advantages over adhesive bonding, such as reduced cycle time, less surface preparation, no shelf life or bonding immediacy issues, and the absence of stress concentrations. However, the heating must be controlled and contained so that the polymeric structure of the components being bonded is not significantly altered, and the parts must be held under pressure until cooled to maintain shape. There are several types of fusion bonding processes used with thermoplastic composites. Some of these processes are outlined as follows.

Induction Welding

Induction and resistance welding are the most commonly used fusion bonding methods for thermoplastics and considered to be the most promising. Both make use of an insert that helps to generate electrical current that then heats the thermoplastic substrates. After cooling down under pressure, a weld is created. In induction welding, the insert is called a susceptor, conductor, or conductive element; in resistance welding it is called a resistive element or welding tape. In both processes, it is usually some type of wire mesh.

Induction welding (also called electromagnetic welding) uses a high radio frequency generator to create an electromagnetic field, which then produces an alternating electric current. The current passes through the conductive element, which heats up, softening and fusing the thermoplastic substrates. The whole process takes place within specially designed fixtures that join the parts together under low pressures and takes only seconds for a typical weld.

Induction welding has been used for 40 years in the plastics industry, mostly with polyolefins. Because the electromagnetic energy precisely targets the susceptor/conductor, induction welding doesn't significantly affect the dimensional stability of the mating components. It also does not distort the material adjacent to the bondline, as is common with welding based on conductive heating or the application of mechanical energy to the joint. Induction welding can easily handle large parts and can also support a fair amount of automation; for example, the susceptor/conductor implants can be installed robotically or even molded into the mating parts. (Figure 11-9)

Resistance Welding

In resistance welding, electric current is passed directly into the insert, which is an electrically resistive element—usually a strip of metal mesh or carbon, also called welding tape. Specially made tooling clamps the welding tape between the two parts at the intended bondline. The metal mesh tape quickly heats as electric current is applied, heating the adjacent part surfaces to the melt temperature of the thermoplastic matrix. The resistive element—which is typically very thin (~0.2 mm/0.01 inch thick) and open—remains in the joint without disrupting the bond.

As with induction welding, heat is introduced into the structure only where it is required, and thus limits the material melt zone, reducing the risk that part shape or dimensional stability will be compromised. The total time to weld is typically a couple of minutes for parts ranging in length from 3 to 4 meters (9.84 to 13.1 feet); therefore, the process adapts well to production and can be automated. (Figure 11-10)

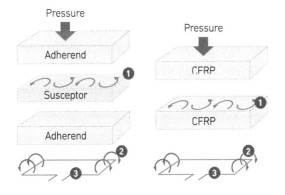

1. Induced currents
2. Alternating magnetic field
3. Induction coil

FIGURE 11-9. Induction welding.

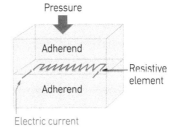

FIGURE 11-10. Resistance welding.

Conduction Welding

Conduction welding uses direct heat conduction through an electric "hot iron" type tool or platen that transfers heat along the weld length, through at least one of the parts to be joined. Since the heat is transferring through at least one part, support tooling is required to maintain the shape of the joined parts. Conduction welding is ideal for spot welding operations and can be adapted to automated machine processes. (Figure 11-11)

Ultrasonic Welding

Ultrasonic welding uses high frequency sound energy to soften or melt the thermoplastic surfaces at the joint. The parts to be joined are held together under pressure between the oscillating horn and an immobile anvil. The joined parts are then subjected to ultrasonic vibrations at a frequency of 20 to 40 kHz at right angles to the contact area. Alternating high-frequency stresses generate heat at the joint interface to produce a good quality weld.

Ultrasonic welding is an easily automated and very fast process, with weld times of less than 1 second. However, because the equipment for this process are expensive, it has traditionally been used for large volume production runs in the plastics industry, and typically limited to small components with weld lengths not exceeding a few centimeters. (Figure 11-12)

Laser Welding

Laser welding uses laser radiation that is directed through an unreinforced laser-transparent part, heating the surface of a laser-absorbent part, resulting in fusion of the materials at the interface. The greatest benefit to laser welding is that it can be performed at a relative high speed ($\approx$24m/min.). The biggest drawback is that it requires on part to be transparent (i.e.: glass, quartz, or no reinforcement). (Figure 11-13)

Joint Design

Many different adhesively bonded joints configurations are possible. "Scarf" joints, with tapered scarf angles and sharp edges, are generally considered to have the best strength. However, tapered scarf angles can be difficult to machine, and achieving proper fit with precured mating parts to be secondarily bonded is challenging. However, tapered scarf cobonded assemblies can easily be accomplished. In general, the more complex the joint design, the harder it is to achieve. (Figure 11-14)

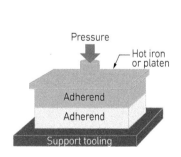

FIGURE 11-11. Conduction welding.

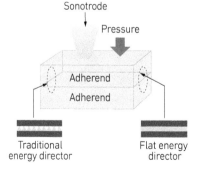

FIGURE 11-12. Ultrasonic welding.

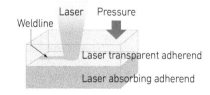

FIGURE 11-13. Laser welding.

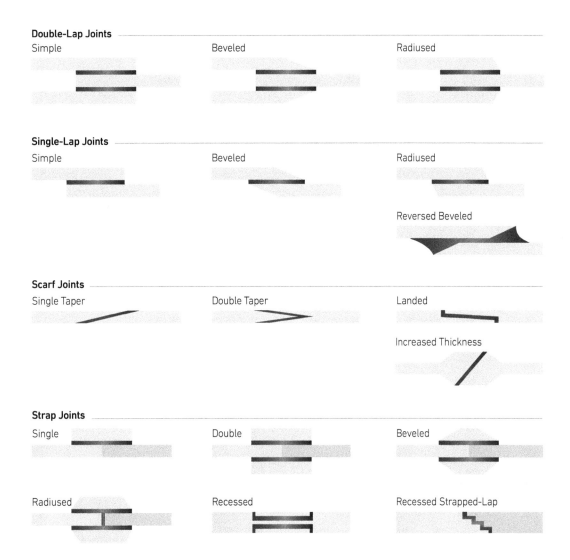

FIGURE 11-14. Bonded joints.

Design alternatives for bonded joints are more numerous than for bolted joints. Illustrated here are various conceptual forms of lap, strap, and scarf joints, along with some less common configurations.

12

Machining, Drilling, and Fastening Composites

CONTENTS

Overview of Machining Methods and Practices

There are two primary methods of machining composites: rotary cutting (sawing, milling, routing, drilling, etc.) and waterjet machining. Rotary cutting can be done manually or by automated machine (CNC or robotic) systems. Waterjet cutting is almost always done in an automated work cell for safety reasons. In today's industry it is common to find both rotary and waterjet capabilities in the same work cell.

FIGURE 12-1. Combined work cell with both waterjet and rotary machining capabilities. [a] CMS work cell at General Atomics facility; [b] waterjet head; and [c] router head. *(Photos courtesy of CMS North America)*

Rotary Cutting

Rotary cutting composites pose many challenges, some of which are associated with either the abrasive nature of the fiber which contributes to tool wear, or the tenacity of the fiber which resists cutting and results in fraying of fibers at the edges of the cut. It is also difficult to uniformly fracture brittle fibers like carbon and glass with any predictability, and aramid fibers are prone to elongate and pull away from the matrix prior to breaking and fraying. These conditions are exacerbated when there is increased void volume in the laminate.

Vibration of the part greatly contributes to poor edge quality. Proper tooling is required to mitigate this issue. The part should be well clamped in a jig or fixture designed to allow for cutter (or drill) clearance, yet with good support near the part net-trim edge.

Friction from the cutting tool can also cause excessive localized heating of the resin matrix, causing damage to the composite and deposition of resin on the cutting tool. Overheating of low-temperature thermoplastic fibers can cause a similar effect. These conditions tend to leave a less than perfect cut. Because of this, machinists often look to coolants to reduce friction, to minimize overheating and to extend tool wear. However, many industry experts will argue dry machining gets better tool wear and that they can get good results without coolant. In some cases, only water is used as a coolant, primarily to mitigate dust and not necessarily to improve the cut.

The risk of contaminating the composite is the major concern, especially when the part will be used in a structural adhesive bonding operation. Even water can be a concern if the part is not dried after machining. For this reason, careful consideration of the type of coolant and whether to use a coolant at all is very important. Many OEMs will specify approved coolants for certain applications if required for machining.

❱ CUTTING TOOLS

Conventional cutting tools made with low-cobalt or tungsten carbide alloys are commonplace in the machining industry. Tools used for cutting composites may be coated with a hard, amorphous diamond coating to improve lubricity and wear resistance over the uncoated counterpart. Each cutting tool is designed for a specific purpose and employed where most applicable.

Some common types of end mill cutters used for routing are listed below:

Straight flute end mill cutters are designed to prevent delamination by applying only radial cutting forces, eliminating the axial forces that are common to a helical-shaped cutting edge. These cutters are good for peripheral edge routing.

Compression cutters consist of a combination of both up and down (split) helixes that merge at the cutting center of the bit. The top helix has right-hand cutting teeth with a left-hand spiral and the bottom helix has right-hand cutting teeth with a right-hand spiral. This creates opposing cutting forces that work towards the middle of the laminate to prevent delamination and fiber fray. Compression cutters work well with aramid fiber reinforced composites, but also may be used to machine glass and carbon laminates.

Chipbreaker cutters are used primarily for roughing and profiling composites that have a high fiber volume. Notch-like "chipbreaker" cutting edges shear the fibers and break up the materials into small chips for better particle evacuation. These tools are less likely to bind up with fibers wrapped around the bit.

Diamond cross-cut cutters typically come in two different geometries; end mill or drill mill configurations, of which the drill mill has a 140° angle point that allows it to be used for penetrating the panel prior to routing. These tools receive their name from the diamond-patterned combination of left-hand and right-hand teeth.

Finish cutters have an optimized geometry for fine edge-finishing. The cutter has a gentle helix and high flute count, allowing for more contact area at the edge of the laminate, resulting in minimal fraying or fracture, thus producing a smoother surface.

Abrasive cutters. Diamond or tungsten carbide grit abrasive coatings can also be applied to a tool blank for router bits, hole saws, band saws, or circular saw blades to make an abrasive cutting tool. Abrasive tools are designed to grind away materials and as a result, are well suited for machining glass and carbon fiber reinforced laminates and not so good for machining aramid fibers. Diamond abrasives tend to last longer than carbide but are generally more expensive.

There are several types of diamond abrasive coating methods as follows:

- **Chemical vapor deposition (CVD):** These tools are made by attaching diamonds to the tool-blank using a chemical vapor deposition method and are generally the least expensive to produce. CVD tools have less abrasion resistance than other coatings and may not be the best choice for high rate production.

- **Polycrystalline Diamond (PCD):** Made by sintering micro-size single diamond crystals at high temperature and pressure. PCD has a low friction coefficient, good fracture toughness, and good thermal stability, making these tools a desirable choice for high-rate production.

- **Abrasive Technology** (Lewis Center, OH) makes a P.B.S.® series of tools where the diamonds are "brazed bonded" with a nickel-chrome alloy, allowing for high diamond exposure and selective spacing on the cutting tool compared to CVD or PCD tools. (Figure 12-3)

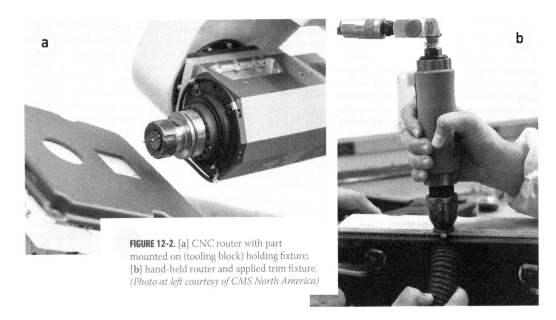

FIGURE 12-2. [a] CNC router with part mounted on (tooling block) holding fixture; [b] hand-held router and applied trim fixture. *(Photo at left courtesy of CMS North America)*

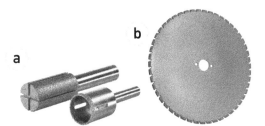

FIGURE 12-3. [a] Brazed-bonded diamond abrasive router bits and [b] circular saw blade. *(Photos courtesy of Abrasive Technology)*

Waterjet Cutting

Typically, waterjet cutting is done with a garnet abrasive fed into a high velocity stream of water directed at the desired edge of part (EOP). An ultrahigh pressure pump generates a stream of water at 60,000 to 94,000 psi. The water is forced through a tiny orifice between 0.030 to 0.040 in diameter, creating the high-velocity stream. Garnet abrasive is pulled into the head via a venturi vacuum system and mixed with the water stream, allowing the high-velocity abrasive waterjet to cut hard materials like metals, ceramics, glass, and composites. (Figure 12-4)

Abrasive waterjets cut by erosive action rather than friction and shearing like a rotary tool, leaving a smooth finished EOP without delamination. This minimizes the need for additional finishing. Benefits to waterjet cutting of composites are as follows:

- High productivity, faster and higher-quality cut without dust.
- Minimal side or vertical forces or thermal impact on laminate.
- No delamination or loose fibers after cutting.
- Minimal kerf—good for close tolerance detail cuts.
- Single-pass trimming, depending on material thickness.
- Reduced tooling costs—can be used with fixtureless setup.

Laser Cutting

Historically, the use of lasers for cutting composites has been limited due to two factors: excessive energy generated by the laser at the cutting edge often resulting in matrix damage, and the loss of a focal point for precision ablation as the laser cuts through the thickness of the material. The latter contributes to a divergence of the beam, resulting in an outward conical shaped cut through the laminate.

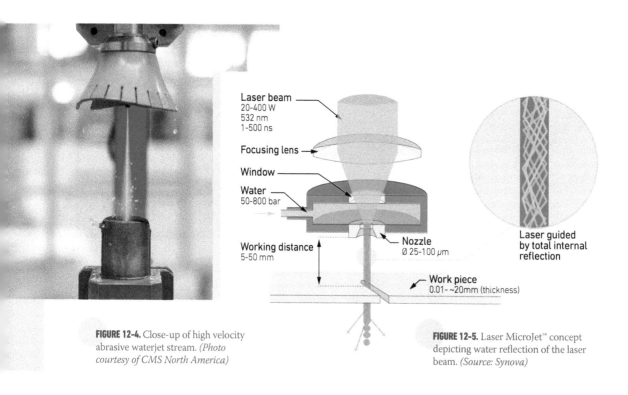

Laser beam
20-400 W
532 nm
1-500 ns

Focusing lens

Window

Water
50-800 bar

Working distance
5-50 mm

Nozzle
Ø 25-100 μm

Work piece
0.01- ~20mm (thickness)

Laser guided
by total internal
reflection

FIGURE 12-4. Close-up of high velocity abrasive waterjet stream. *(Photo courtesy of CMS North America)*

FIGURE 12-5. Laser MicroJet™ concept depicting water reflection of the laser beam. *(Source: Synova)*

To overcome these issues, a fairly new technology called the Laser MicroJet™ (Synova, Duillier, Switzerland) uses a micro-stream of water around the laser beam, of which the beam is completely reflected at the air-water interface, using the refractive indices of the air and water to maintain the focal point of the laser as it penetrates the material. The result is that the laser is completely contained within the water jet as a cylindrical beam, comparable in principle to the function of an optical fiber. (Figure 12-5)

The Laser MicroJet (LMJ) process works in two stages: the energy from the pulsing laser vaporizes the material while the water simultaneously cools and cleans the surface between pulses. Through a continuous scanning process, a trough is formed that becomes deeper with each sequential pass.

Unlike conventional lasers that are limited to only a few millimeters in the depth due to the beam divergence, the water encompassed laser cuts with a parallel beam that has a working distance of up to several centimeters. Additional advantages to the LMJ system are as follows: there is no need for constant focal adjustment of the beam to obtain parallel kerf edges in the cut; a minimal heat-affected area due to water cooling; lastly, there is a high rate of cut and the debris is flushed away from the part with the water. (Figures 12-6, 12-7)

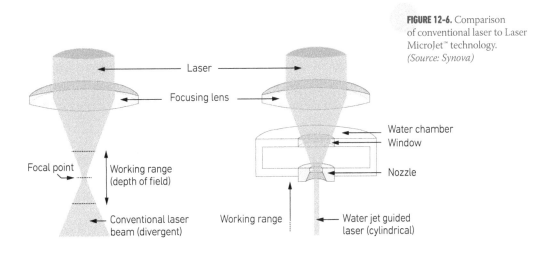

FIGURE 12-6. Comparison of conventional laser to Laser MicroJet™ technology. *(Source: Synova)*

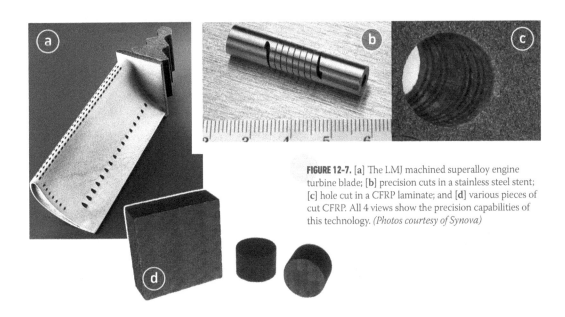

FIGURE 12-7. [a] The LMJ machined superalloy engine turbine blade; [b] precision cuts in a stainless steel stent; [c] hole cut in a CFRP laminate; and [d] various pieces of cut CFRP. All 4 views show the precision capabilities of this technology. *(Photos courtesy of Synova)*

Drilling Tools and Techniques

The advanced composites industry is rife with special tools to cut, machine, and drill composite materials. Equally impressive is the voluminous quantity of text concerning techniques for drilling and machining composites that has been published over decades. Since each fiber type has its own set of unique machining characteristics, and these characteristics may change, perhaps dramatically, depending on the volume ratio and properties of the matrix in which they are embedded.

❱ SPEED AND FEED RATE

For years, the rule of thumb in drilling composites has been high speed – low feed. In other words, the tool's cutting edge(s) should be traveling as fast as is practical for the operation, and the operator should allow the tool time to do its job by not using excessive feed rates.

At lower RPM, even drill bits especially designed for use in composites tend to generate more delamination at the surface plies and fiber breakout on the far side of the laminate as the drill exits the work. In general, only high RPM drill-motors capable of turning at between 5,000 and 22,000 RPM, with enough torque to maintain these speeds, should be used for drilling composites.

Feed rates vary tremendously with the type of laminate being drilled. It is important to never force the drill to cut faster than its design or condition allow as this will usually translate into excess heat within the laminate and may compromise the matrix material. Assuming the proper tool has been selected for the material being drilled, common feed rates vary from 15 up to 60 inches per minute.

❱ CONTROLLING ANGLE AND FEED RATES

As mentioned earlier, successful installation of fasteners in composite materials relies on strict adherence to the manufacturer's specifications and tolerances. Chief among these are the ones concerning installation holes. Not only must the holes be of the correct diameter, but they must be perpendicular to the surface being drilled.

Delamination on the far side of a hole is probably the most common difficulty when drilling. Drill design, speed and firmly backing up the work are important, but sometimes more is required to get satisfactory results. Often, a controlled-feed drill will provide the extra level of control necessary. As the name suggests, a controlled-feed drill restricts the speed at which a drill bit penetrates the work. This eliminates any tendency a bit might have to "screw" its way through the laminate and generate delamination.

Some manufacturers, such as Monogram, offer a kit with a controlled-feed drill motor and a fixture designed to ensure that the drill remains normal to the surface throughout the drilling process (*see* Figure 12-8). In other cases, it might be necessary to use a conventional drill guide such as the one shown in Figure 12-9.

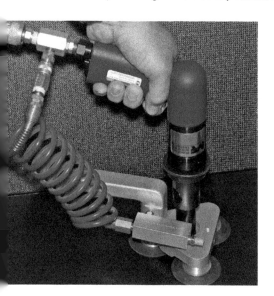

FIGURE 12-8. Monogram controlled feed drill and fixture.
(Photo courtesy of Monogram Aerospace)

❱ DRILLING CARBON AND GLASS FIBER REINFORCED COMPOSITES

Carbon fibers are relatively stiff and brittle, while glass fibers are more flexible by comparison. Despite this, they have a lot of features in common when it comes to drilling and machining. Therefore many tools that are designed to cut, machine, or drill carbon fiber composites also work well on fiberglass laminates.

Carbon and glass laminates are extremely tough on cutting tools, able to quickly round off the cutting edge on any high-speed steel (HSS) tool. The solution here is to manufacture the drill (or countersink) from a material hard enough to maintain a sufficiently sharp edge against carbon and glass. Two such materials are carbide and diamond.

Carbide drills designed for use in composites usually have a positive (often extreme) rake angle, a fine chisel edge, and a gradual taper up to the final drill diameter. An acute positive rake angle encourages the fibers to be drawn into the cutting area and sheared between the cutting edge and the uncut material. A positive rake removes more material in a shorter amount of time than a negative rake, but it also makes the cutting edge more sensitive and fragile.

The penetration rate, and thus the production rate, of a drill bit can be enhanced by incorporating a small chisel edge at the point. Because of the proportionately high-penetration rate, the optimum chisel edge for drilling composites would be as close to a sharp point as possible.

These features conveniently harmonize with a gradual taper in the bit that is often incorporated into drill designs. The gradual taper minimizes cutting pressures at the edge of the hole, gradually opening it up to its finished dimension. Additionally, this feature helps to minimize breakout and delamination on the far side of the laminate that can occur with standard twist-drill configurations. Regardless of the drill design, supporting the backside of the composite panel with wood or suitable fixturing is recommended to minimize backside-fiber breakout.

Diamond-coated carbide tools are highly effective for cutting and drilling carbon fiber and fiberglass structures. With these tools, small diamonds of a specific grit size are deposited on the cutting edge and are embedded in a metal matrix material. Ideally, as the sharp edges of the exposed diamonds are worn away, new diamond edges begin to appear as the matrix is worn away. This provides additional longevity of the cutting tool.

Diamond coating technology is extremely versatile and can be used for drill bits, hole saws, cutting wheels, countersinks, and a myriad of other composite machining applications.

❱ DRILLING ARAMID FIBER REINFORCED COMPOSITES

Aramid fiber's incredibly high tensile strength and durability has been leveraged for years by engineers looking to design abrasion-resistant, lightweight structures. Unfortunately, that very same toughness and durability become liabilities when machining aramid composites.

Conventional twist drills tend to produce rough, fuzzy edges in aramid fiber laminates that might be considered unacceptable in most applications. This is because these drills tend to grab the fibers and pull them in tension until the fibers break, rather than shearing them neatly.

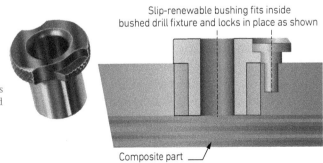

FIGURE 12-9. Conventional drill guide—uses a slip-renewable bushing *(left)* in a pressed bushing in a drill fixture or template.

Slip-renewable bushing fits inside bushed drill fixture and locks in place as shown

Composite part

While the resin content of all composite laminates affect the machinability, this is especially true of aramid-reinforced laminates. Resin-rich laminates tend to make for cleaner holes, while resin-lean laminates tend to have more fuzz at the edges.

One of the better drilling solutions for aramid composites is the brad-point carbide twist drill. This self-centering, modified, two-flute twist drill features a sharp center-point surrounded by two sharp peripheral cutting edges that draw the fibers in tension and then shear them off. This results in a nearly fuzz-free hole through most aramid laminates.

DuPont, who makes Kevlar® aramid fiber, recommends the use of brad-point drills at a speed between 6,000 and 25,000 RPM and with a slow, controlled feed rate. They also suggest that, unlike other composites, RPM should be decreased as the size of the drill bit increases. (Figure 12-10)

FIGURE 12-10. Brad-point drill for drilling aramid fiber laminates. *(Photo courtesy of Traditional Woodworker)*

❭ DRILLING COMPOSITE-METAL STACKS

Drilling two or more dissimilar materials such as a CFRP-titanium (Ti) stack presents a special challenge in that each material requires a different drill speed and feed rate along with a common bit design that will adequately cut each material.

The conventional approach to stack drilling involves using a semi-automated drill motor with controlled feed, speed and force, to drill the CFRP laminate at a higher rotational speed and feed rate than that of when the Ti plate is reached. To achieve the best results, the drill is programmed to first drill through the laminate, "peck" the surfaces of the Ti, back up and clear debris, then return to drill the Ti at the slower rotational speed and feed. Often, after drilling, the stack is disassembled, and all pieces are cleaned of debris before reassembly. Care must be taken with bonded stacks so as not to overheat the materials during drilling.

❭ ORBITAL DRILLING

Orbital Drilling™ by Novator (Spånga, Sweden) combines three motions in one tool: feed, spindle rotation and orbital rotation. This works by rotating a cutting tool around its own axis and simultaneously around an offset (hole center) axis while penetrating the laminate. This produces a helical cut through the material using an end mill that is smaller than the hole diameter.

The drilling process generates less heat and a far smaller axial force on the materials being cut, resulting in high quality and mostly defect-free holes. NovaGrip™ nose bushings have built in RFID antennae used to work with and RFID tag embedder in the drill fixture, which identifies the hole and assigns the proper drilling specifications (speed, feed, and force) to the drill motor. This technology is ideal for drilling stacks of dissimilar materials. (Figures 12-11, 12-12)

Mechanical Fastening Considerations

As mentioned in an earlier chapter, adhesive bonding is preferred to mechanical fastening for thin composite structures, as the loads are distributed more uniformly across the joint. However, in thick, heavily-loaded composite structures, the use of mechanical fasteners is preferred over adhesive bonding. This preference is based on two considerations. First, thinner composite laminates have a low tolerance for edge-loading; second, they are easily damaged by the bearing loads imposed on them by mechanical fasteners.

Unfortunately, adhesive bonding is not always practical for other reasons too. For example, access to the internal structure may be required for manufacturability or maintenance purposes. In these cases, mechanical fasteners have long been employed to facilitate the joining of structures for permanent or semi-permanent applications as well as for frequent removal. These fasteners have evolved over time to meet the specific needs of a host of different materials and applications. As a result, manufacturers have designed a variety of fasteners that provide the desired structural integrity while also accounting for certain unique requirements of advanced composite materials.

Traditional cold-worked solid rivets should never be used in composite applications. Not only is there a high potential for damage to the laminate by the rivet gun or bucking bar, but the fundamental installation process of solid rivets causes the shank to swell remarkably. This causes serious delamination around every rivet hole, compromising the structure's integrity. Thus, in any discussion involving composite materials, when the term "rivet" is used, it can generally be assumed to refer to a blind fastener like the ones seen in Figure 12-13.

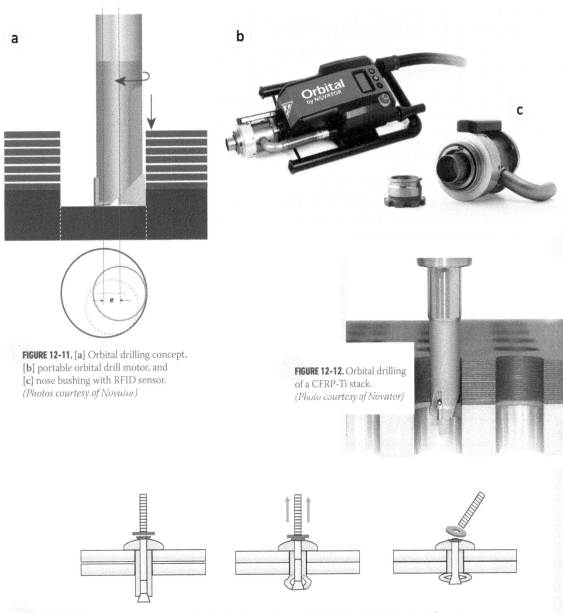

a   b   c

FIGURE 12-11. [a] Orbital drilling concept, [b] portable orbital drill motor, and [c] nose bushing with RFID sensor. *(Photos courtesy of Novator)*

FIGURE 12-12. Orbital drilling of a CFRP-Ti stack. *(Photo courtesy of Novator)*

FIGURE 12-13. Blind rivet pull-up sequence *(left to right)*: place fastener in the hole, pull up on stem, and stem breaks off flush with head.

The term *blind* is used to denote that access to the far side of the joint is not required for installation. Generally speaking, blind rivets are either hollow or made from materials not well suited for composite use for the reasons mentioned above. Therefore, the vast majority of blind fasteners seen in structural composite applications are actually blind bolts rather than blind rivets.

The difference in terminology is important to note as blind bolts have a significantly higher shear strength and clamp-up properties than do blind rivets. Put simply, they perform more like a traditional bolt than a rivet. Good clamp-up is essential to the longevity of a mechanically fastened joint. However, care must be exercised in selecting a fastener whose clamp-up pressure at installation does not exceed the compressive strength of the composite matrix. Most traditional laminates have little or no fiber reinforcement along the longitudinal axis of the fastener. Therefore, the pressure applied to the laminate during installation is born largely by the laminate's matrix system.

❭ COMPOSITE-SPECIFIC CONCERNS

Historically, organizations that have transitioned from making metallic structures to making composite structures have learned that composite materials have certain peculiarities that cannot be ignored when discussing mechanical fasteners. The following list comprises the general, overarching issues to be considered for any mechanical fastening application involving composite structures. Prudent consideration of these issues in the early design phases can result in significant cost-savings during subsequent operations and sustainment phases.

Materials

Most fastening systems specifically designed for composite applications tend to use metals from the cathodic end of the galvanic scale. (*See* Table 6.5, the galvanic series chart, on page 171.) Unlike aluminum, magnesium, and cadmium, metals such as titanium, Monel, and certain alloys of stainless-steel (for example, A-282 and 18-8) exhibit little or no tendency to corrode when exposed to carbon fiber.

While not every composite application of mechanical fasteners involves carbon fiber, fastener manufacturers tend to make composite-specific structural fastening systems primarily from these alloys. Ostensibly, this is because fastener manufacturers have no control over the end-use of their products and wish to both enhance economy of scale as well as minimize liability. A few aluminum fastener exceptions can be found in the industrial roofing and siding industries that use fiberglass almost exclusively.

Footprint

Structural fastening systems designed to secure laminates together through the Z-axis like screws, bolts, and rivets generally have a larger manufacturer's head and shop head than their traditional counterparts. Since mechanically fastened joints are typically loaded in shear, a larger diameter head helps the fastener resist rotation, or "tipping," particularly under heavily-loaded conditions. (Figure 12-14)

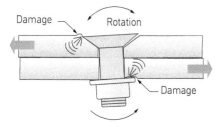

FIGURE 12-14. Fastener tipping under shear loading.

Each tipping event inflicts a certain amount of damage to the edge of the laminate as well as the area immediately under the head of the fastener. This damage is cumulative and ultimately results in a working fastener that no longer carries loads. These larger heads are generally not necessary in metallic structures as the isotropic nature of metals makes them better able to accommodate the tipping forces generated by shear loads. The head of protruding head fasteners specifically designed for composite use (in shear) is between 10% and 30% larger in diameter than their counterparts designed for metallic structures. (Figure 12-15)

Flush head, or *countersunk* fasteners also require a greater surface area to support the reaction loads and prevent tipping. For decades the standard countersink angle for load-bearing aerospace fasteners was 100°. This was found to be insufficient for composite applications in terms of surface area, leading to the invention of the 130° countersunk head.

Edge-Loading

Employing mechanical fasteners typically involves drilling a hole through the laminate in which the fastener is being installed. This exposed edge of traditional fiber reinforced laminates represents one of the more difficult issues associated with mechanical fasteners in composite applications, particularly for thin laminates (< .100 inch).

For example, if a fastener is installed in a hole that is too large (a **clearance fit**), it will begin to tip almost immediately when the joint is loaded, damaging the edge and elongating the hole further. If a fastener is installed in a hole that is too small (**interference fit**), the force necessary to seat the fastener in the hole will generate enough interlaminar shear to delaminate the area around the fastener, thus damaging the laminate before the fastener is even fully installed. Indeed, even properly installed fasteners designed for composite materials often do some damage to the hole on installation. (Figure 12-16)

FIGURE 12-15. [a] Standard blind fastener for metallic applications and [b] large footprint blind fastener for composite applications.

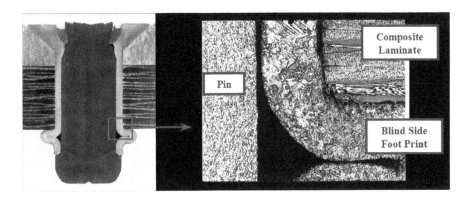

FIGURE 12-16. Fastener in a composite structure. *(Photos courtesy of Arconic, Inc.)*

This delamination is often more pronounced at the far side (exit) of the hole, where it may escape the attention of the technician and inspectors entirely. From a design standpoint, one measure that can be taken to make these exposed edges more robust is to gradually build up laminate thickness toward the edge of the part to increase its ability to resist tipping and increase the bearing area of the laminate-to-fastener interface.

While it is difficult to achieve, an interference fit is still highly desirable for both the ultimate strength of the mechanically fastened joint as well as for increased resistance to fatigue. For this reason, manufacturers have designed innovative fasteners that incorporate a thin-walled sleeve that expands up to .006" radially as they are installed, thus achieving an interference fit while avoiding damage to the hole. (Figure 12-17)

❱ EDGE DISTANCE AND SPACING

Full bearing strength is developed when the edge distance is equal to or greater than 3.0d (d = fastener diameter). The pitch distance (one fastener center to the next in the same row) and transverse pitch are equal to or greater than 4.0d. Hole spacing is based upon load calculations, which are partly dependent upon lamina form (uni-tape or fabric), orientation, and laminate thickness.

❱ HOLE TOLERANCES

Another contributing factor to joint fatigue is the quality of the installation hole. The hole must be normal to the surface and must be the proper diameter. Unlike their metal counterparts, composite laminates are rather fragile edge-on, and are unable to handle edge-loading inside the fastener hole, unless the fastener fits tightly during installation and remains so throughout its service life.

Absent specific engineering or process tolerances, best practices in typical aerospace applications are as follows:
- Surface roughness (Ra) < 4.8 μm.
- Delamination < 1 mm over the diameter.
- No splintering.

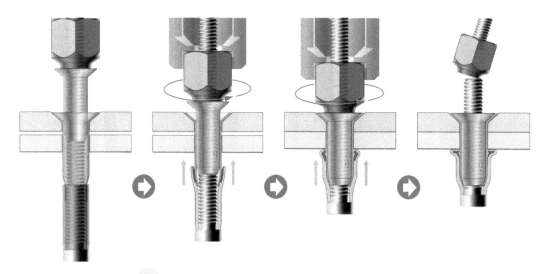

FIGURE 12-17. Monogram's Radial-Lok® fastener has a sleeve designed to expand radially on installation, for an interferance-fit without delaminations.
(Illustrations ©2019 Monogram Aerospace Fasteners, Inc. - Any portion of this image can only be reproduced or reused with the express written permission of the copyright owner.)

A net-diameter hole is somewhat difficult to achieve in high fiber volume composite structures due to the nature of the materials involved; the ends of the fibers tend to move away from the drill or ream during machining and then "rebound" back into place afterwards. This results in slightly smaller diameter holes and a tendency for the fastener not to fit. Because of this, the importance of proper tool selection and the use of proper techniques cannot be overstated.

❯ TWO-PART FASTENERS

Hi-Lok Fasteners

If there is access to both sides of a fastened joint, a variety of composite-compatible fasteners exist. Most of these are variations of fasteners traditionally used for decades on metallic structures. The Hi-Lok line of fasteners for composite structures from Lisi Aerospace comprises a variety of material and head configurations to suit a broad range of applications. However, the basic installation process has changed little in decades (*see* Figure 12-18). Once the countersunk or protruding head pin is installed in the substrate, a threaded collar is installed on the shop-head side. The fastener is installed by applying torque to the wrenching element, and at a predetermined torque value, the wrenching element shears away leaving the collar behind.

Lockbolts

Lockbolts such as the Huck LGP lockbolt work by swaging the collar into annular grooves on the pin rather than threading on (*see* left view of Figure 12-19). During installation, the tool draws the pin into the gun thus swaging the collar into the grooves. At a predetermines tension, the pin-tail breaks away leaving the properly seated fastener. This type of installation provides excellent clamp-up pressure and is also available in a sleeved version providing an interference-fit (right view of Figure 12-19).

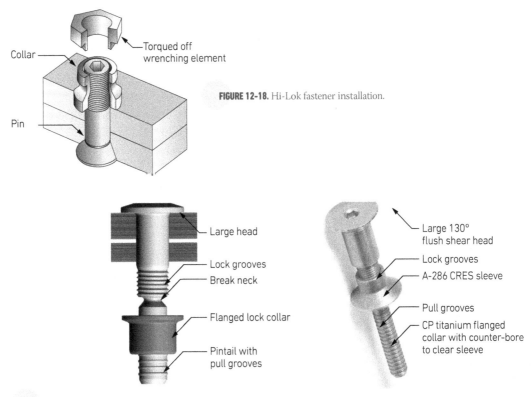

FIGURE 12-18. Hi-Lok fastener installation.

FIGURE 12-19. Huck LGP lockbolt installation.

Adjustable Sustained Preload Fasteners

This unique fastener system is available from Howmet (ASP) as well as Monogram (MAF). Its revolutionary design allows a fastener to be installed through thin-skinned sandwich panels without crushing the skins on either side (*see* Figure 12-20). This is accomplished via the fastener's unique two-piece design: the threaded shank (A) is installed from one side of the panel while the sleeve (B) is threaded down onto the shank from the opposite side. After installing a locking ring onto the pin-tail, the installation tool is used to apply tension to the pin-tail, while simultaneously seating the locking ring into the sleeve.

As with other tension-actuated fasteners, the pin-tail breaks away at a predetermined tensile value. What is so different about this fastener design is once the sleeve is threaded onto the shank and torqued (typically 25 in/lb), 100% of the tension from the installation tool is reacted in the fastener and not allowed to translate to the skins.

Bonded Fasteners and Inserts

As we've seen, using mechanical fasteners in advanced composite materials poses some unique problems. Unlike metal structures, composites are tremendously strong and light-weight, yet can be fragile and easily damaged. These peculiarities must be considered when considering what type of fasteners should be used in a given composite structure. When blind rivets/bolts or lockbolts are impractical or simply not appropriate to the application, potted inserts or adhesively-bonded fasteners may be the solution to the fastening problem.

❯ INSERTS

The need for an effective means to attach components to honeycomb sandwich structures was realized shortly after the first such panel was built. The relatively delicate, light-weight honeycomb core material adhesively bonded to two thin skins was ill-suited to accommodate significant loads in any direction. For decades, potted inserts provided the capability to attach components and sub-assemblies to such sandwich structures. Brand names like Rosan® Delron®, and Shur-Lok® offer a wide variety of types, styles, materials and finishes for sandwich panel fastening.

Potted fasteners are typically available in two forms: threaded inserts often installed from one side, and matched bushings installed from both sides creating a hard-point in a sandwich structure for through-fastening applications. (Figure 12-21)

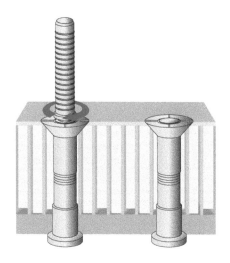

FIGURE 12-20. Monogram's MAF fastener installation for use in structures sensitive to high clamping loads.
(Illustrations ©2019 Monogram Aerospace Fasteners, Inc. • Any portion of this image can only be reproduced or reused with the express written permission of the copyright owner.)

❱ BONDED FASTENERS

In cases where threaded studs, sleeves, or other specific hardware must be installed, adhesively-bonded fasteners may be the best option. Click Bond, Inc. (Carson City, NV) has developed the broadest array of solutions to these fastening challenges for use on advanced composite structures.

Adhesive bonding of fasteners and/or other hardware has several advantages over previous installation methods. For example, installing an adhesively-bonded nut-plate requires that only one hole be drilled in the structure since the need for the retention rivets is eliminated. Additionally, it takes approximately 80% less time to install these fasteners compared with riveting them on. Many of Click Bond's solutions include a temporary fixture to align the fastener and apply pressure to the bond line. A few examples are shown in Figures 12-22 and 12-23.

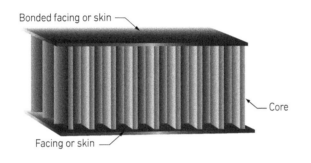

Bonded facing or skin

Core

Facing or skin

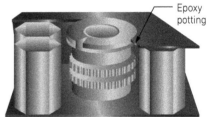

Epoxy potting

Typical Series 400 SF Snap-in, floating nut insert installed in honeycomb sandwich panel. Insert is held in place by a cured epoxy compound.

FIGURE 12-21. Fasteners for sandwich panels.

FIGURE 12-22. Adhesively-bonded nut-plates.

[a] Metallic base and [b] glass fiber and carbon fiber bases. When pulled through a properly sized hole in a composite panel, the silicone rubber fixtures keep the assembly centered and maintain pressure on the bondline as the adhesive cures. *(Photos courtesy of Click Bond, Inc.)*

FIGURE 12-23. [b] Adhesively-bonded cable-tie mount and [a] adhesively bonded threaded stud. Shown with acrylic installation fixtures that are removed after the adhesive cures. *(Photos courtesy of Click Bond, Inc.)*

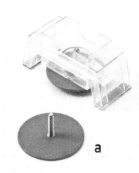

13

Repair of Composite Structures

CONTENTS

Repair Design Considerations

Designing a composite repair can be a complex process. Some of this can be seen in Table 13.1, which lists some of the considerations, practicalities and parameters of composite repair. In addition, engineering analysis of the structure (and repair) may be required for large or more critical repairs.

TABLE 13.1 Composite Repair Design Considerations

Design Considerations	Practicalities	Parameters
• Temporary or permanent	• Time	• Strength
• Bonded or bolted	• Environment	• Stiffness
• Flush repair or external doubler	• Equipment	• Weight
• Single-sided or double-sided	• Skills	• Shape/Contour
• Wet layup or prepreg	• Materials	• Surface finish
• Appearance desired	• Regulatory requirements	• Service environment
• Access to damaged area	• Documentation	
• Complexity of repair		
• Primary, secondary, or tertiary		

If replacement is not possible, then the ideal repair is to match all original design parameters (e.g. materials, fiber orientation, curing temperature, etc.). This is rarely possible, and compromises are inevitable. However, the goal of the repair design is to return the structure, as much as possible, to an acceptable strength, stiffness, shape and surface finish.

Note: If the extent of damage or overall damaged area exceeds allowable repair limits in applicable repair instructions, then specific engineering support is required before proceeding with the repair. (A specifically designed repair is necessary.) Since it is such a complex subject, specialized knowledge/training in composite repair design and analysis is required.

❱ STRUCTURE TYPES

Repairs to lightly loaded structures might not require as much detailed analysis as those to heavily loaded primary structures (*see* Table 13.2). However, repair design is still required in order to ensure that the repaired structure will not fail in service.

TABLE 13.2 Composite Structure Types

Structure Type	Description
Primary	Heavily loaded or safety-of-flight structures. Example: wing spar, fuselage, boat hull
Secondary	Intermediate in load and safety criteria. Example: aerodynamic fairing, cowling
Tertiary	Lightly loaded or non-critical structures. Example: interior sidewall panel

Note that in some cases, a structure is designated as secondary or tertiary from a design load standpoint, but it might be treated as primary structure when repaired, due to the critical nature and location of the component. An example would be a lightly loaded fairing located in front of the intake of a jet engine. Failure of the fairing itself would not be structurally critical. However, if it gets sucked into the engine, the resulting engine failure could be catastrophic.

Types of Damage

Most composites are damaged in service from overloading, impact, exposure to excessive heat, or other environmental effects. Parts may also need to be repaired due to a manufacturing defect that would make a composite structure more susceptible to in-service failure.

❱ HOLES AND PUNCTURES

Holes and punctures are classified as visible impact damage (VID) and are usually caused by high or medium velocity impacts. They can be severe but are usually easily detected. (Figure 13-1)

FIGURE 13-1. [a] Hole and puncture damage to the carbon fiber/aluminum honeycomb nose of an F3 race car monocoque; and [b] its appearance after it has been prepared for repair by removing primer and paint.

❭ DELAMINATION

Perhaps the most common of all damage types, **delamination** is often caused by impacts such as a tool drop or a glancing bird strike on an aircraft. They can sometimes be visible if they are near the surface. Delamination can also be caused by manufacturing defects, matrix damage, low resin content, or **moisture ingression**. (Figure 13-2)

❭ DISBONDS

A **disbond** indicates a failure between materials. This is usually seen as a face sheet disbonding from an underlying sandwich core material. The terms *disbond and delamination* are often used interchangeably, however there is a subtle, but important difference. Whereas delamination refers to a condition where plies of the laminate are no longer bonded to each other, a *disbond* is a condition where a laminate is no longer bonded to another constituent of the composite such as a core material. Both conditions can exist in close proximity, particularly as a result of an impact or excessive heat. (*See* Figure 13-3.)

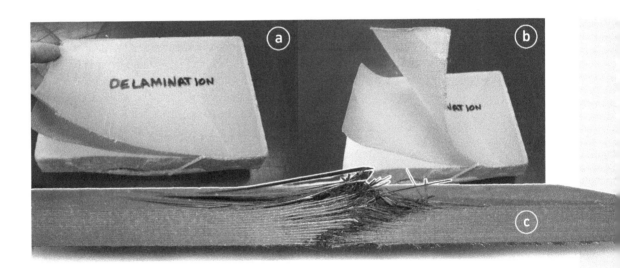

FIGURE 13-2. Delamination damage.

[a, b] This aramid-skinned/foam-cored panel illustrates how the plies of a laminate can actually come apart, or "de-laminate." [c] Carbon fiber sample shows how a localized impact may cause delamination within the laminate across a significant area.

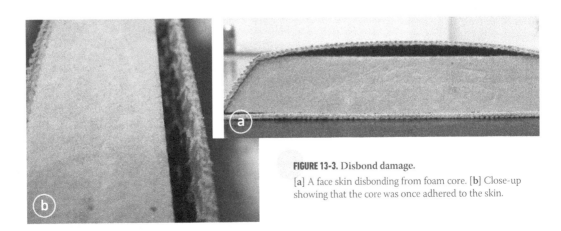

FIGURE 13-3. Disbond damage.

[a] A face skin disbonding from foam core. [b] Close-up showing that the core was once adhered to the skin.

❱ CORE DAMAGE

Core damage can occur with any type of core and is typically caused by improper handling in manufacturing, or impacts or fluid ingress while in-service. Fluid ingress is a common problem with honeycomb core panels. This can facilitate extended damage from freeze-thaw cycling in aircraft parts. (Figure 13-4)

❱ RESIN DAMAGE

Resin damage is caused by many factors, such as BVID/VID impact, exposure to excessive heat, UV exposure, paint stripper, over-stressing the structure, or anything else causing degradation of the resin properties. This type of damage may be the most difficult to detect in the field, and it is especially difficult to quantify the effect on the structural integrity of the part. As a rule, resin damage leads to a greater loss of compressive (interlaminar shear) strength than of tensile strength.

❱ WATER INGRESSION OR INTRUSION

Water ingress is a problem with exposed laminates and within honeycomb core, causing weight gain, corrosion (in aluminum honeycomb), and skin disbonds if the water freezes and expands. Trapped or residual water can be a problem in high-temperature repairs; e.g., the heat of curing the repair causes trapped water vapor to expand, often disbonding face sheets around the repair and ultimately converting a small area of damage into a large area of damage. (Figure 13-5)

 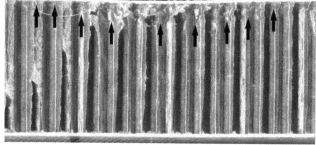

FIGURE 13-4. Core damage to honeycomb core caused by impact. *(Photos courtesy of The University of Auckland/LUSAS)*

 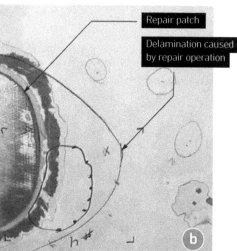

FIGURE 13-5. Water ingression damage.

[a] Note the discoloration of damaged cells in the honeycomb core damaged by water ingression.

[b] A trailing edge flap from an F/A-18 Navy fighter jet illustrates how moisture-induced disbond damage can be caused during repair. *(Photo courtesy of Austal USA-ElectraWatch, Inc.)*

❭ LIGHTNING, FIRE AND HEAT DAMAGE

Lightning damage ranges from simple surface abrasions to large holes, and includes thermal damage to the resin. Fire or excessive heat from a lightning strike may char or burn resin and fibers. It may be difficult to determine the extent and depth of this type of damage, but it is extremely important to do so. (Figure 13-6)

Damage Detection

Damage to a composite is often hidden to the eye. Where a metal structure will show a dent or "ding" after being damaged, a composite structure may show no visible signs, yet may have delaminated plies, broken backside fibers, or other damage within.

Impact energy affects the visibility, as well as the severity, of damage in composite structures. High and medium energy impacts, while severe, are generally easy to detect (VID). However, high energy, wide-area blunt impact and some low energy local impacts can easily cause *hidden* damage. This is usually classified as barely visible impact damage (BVID). (Figure 13-7)

There are a variety of non-destructive inspection techniques available to help determine the extent and degree of damage. *See* Chapter 10 for illustrations and in-depth descriptions of each.

Laminate and Ply Determination

Ideally, composite components should be fully identified before a repair is performed. Laminate details such as material specifications, number of plies, ply orientation, core ribbon direction, ply buildups and drop-offs, and numerous other details need to be clarified and understood before beginning a repair.

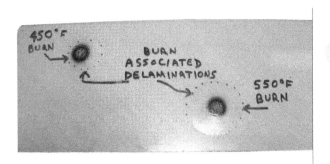

FIGURE 13-6. Burns cause resin damage, which is easy to see here in the charred circles. However, note the larger areas of delamination caused by the same overheating, which is much more difficult to detect visually.

Medium Energy Impact

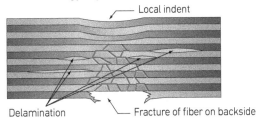

Local indent

Delamination   Fracture of fiber on backside

Low Energy Impact

Pyramid pattern matrix crack from impact

FIGURE 13-7. Difficulties in damage detection—differences in damage visibility and severity resulting from medium and low energy impacts.

For aircraft, this type of information is often available in the structural repair manual (SRM) or equivalent documents. Certain boat manufacturers will supply specific repair instructions, and some may provide repair kits. If an SRM or equivalent document is not available, this type of information is still needed for proper repairs and determination of these details can be performed by the operator if necessary.

Determination can be done by careful taper sanding through a small sample of the damaged part, reading the information directly from the composite itself. However, a thorough understanding of composite materials—including fiber type, weave patterns, and ply orientation concepts such as balance, symmetry, nesting, etc.—is mandatory for this type of analysis. (Figures 13-8, 13-9)

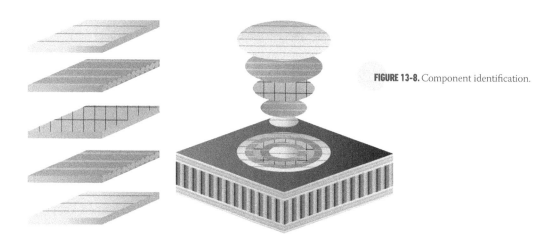

FIGURE 13-8. Component identification.

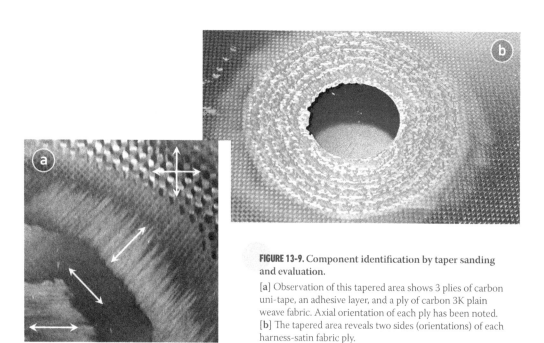

FIGURE 13-9. Component identification by taper sanding and evaluation.

[a] Observation of this tapered area shows 3 plies of carbon uni-tape, an adhesive layer, and a ply of carbon 3K plain weave fabric. Axial orientation of each ply has been noted. [b] The tapered area reveals two sides (orientations) of each harness-satin fabric ply.

❯ REPAIR MATERIALS

Ideally, a repair would be performed using the same materials, fiber orientation, core ribbon direction, stacking sequence, and cure temperature that was used for the original fabrication of the part. However, sometimes this is not feasible, especially in the case that an original structure was manufactured in an autoclave and the repair is being performed without an autoclave. Therefore, material or process substitutions may be allowed or even required by the repair instructions. (For temporary repairs, more flexibility in material substitution may be allowed than for permanent repairs.)

Nevertheless, the goal is to return the structure to an acceptable strength, stiffness, shape and surface finish. For any composite repair the following types of repair materials need to be considered and evaluated before the repair is begun, ensuring that the repair is structurally sound:

Matrix Resins

- Resin/adhesive systems—wet layup versus prepreg, low-temperature versus high-temperature.
- Cure cycle requirements and available equipment.

Fibers/Fabrics

- Fiber reinforcement—type of fiber (fiberglass, aramid, carbon, etc.).
- Fiber reinforcement form (unidirectional tape, woven cloth, weave style, etc.).

Core Materials

- Core type and density—Nomex® honeycomb, aluminum honeycomb, foam, balsa, etc.
- Honeycomb ribbon direction—different properties in the ribbon versus transverse direction.
- Core adhesive or potting compound.

Lightning Strike Materials

- Mesh or fiber materials—copper or aluminum mesh, conductive fibers, etc.
- Coatings—conductive coatings, paint thickness.
- Grounding strips (often bonded to external surfaces).

Sealants

- Polysulfide (type of coating used on older aircraft parts).
- Polyurethane—newer systems and hybridized coatings that are less brittle, more damage tolerant.
- Rubberized coatings (variety of products to help prevent damage from erosion, etc.).
- Others—boats use gel coats and water-resistant paints; pipes often use thermoplastic coatings for UV and abrasion resistance; sporting goods use a variety of unique finishes.

Paint Removal Methods

The first step is to remove paint or outer coatings. Chemical paint strippers must **not** be used, unless it is certain they are specifically designed for composite structures. Most paint strippers are based on methylene chloride and will attack cured resins. Paint and coatings may be removed by:

- Hand sanding.
- Bio-based starch media blasting (wheat, corn, corn hybrid polymers).
- *Careful* grit blasting or plastic media blasting.

It is important to check any repair instructions or guidelines offered by manufacturers, and to make sure all health and safety requirements are met.

❱ HAND SANDING

Hand sanding is widely used as a paint removal method for composites. No expensive equipment is needed, and the sanding pressure can easily be controlled to avoid laminate damage. Hand sanding is recommended for paint removal in a small area (e.g., around the area to be repaired). One must be careful not to damage fibers in the surface ply.

The obvious disadvantage is the high labor costs, especially on large parts. However, as an option, nylon abrasive pads (Scotch-Brite™) attached to a die grinder can be used to remove paint. This may provide faster paint removal operation but at a slightly higher risk. (Figure 13-10)

❱ MEDIA BLASTING

Media blasting is possible with composites but must be done very carefully, as it is highly likely that fibers on the top surface will be damaged if extreme care is not taken.

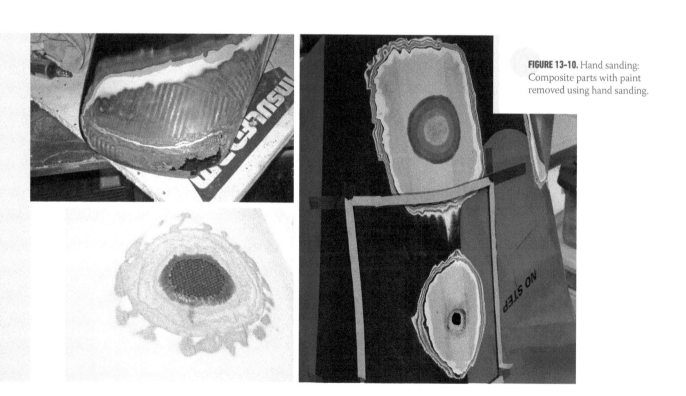

FIGURE 13-10. Hand sanding: Composite parts with paint removed using hand sanding.

Plastic Media Blasting

This method is less aggressive than sand (silica), garnet, or aluminum oxide media blasting, and is used often with composites. (Figure 13-11) While effective, this technique has shown some difficulties in the field. The plastic media is most often cleaned and re-used in such systems. This can lead to two problems:

1. The cleaning process is largely designed to remove paint chips and solid matter. If the plastic media becomes contaminated with oil, grease, fuel, etc., these contaminates can be driven back into the composite surface, causing paint adhesion problems.

2. The plastic particles become dull with reuse, and a poorly trained operator may turn up the air pressure to compensate. This can lead to damage of the composite surface, or even blowing a hole through these rather brittle materials.

As a result, fresh media is recommended for each new paint removal operation.

Starch Media Blasting

This method uses starch media to blast paint from parts. Many OEMs have approved this method for paint removal from composites. As with plastic media, there can be a contamination problem as the starch is reused. This is typically mitigated by limiting the number of reuse cycles.

Starch is much more aggressive when used at higher pressures (over 30 psi). The particles fracture upon impact into smaller particles with increased surface edges. These are even more effective at removing coatings, but they can also damage composite substrates. A lower blasting pressure and smaller angle of impingement are recommended for composites, and the media flow range and blasting standoff distance should be carefully selected for the part being treated.

FIGURE 13-11. Blast media.

[a] Both plastic media and bio-based starch media are used to remove coatings from composite structures. *(Photo courtesy of Archer Daniels Midland Company)*

[b] Various blast medias for different applications are shown with Plasti-grit® (made from Urea) represented in the middle, as well as the two white piles (made from Clear Cut and Acrylic). The brown-colored pile is walnut shell, the tan colored is corn cob, and the red piles are Nylon and Poly-Carb plastic media. *(Photo courtesy of Composition Materials Co., Inc.–www.compomat.com)*

With starch media, there is no chemical attack to worry about, and it is an environmentally benign and biodegradable material that can easily be disposed of. Although very rare and limited to specific areas within the process, dust generated from wheat starch could possibly explode. This risk is reduced when standard blast protection procedures and proper ventilation are used. (Figure 13-12)

❱ OTHER METHODS

Laser Ablation

The technology behind laser ablation of coating systems was first explored by the U.S. Air Force in 1991. Environmental concerns, worker safety, and recent advances in robotics technology have thrust this technology to the fore once again. General Lasertronics has developed a neodymium-YAG laser paint removal system currently used on the Sikorsky CH-53 heavy-lift helicopter. Concurrent Technologies produces the Advanced Robotic Laser Coating Removal Systems (ARLCRSs) for the U.S. Air Force. This CO_2 laser system is currently in use to strip metallic and composite structures on the C-130 and F-16 aircraft. Additionally, Netherlands-based LR Systems is currently building a system that is 52 feet tall with a reach of 85 feet to appeal to the commercial airlines.

There are many other methods, such as dry ice blasting, Xenon flash lamps and scraping, baking soda blasting, UV nanosecond pulsed laser, etc. Work in this field is continuing, and better automated methods are currently being researched by many organizations.

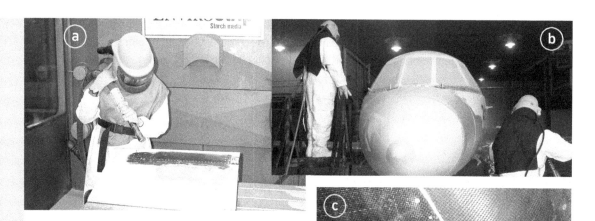

FIGURE 13-12. Starch media blasting.
[a] ADM has engineered ENVIROStrip® and eStrip® dry starch-based media to remove multiple tenacious aircraft paint and primer coats from sensitive composite substrates. [b] The dry media process can be used with portable or booth-based blasting systems. [c] With some types of primers, ENVIROStrip® can remove coatings down to the final layer, leaving the substrate primer intact. *(Photos courtesy of Archer Daniels Midland Company)*

Traditional round nozzle path with hot spot

Preferred fan-shaped blast path of a flat nozzle

FIGURE 13-13. Flat vs. round nozzle. *(left)* Traditional round nozzle path with hot spot. *(right)* Preferred fan-shaped blast path of a flat nozzle.

Regardless of the blasting system used, one development that is universally accepted is the preference for a flat nozzle with a fan-shaped blast path. This avoids the damaging hot spot in the blast path center of traditional round nozzles. (Figure 13-13)

Damage Removal

After paint removal, additional damage assessment is performed because visible damage now becomes more apparent. All damaged material must be removed, including anything contaminated by oil or other synthetic fluids, and anything metal showing signs of corrosion. Water should be removed, and any residual moisture properly dried (*see* "Drying" on page 351).

Aluminum honeycomb should always be inspected for corrosion. Beginning signs of corrosion include a dull instead of a shiny appearance, and white or gray-colored areas. With advanced stages of corrosion the honeycomb becomes brittle and tends to flake.

❱ DAMAGE REMOVAL SCENARIOS

The following scenarios apply to both solid laminate and sandwich structures. The concepts depicted are representative of the damage removal process that one might use for repairing sandwich structures; however, the basic approach is the same for solid laminate structures that are to be repaired using taper-scarf preparation. Damage removal can be performed using hand-held tools as described, or via automated machining methods.

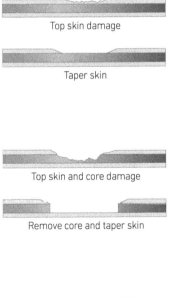

Top skin damage

Taper skin

Top skin and core damage

Remove core and taper skin

Damage through the laminate

Remove core and taper both skins

Remove core and taper both skins from one side

Top Skin Damage Only

One skin is damaged. This situation is usually caused by low-energy impacts and results in matrix resin cracks, fiber breaks, and delamination of the top skin. In a sandwich structure, disbonding of the skin from the core is possible and should be assessed.

Remove the damaged skin plies as needed by taper sanding to the depth of the damage and the surrounding area at the desired scarf angle. (*See* "Tapered Scarf Repair," below.)

Top Skin and Core Damage

Damage extends through one skin into the core material. This damage results from greater impact and may produce disbonding and/or delamination in the bottom skin, so be sure to perform a thorough inspection. (If the bottom skin has been damaged, then see the through-damage case below.)

Remove all damaged skin and core. Surrounding core material must be inspected to ensure it is not contaminated. If it is contaminated, it must be dried or removed. Taper sand the surrounding area to the desired scarf angle.

Damage Through the Laminate

In the case of a hole or puncture through the laminate, both skins will require damage removal.

Remove the damage to both skins and damaged core material. Taper sand the surrounding undamaged area on both sides to the desired scarf angle. The tapered areas for the top and bottom skins do not necessarily have to match. They should be based on the size of the damage in each skin.

Note that it is also possible to remove a larger area of one skin to access the inside of the opposite skin, in order to taper scarf and perform a repair to both skins from one side.

❯ AVOID SHARP CORNERS

Damaged composite skin plies may be removed by careful routing or grinding through the damaged surface in circular or oval shapes, without making sharp corners. If an irregular shape is required, then radius each corner to minimize stress concentration. (Figure 13-14)

❯ ROUTING

This is the method of choice for removal of damage through the (skin) laminate thickness. Diamond coated or carbide-tipped router bits give a cleaner cut and reduce splintering in glass and carbon fiber reinforced laminates. For repair of aramid fiber reinforced laminates that tend to fuzz when cut, a tungsten carbide split-helix router produces the best results.

❯ GRINDING

A 90-degree angle grinder with a mandrel and an abrasive disk works well for grinding and scarfing composites. Carefully sand away damaged plies until undamaged plies or the top of the core is reached. The trick with using an angle grinder on a composite is to "feather" the materials using the trailing edge of the disk rather than grinding with the leading edge of the disk, as with metals. If the leading edge is used, the disk will dig into the plies and create a larger damage area.

❯ CORE REMOVAL

Cutting and Peeling

This is commonly used with light density honeycomb core of any type. Use a sharp square-edged X-acto knife, putty knife, or flat chisel to make vertical cuts along the edge where the skin has been removed. Be careful not to penetrate or cause damage to the far-side skin. Next, slide the knife-edge between the core and the far-side skin to make a horizontal cut, and then remove the core along its adhesive bondline. (Figure 13-15)

Routing and Grinding Core

Routing may be preferred for removal of cores of higher density. A router may be used to remove the top skin and core; however, care must be taken not to penetrate the far-side skin. A high-speed grinder can be used to carefully remove the remaining core from the far-side skin. (Figure 13-16) Extreme care must be taken not to damage the plies in the far-side skin.

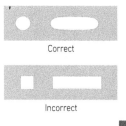

Correct

Incorrect

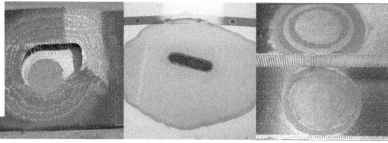

FIGURE 13-14. Damage removal using "radiused" shapes.

Important: This may mean that there will still be a slight imprint of the core cells. The objective is to have a smooth surface for which to bond the replacement core plug, not to remove all evidence of the previous core.

CONTAMINATION

Damaged composite structures may be contaminated by fuel, hydraulic fluid, oil, etc. Contaminated surfaces must be removed or treated in a way that will enable resins and adhesives to bond. (Figure 13-17)

Solid laminates contaminated with fuel, oil, grease, etc., may first be treated by wiping thoroughly with a solvent and multiple clean wipes, then using reagent grade solvent for the final wipe. It is important to know what the composite is made of, what the contaminant is, and to check all suggested procedures to ensure the solvent being used can dissolve the foreign fluid without further damaging the composite structure.

If the core in a sandwich structure is contaminated, replace the affected material. For water contamination a proper drying operation is recommended.

DRYING

Damaged laminates will absorb water and must be addressed to achieve a successful composite repair. All affected composite materials must be dried before an effective repair is possible: cured resin, fibers, and honeycomb core can all retain moisture that must be removed. Damage removal and (taper scarf) preparation should be conducted prior to drying.

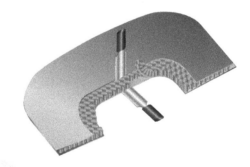

FIGURE 13-15. Core removal using a putty knife or chisel.

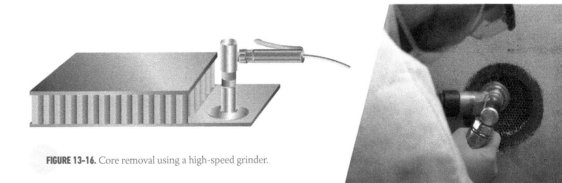

FIGURE 13-16. Core removal using a high-speed grinder.

FIGURE 13-17. Removal of contaminated core.

Sandwich structures containing Nomex® or glass honeycomb core, will almost always be hydrated. Trapped, standing water in honeycomb core is common and can be removed with a wet and dry vacuum or other non-destructive means prior to drying. Refer to the drying cycles specified in repair instructions. It should be noted that one hour of drying time is usually not enough. Normally, many hours or even days of drying time is required.

Drying is typically done with a heat source, breather layer and vacuum bag, using heat and vacuum to convert the moisture to vapor and to draw it out. Lower temperatures and vacuum over a longer period of time provides the least risk of disbonding skins and generally produces the best results. (Figure 13-18)

Guidelines include:

- Drying temperatures vary but should not be too high: 150°F–170°F is common.
- Always use a drying temperature below the wet T_g of the matrix resin.
- Slower drying using lower temperatures for longer time under vacuum is best.
- Do not exceed 5°F per minute heat ramp rate.

One trick for telling when the structure is dry is to exhaust the vacuum pump through a desiccant that will change color with a change in moisture content. (Figure 13-19)

FIGURE 13-18. Drying moisture from cored structures.

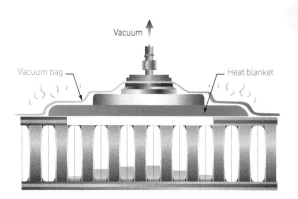
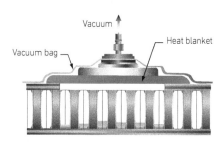
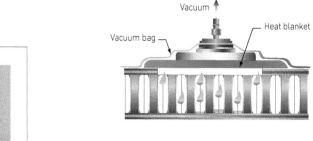
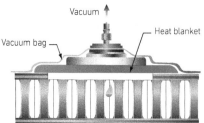

FIGURE 13-19. Change in desiccant color as it gains moisture—this blue silica gel turns pink when saturated. *(Photo courtesy of AGM Container Controls)*

A moisture meter (Figure 13-20) may also be used to determine the dryness of composite skins and non-metallic honeycomb core. Moisture meters work best with non-conductive materials; thus, they may not produce good results with carbon fiber composites or metallic cores. Another important consideration is that moisture meters do not give a reliable calibrated moisture content reading—i.e., an absolute measurement. However, they can be used for relative readings, such as "is it drier now than it was yesterday?" Note that the readings may vary widely between instruments from different manufacturers.

Types of Repair

❭ COSMETIC

Superficial, non-structural epoxy-based filler is sometimes used to restore a surface. Sometimes this is done with a single ply of glass over the filled area to keep fluids out and/or hold the filler in. This type of repair does not regain strength and is used only for small damage areas to secondary or tertiary structures. (Figure 13-21)

Many non-structural filling compounds are made from polyester resin (auto-body filler). They are cheap, readily available and easy-to-use, but they simply may not be the right material for composite aircraft repair. Therefore, epoxy-based fillers (syntactics) that require precise mixing, and may need elevated temperature to cure, are the material of choice for cosmetic repairs—simply because they stick, and they remain stuck. Epoxy-based fillers have higher elongation and much lower shrinkage than polyesters, enabling them to resist cracking and perform better over the long-term.

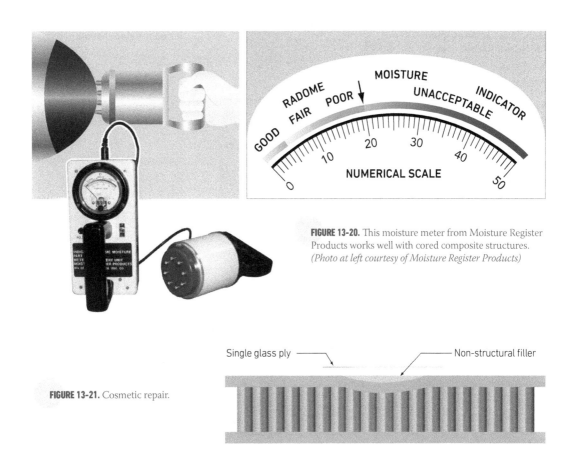

FIGURE 13-20. This moisture meter from Moisture Register Products works well with cored composite structures. *(Photo at left courtesy of Moisture Register Products)*

Single glass ply ⸻ Non-structural filler

FIGURE 13-21. Cosmetic repair.

❱ RESIN INJECTION

This type of repair is rarely recommended. It is accepted practice for repair of very limited delamination at the edge of a composite part. (Figure 13-22)

For any significant area of delamination or disbond, resin infusion is preferred over resin injection for repair. The atmospheric pressure created by the vacuum—used in resin infusion— moves the resin into cracks and delaminations, re-bonding the laminate. (*See* "Resin Infusion Repairs" on page 357.) Resin injection, however, uses hydraulic pressure to push resin into the laminate. Traditionally, resin injection repair techniques specified that vent holes be drilled peripherally around the damaged area to prevent further delamination. Even with this measure, the risk is still great that excessive pressure will force cracks open, grow the delamination, and increase overall damage.

❱ MECHANICALLY FASTENED COMPOSITE DOUBLER

Bolted composite doublers are typically used for thick, heavily loaded solid laminate structures. This is often the only practical means of repairing such structures. These patches are often fabricated on transfer tooling that controls the contour of the area to be repaired. A liquid shim (paste adhesive) or polysulfide sealant is often required between the patch and the structure to maximize contact and seal against fluid ingress.

Bolted titanium patches are commonly used on thick carbon fiber solid-laminate primary structures because the titanium can easily match the strength and stiffness requirements to transfer loads around the damage. Also, titanium is not susceptible to galvanic corrosion in direct contact with the carbon. Such repairs are not flush and are not specified for critical aero-

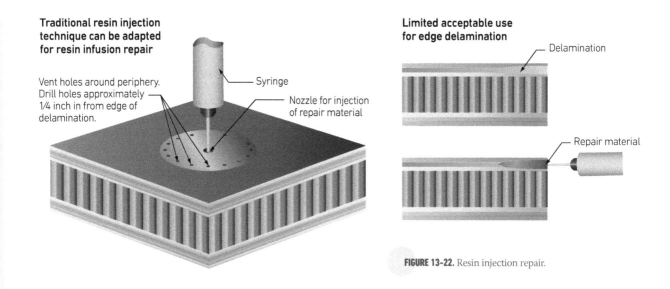

Traditional resin injection technique can be adapted for resin infusion repair

Vent holes around periphery. Drill holes approximately 1/4 inch in from edge of delamination.

Syringe

Nozzle for injection of repair material

Limited acceptable use for edge delamination

Delamination

Repair material

FIGURE 13-22. Resin injection repair.

FIGURE 13-23. Mechanically fastened composite doubler repair on an external aircraft structure.

dynamic or low observable surfaces. The edges of metallic and composite doublers are always beveled and the corners are rounded to lessen the peak shear-stress concentration around the edges. (Figures 13-23, 13-24)

This type of repair can be quicker and may cause less damage to the original structure than a scarfed skin repair. It is also possible to perform while the part is still assembled (with access to only one side of the part). Knowledge of underlying structure and systems is required before cutting or drilling for doublers. These type or repairs are considered permanent and may require a filler plug in the cleaned-up damage area to satisfy the repair design.

❯ STRUCTURAL ADHESIVELY-BONDED DOUBLER REPAIRS

Bonded external doublers are sometimes used for repairs to lightly-loaded thin laminate structures. They may require ambient or high-temperature cure cycles, depending on the adhesive system used for repair. Bonded doubler repairs can regain a significant portion of the original strength of the structure but with a significant stiffness and weight penalty in many cases.

Tooling may be required to fabricate the doubler (repair patch) so that it fully conforms to and restores the part's shape and surface. This type of repair is generally easy, relatively quick, and does not require the highly developed skills needed for flush structural repairs.

❯ FLUSH STRUCTURAL REPAIR (TAPERED OR SCARF REPAIR)

This type of repair restores axial load properties by forming a joint between the prepared repair area and the repair patch, where each ply in the patch overlaps the original ply in the structure, transferring load on axis. The repair patch is made by replacing each ply of the original laminate that has been removed from the damage area. (Figure 13-25)

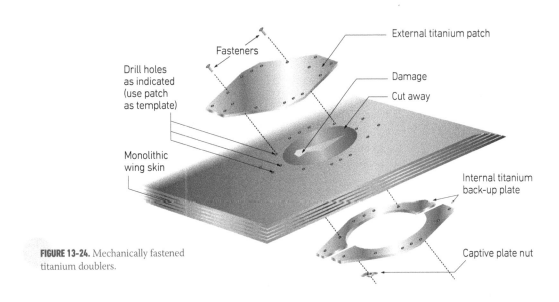

FIGURE 13-24. Mechanically fastened titanium doublers.

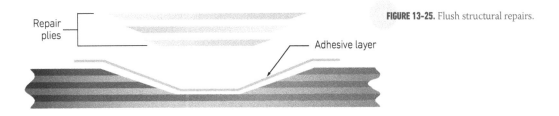

FIGURE 13-25. Flush structural repairs.

The size of the repair patch should fit closely the area prepared for repair, including the additional area for an outermost, extra-structural repair ply (or plies) and/or a final cosmetic or sanding layer that may be specified for the repair, as outlined in the repair instructions. (*See* "Tapered Scarf Repair" on page 358, for more details.)

Flush structural repairs are commonly called *tapered-scarf* or *scarfed* repairs, because they require tapering the repair area around the removed damage in preparation for bonding a cobonded repair patch. They are common with thin solid laminates and with sandwich structures and are currently considered the best type of repair in terms of providing the desired restoration of strength and stiffness with minimal weight increase. They can also be perfectly flush with the original surface if performed carefully. Scarfed repairs are commonly specified by most aircraft manufacturers in their SRMs and may be recommended by boat companies and other original equipment manufacturers (OEMs).

Tapered scarfing is a vital skill to be perfected for this type of repair, because if done incorrectly, it can increase the damage to the part.

Achieving an intimate bond between the repair patch and the prepared repair area is critical to completing a successful scarfed repair. One of the basic requirements is careful removal of material in a smooth fashion and at a precise angle relative to the surface contour, all around the area where the damage has been removed. This scarfing or *taper sanding* is done with a compressed-air-powered high-speed angle grinder with a mandrel and abrasive disk attached. Practice is the only way to obtain the necessary skill in scarfing required to perform this type of repair. (Figure 13-26)

After scarfing, the repair materials are applied to the prepared area, vacuum bagged and cured. Much of the remainder of this chapter will discuss surface preparation, curing and vacuum bagging of scarfed structural repairs.

❱ DOUBLE VACUUM DEBULK REPAIRS

Double vacuum debulk (DVD) processing is an out-of-autoclave technique used to remove trapped air and gas from a repair patch prior to applying the patch, or prior to curing the patch on a mold, in order to create a low-void doubler. The idea is to perform a warm debulk under vacuum without atmospheric pressure compacting the repair laminate. This is done by creating a pressure differential between a box sealed to the vacuum bag and the vacuum bag itself. The differential is created by pulling slightly more vacuum on the box then the bag. (Figure 13-27)

By pulling vacuum inside the box, the pressure differential is created, thus the layers under the bag are no longer under pressure yet are still under vacuum. This allows for the repair

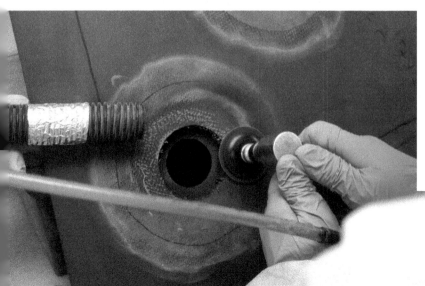

FIGURE 13-26. Scarf repairs.

laminate to "degas" as if it were in a vacuum chamber. Heat is applied during the process to lower the resin viscosity and promote the degassing event. (Figure 13-28)

After enough time the box is vented, and the layers are compacted within the initial vacuum bag. Venting the box allows the inner vacuum bag to pull down tight against the repair laminate effectively consolidating the freshly degassed layers against the tool surface. The resulting repair layup is now ready for cobonding to a prepared structure or molding on a tool to create a doubler patch.

❱ RESIN INFUSION REPAIRS

NOTE: The information in this section was contributed by André Cocquyt, president of GRPguru.com and author of *VIP: Vacuum Infusion Process*.

Resin infusion uses vacuum/atmospheric pressure to move liquid resin into dry reinforcements. Although infusion can be used with 1-sided or 2-sided repairs, it is essential for the mold, tooling or back surface of the repair to hold vacuum. Even the tiniest of leaks will bring air into the repair, potentially destroying its integrity and strength.

The basics of resin infusion repair are the same as with a structural scarfed repair: assess and remove damage, design a repair scheme, and prepare the repair area for bonding. However, in this case repair materials are laid in dry instead of being pre-wet with resin. Part of the repair design strategy is to create pathways for channeling the resin and to make sure it flows completely through the damaged area. The repair area is then sealed, and vacuum is applied. Be sure to use epoxy resins formulated for infusion (lower viscosity) and with slow

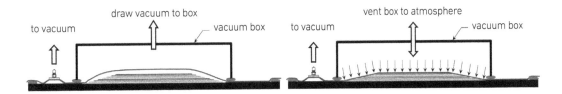

FIGURE 13-27. Double vacuum debulk concept.
Drawing slightly more vacuum on the box *(left)* creates the pressure differential that allows the bag to relax and resin in the repair plies to degas. After degassing, the box is vented to the atmosphere and the patch is compacted for a period until the layers are consolidated *(right)*.

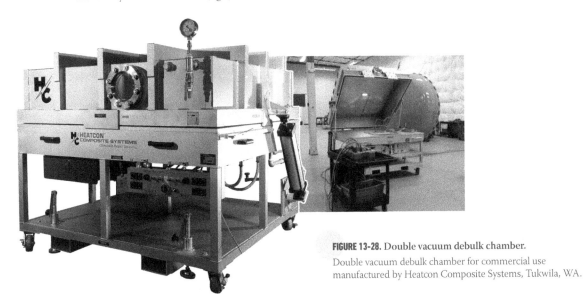

FIGURE 13-28. Double vacuum debulk chamber.
Double vacuum debulk chamber for commercial use manufactured by Heatcon Composite Systems, Tukwila, WA.

cure times. This enables the resin to completely penetrate the confined spaces of the repair area before it hardens. At the same time, it is important to stay with a system that matches the properties of the original laminate.

Resin infusion is especially useful for repair of large structures (e.g., boat hulls) and with sandwich structures using balsa or foam core. It can significantly reduce the repair time and eliminates the mess of the traditional wet layup repair. However, resin infusion repairs require careful planning and significant experience to successfully balance the many variables involved—vacuum, resin pathways, resin cure time, part configuration, etc. (Figure 13-29)

Tapered Scarf Repair

After completing initial damage removal, the area around the repair must be prepared. The corners of the damage removal hole must be rounded off and the area beyond this should be tapered to provide the best load transfer when the repair patch is bonded in. **Scarfing**, or **taper sanding**, is usually achieved using a compressed-air-powered, high-speed grinder. This is a gentle process, which prepares the damaged area for application of a repair patch. It is imperative to follow all repair manual guidelines, and significant skill and practice on the part of the repair technician is mandatory.

Note: If the damaged area exceeds the allowable repair limits in the governing repair manuals, then specific engineering support is required to proceed with the repair.

Tapered (scarf) repairs are typically circular or oval but can be any shape. There are two ways to specify how much area should be tapered around the damage: (1) tapered-scarf distance per ply, or (2) tapered-scarf angle.

❯ TAPERED SCARF FIXED DISTANCE PER PLY

A rough rule-of-thumb is to taper sand approximately ½-inch of area per ply of composite laminate. In the example below, following this rule would result in scarfing ½-inch for each of 4 plies to produce a scarf distance of 2 inches. Scarf distance is measured from the inner edge of the scarf to the outer edge of the scarf. The inner edge of the scarf begins where the damage has been removed. Counting the hole diameter of 0.5 inch, the scarf diameter would be 4.5 inches. The scarf diameter is the diameter of the whole scarfed area. Typical scarf distances range from 12 to 120 times the thickness of the removal. (Figure 13-30)

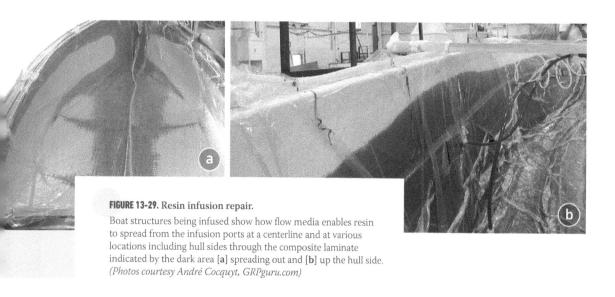

FIGURE 13-29. Resin infusion repair.
Boat structures being infused show how flow media enables resin to spread from the infusion ports at a centerline and at various locations including hull sides through the composite laminate indicated by the dark area [a] spreading out and [b] up the hull side.
(Photos courtesy André Cocquyt, GRPguru.com)

❱ TAPERED SCARF ANGLE (SCARF RATIO)

The tapered or scarf area may also be specified by a scarf angle, rather than distance per ply. Scarf angle is defined as the scarf distance divided by the thickness of the laminate and is always expressed as a ratio (e.g., 12:1, 20:1, 40:1, etc.), where "1" equals the laminate thickness. In the Figure 13-31 example, a 0.25-inch thick laminate is to be tapered at a 20:1 scarf angle ratio. To calculate the scarf distance required, simply multiply the laminate thickness by the scarf angle (0.25-inch x 20 = 5 inches).

The steeper the scarf, the less undamaged material is removed. Lightly loaded structures may be able to tolerate a smaller, steeper scarf. A typical scarf angle for lightly loaded non-aerospace structures is 12:1 (~5°). The flatter the scarf (more area per ply), the larger the adhesive bond is and the lower the load per square inch on the bond. Heavily-loaded structures usually require a larger, gentler scarf. For example, aerospace structures require a scarf angle of at least 20:1 (~3°), and thin, heavily-loaded primary structures may require a scarf angle of 50:1 (~1.3°).

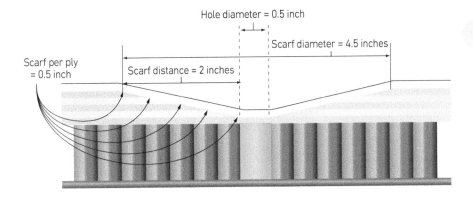

FIGURE 13-30. Tapered scarf distance.

The amount of material removed during scarfing may be specified by scarf distance per ply. Scarf distance is measured from the edge of the removed damage to the outer edge of the scarf, while scarf diameter is measured across the entire scarf area.

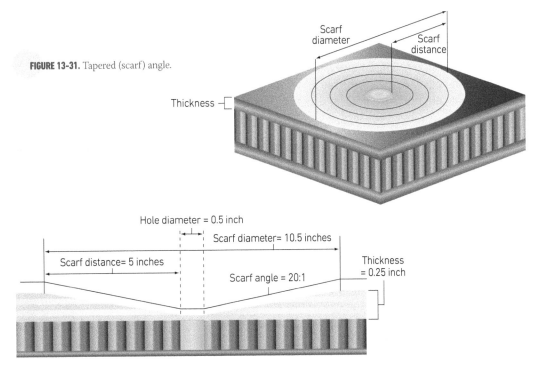

FIGURE 13-31. Tapered (scarf) angle.

Scarf distance = 0.25 inch (laminate thickness) × 20 = 5 inches

❱ TAPERED SCARF VS. STEPPING

Stepping is an alternate but less desirable method for removing material in preparation for a repair patch. Stepping is achieved by removing "steps" of material at the depth of each ply of the original composite laminate. The result is a "terraced" slope as opposed to a tapered slope in a tapered scarf. Stepping is more difficult to perform and less forgiving, in that each repair ply must fit perfectly into the recess without gaps or overlaps of the step.

The process of machining the step presents a high risk of damage to underlying plies because it is very difficult to cut cleanly through one ply without damaging fibers in the adjacent ply below. Also, great care must be taken not to extend delamination as each ply is dug out or peeled away. The sharp edges of the steps also lead to abrupt stress concentrations in the repaired laminate, increasing the possibility of failure. This has led to reduced use of this technique and a preference for tapered scarf repairs. (Figure 13-32)

❱ AUTOMATED SCARF REMOVAL

In recent years automated equipment has emerged as an alternative to manual taper-scarf and step removal operations. Both 5-axis machines and robotic adaptions have been deployed that show promise in providing precise material removal with less risk of damage by operators to the surrounding structure. (Figures 13-33, 13-34)

In addition to machining, automated kitting or placement of patch materials are possible. Work in this area continues and will likely become mainstream in the future.

The Repair Patch

Plies for the repair patch are precisely cut to fit the prepared repair area. Each ply in the repair patch replaces each ply in the original laminate, on axis, restoring fiber loads in the repaired structure. Thus, the number of plies and orientation must match, layer for layer, to that of the original structure. Each ply in the repair patch overlaps the corresponding layer in the scarfed repair area along the tapered surface. (Figure 13-35)

An exact ply-for-ply replacement to the repaired structure is typically about 70–80% as strong as the original undamaged structure. Is it possible to make a repaired structure as strong as the original? Yes. However, extra repair plies can be added to compensate for the loss of strength. This means the repair will not be perfectly flush, and that the repair will be somewhat stiffer than the original structure.

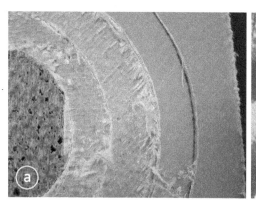

FIGURE 13-32. Scarfing vs. stepping. [a] Step-removal on Kevlar skin. **[b]** Tapered (scarf) on carbon skin.

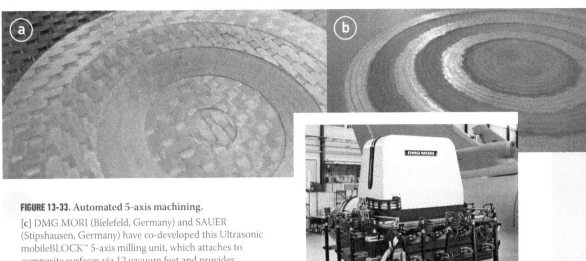

FIGURE 13-33. Automated 5-axis machining.

[c] DMG MORI (Bielefeld, Germany) and SAUER (Stipshausen, Germany) have co-developed this Ultrasonic mobileBLOCK™ 5-axis milling unit, which attaches to composite surfaces via 12 vacuum feet and provides multiple functions, including laser surface scanning, ultrasonic milling and plasma surface treatment. Examples of [a] a step and [b] a tapered scarf produced by this unit. *(Source: DMG MORI/SAUER)*

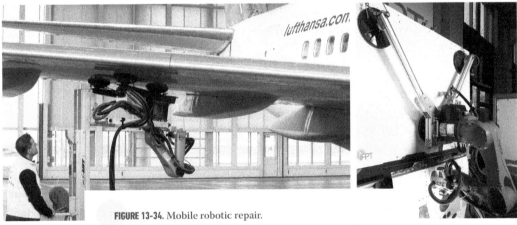

FIGURE 13-34. Mobile robotic repair.

Lufthansa Technik's mobile robotic repair system reportedly cuts repair time by 60% while enabling bonded patch repairs previously not possible or simply too risky to attempt with conventional manual methods. *(Photos courtesy of Lufthansa Technik AG)*

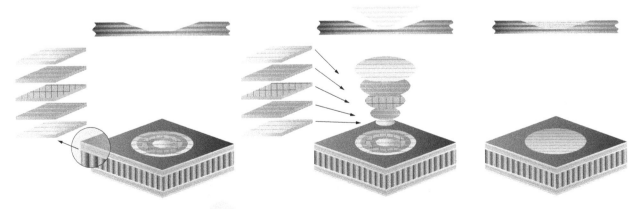

FIGURE 13-35. The repair patch must match the original structure, ply for ply.

Is the extra stiffness a problem? If the original structure is not stiffness-critical and is primarily loaded in straight tension or compression, then a stiffer repair will most likely be fine. However, if the structure flexes significantly under load, the stiffened area of the full-strength repair may contribute to failure of the repair. Some repairs may therefore need to be deliberately under-strength to match the stiffness of that original structure. (Figure 13-36)

Vacuum Bagging Materials for Composite Repair

How to sequence and apply the variety of materials used in vacuum bagging is one of the most challenging aspects in performing composite repair. The sequence of materials is called the vacuum bagging schedule (or simply, bagging schedule).

❯ VACUUM BAGGING REQUIREMENTS FOR REPAIRS

To obtain a good vacuum, certain requirements must be met:
- The vacuum bag must hold a prescribed minimum vacuum.
- Leaks must be eliminated or reduced within a specified range.
- The vacuum pump must accommodate the volume of the bagged area.
- A vacuum gauge should be used to check that the required vacuum is achieved.

The **breather** is another important factor in achieving good vacuum. The basic rules are:
- There should be a continuous breather layer across the bagged area (*see* Figure 13-37).
- The breather should not be allowed to fill up with resin or seal-off flow.
- The bleeder and breather layers should make contact at the edges or through perforations in the separator layer of film.

If both the bleeder and breather layers fill up with resin, the dynamics beneath the vacuum bag change. Instead of having atmospheric pressure exerting down-force on the laminate consolidating the plies, the result is a mass of liquid resin under pressure (a **hydrostatic** state).

FIGURE 13-36. Extra stiffness—when the repaired structure flexes significantly under load, a repair that is stiffer than the original structure may fail at the edges.

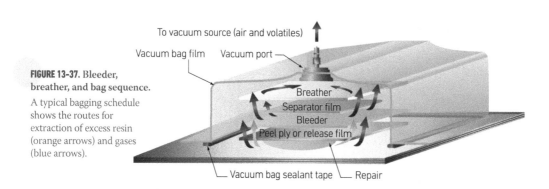

FIGURE 13-37. Bleeder, breather, and bag sequence.
A typical bagging schedule shows the routes for extraction of excess resin (orange arrows) and gases (blue arrows).

To vacuum source (air and volatiles)

Vacuum bag film   Vacuum port

Breather
Separator film
Bleeder
Peel ply or release film

Vacuum bag sealant tape   Repair

In this case, the plies are free to float within the resin mass, rather than being consolidated. For this reason, a separator layer is used between the bleeder and breather layers, so that the excess resin being pulled into the bleeder layer does not saturate the breather.

❯ PEEL PLY

Peel ply is a lightweight fabric—typically Nylon or polyester—that is applied on top of the repair patch. After the repair has cured, the peel ply is removed, leaving a smooth surface which is then easy to prepare for painting or bonding. If the repaired area is to be prepared for subsequent secondary bonding, it is important to choose a peel ply fabric that leaves a clean, energized surface.

Multiple layers of peel ply can also be used as an adjustable, absorbent bleeder stack, with or without additional layers of glass fabric. (Figure 13-38)

❯ BLEEDER LAYER

The bleeder layer is used to absorb resin from the laminate either through a porous peel ply, or a perforated release film as described below. The bleeder layer is usually a single (or multiple) layer(s) of dry 120-style or 1581-style glass fabric. Multiple layers can be added to accommodate heavier resin bleed requirements. The bleeder layer extends beyond the edge of the repair lay-up and is secured in place with flash-breaker tape as required. (Figure 13-39)

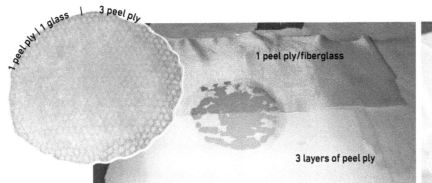
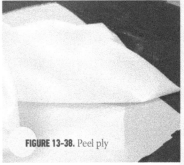

FIGURE 13-38. Peel ply

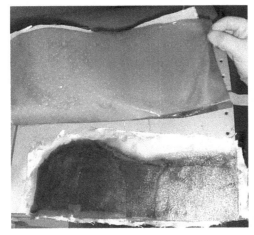
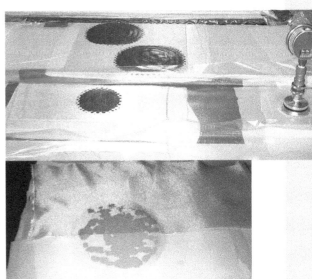

FIGURE 13-39. Bleeder.
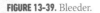

❯ SEPARATOR FILM LAYER

The separator layer is used between the bleeder layer and the subsequent breather cloth to restrict or prevent resin flow. This can be either a solid or perforated release film that extends to the edge of the layup but stops slightly inside the edge of the bleeder layer, to allow a gas path to the breather and vacuum ports. (Figure 13-40)

❯ PERFORATED FILM

This is a release film that has been perforated with a uniform hole pattern. It can be used as an alternative to solid release film as a separator film. It may also be used as a filter between multiple bleeder layers to slow the movement of resin away from the repair. Perforated film is available off-the-shelf with a variety of hole sizes and distance between holes. (Figures 13-41, 13-42)

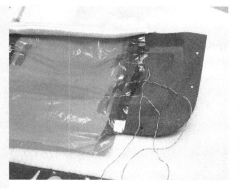

FIGURE 13-40. Separator film layer.

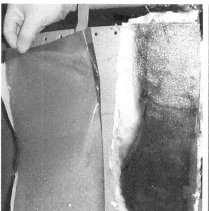

FIGURE 13-41. Perforated film.

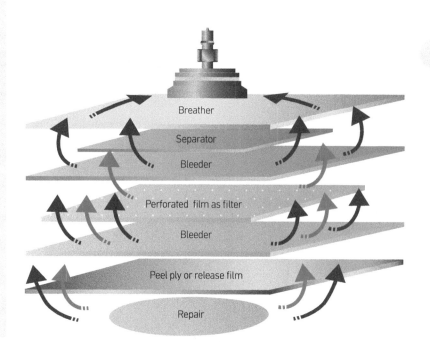

FIGURE 13-42. Perforated film as filter.
Perforated film can be used as a filter between two bleeder layers, in one of many bagging sequence variations. The orange arrows show resin flow; the blue arrows show extraction of air and volatiles. This is just one example of the numerous bagging schedule variations that are possible, each variation being used to achieve specific processing and cure objectives.

❱ BREATHER LAYER

The breather layer is used to maintain a "breather" path throughout the bag to the vacuum source, so that air and volatiles can escape. It also enables continuous pressure to be applied to the laminate under the bag. Typically, synthetic fiber materials or glass fabric is used for this purpose.

The breather layer extends past the edges of the repair layup and heat blanket (if applicable) so that the edge band contacts the bleeder ply around the edges, or through small holes in the separator film. The vacuum ports are connected to the breather layer adjacent to the repair area. It is especially important that adequate breather material be used in the autoclave at pressure. (Figure 13-43)

❱ BAG FILM AND SEALANT TAPE

The bag film is used as the vacuum membrane that is sealed at the edges to either the surface of the part being repaired, tool surfaces, or to the bag film itself when an envelope bag is used. A rubberized sealant tape is used to provide the seal. The bag film layer is generally much larger than the area being bagged, as extra material may be required to form pleats at all the inside corners and about the periphery of the bag as required to prevent bridging.

Bag films are made of Nylon® (or Kapton® for high temperatures). They come in a variety of thicknesses, elongations and temperature ratings. (Figure 13-44)

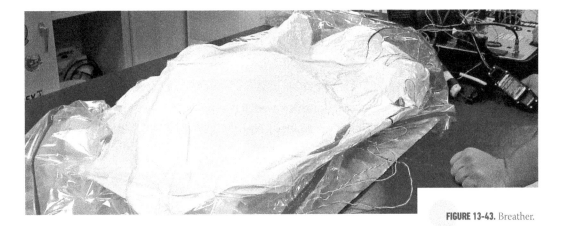

FIGURE 13-43. Breather.

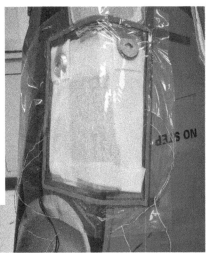

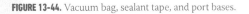

FIGURE 13-44. Vacuum bag, sealant tape, and port bases.

❱ VACUUM PORTS

Vacuum ports are used to connect the vacuum bag to the vacuum source. Quick-disconnect fittings attach to the vacuum ports, enabling vacuum hoses to be easily connected and disconnected without losing vacuum in the bag. The ports are installed as follows:

1. Place the base plates on top of the breather cloth at desired locations. Cover area with bagging film, cutting a small slit through the film directly over each base plate.
2. Lock the port assemblies into each base plate through the slits in the bagging film. The gasket in between enables an air-tight seal through-hole in the bagging film.
3. Connect the vacuum hoses into the quick-disconnect fittings and turn on vacuum pump.

A general rule of thumb is to use a minimum of two vacuum ports and an additional port for each additional square yard of surface area being vacuum bagged. More ports allow for greater volume extraction of air and volatile gas from the repair patch.

❱ VACUUM LEAKS

It is recommended to check for the leaks in the bag by pulling at least 20 inches Hg while watching the vacuum gauge for 2 to 5 minutes as the leaks are sealed prior to curing the repair. An ultrasonic leak detector with headphones is a relatively cheap way to double-check vacuum integrity and track down leaks—a few hundred dollars can save thousands in a failed repair (*see* Figure 13-46). This time-saving device amplifies the ultrasonic frequencies of leaks while filtering out background noise. Sweep the perimeter of the bagged area with the device probe, listen for the rushing sound of leaks through the device headphones, and check the LED display which indicates leak strength. It may be impossible to have a leak-free bag but minimizing the leak rate is very important and can be accomplished with persistence.

❱ VACUUM GAUGES/MONITORS

Vacuum gauges show the actual pressure achieved at the repair. The ports for these should be located adjacent to the repair, but on the opposite side from the vacuum port. Vacuum gauges are typically removed during high-temperature cures, as they can be damaged by the high temperatures. However, vacuum monitors with hoses to the hot bonder can run along with the cure cycle. Take caution not to overheat heat the hose. (Figure 13-47)

FIGURE 13-46. Checking for vacuum leaks—ultrasonic leak detectors help to pinpoint vacuum bag leaks.

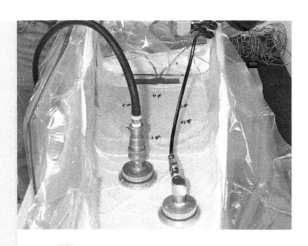

FIGURE 13-45. Vacuum and monitor ports and quick disconnect fittings.

FIGURE 13-47. Vacuum gauges should be located on the opposite side from the vacuum port *(left).*

Curing Methods and Equipment

❯ HOT BONDERS

Portable repair equipment, or "hot bonders," are produced by a variety of manufacturers. Most hot bonders are suitcase-sized or smaller and are used to control (and document) the application of heat, and sometimes vacuum, to a composite repair. They are especially useful for field repairs, in situations where it is not possible to remove the damaged part for repair. They also can be used to monitor temperature for oven-cured repairs.

Hot bonders are most commonly used to control heat blankets, but can also be used to control heat lamps, hot air machines, heat guns, radiant heaters, or other heat sources. Most can be programmed to store multiple cure cycles in memory. They require an electrical power source and control the heating through a thermocouple feedback loop. (Figure 13-48)

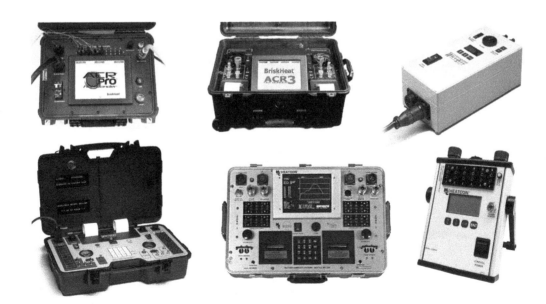

FIGURE 13-48. Hot bonders.
BriskHeat ACR3 dual zone and Mini-Pro single zone hot bonders *(top left and center)*; WichiTech HB-1 single zone *(top right)* and HB 2 dual zone *(bottom left)* hot bonders; Heatcon Composite Systems dual-zone HCS9200B *(bottom center)* and HC8806 Micro Bonder *(bottom right).*
(Photos courtesy of BriskHeat Corporation, WichiTech Industries, Inc., and Heatcon Composite Systems)

Hot Bonders Rely on Thermocouples

A thermocouple consists of two wires of different metals that are welded together at one end and joined at the other end by a plug. The wires at the plug end are held at a fixed lower temperature. When the wires at the welded end are heated, an electromotive force (EMF) is produced, which varies according to the temperature difference between the welded (hot) and the plug (cold) ends of the circuit. A voltmeter in the hot bonder translates this voltage into a very accurate temperature reading—up to 700°F for the "J"-type thermocouples most commonly used with hot bonders.

Thermocouples are placed around the repair area to monitor the temperature of the repair. Based on those temperature readings, the hot bonder's controller processor determines whether to turn the heat source (e.g., heat blanket, oven, heat lamps, etc.) ON or OFF to adjust the temperature at the thermocouples and maintain the programmed cure cycle parameters. (Figures 13-49, 13-50, 13-51)

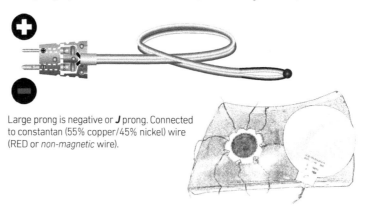

Small prong is positive. Connected to iron wire (WHITE or *magnetic* wire).

Large prong is negative or **J** prong. Connected to constantan (55% copper/45% nickel) wire (RED or *non-magnetic* wire).

FIGURE 13-49. Thermocouples.

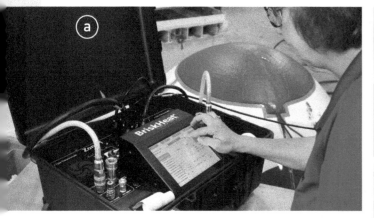

FIGURE 13-50. Heat lamps and blankets.

[a] Custom molded heat blankets are available for use on compound or complex geometries.

[b] A composite repair using a hot bonder and heat lamps.

[c] Heat blankets come in a variety of different shapes and sizes.

(Photo [a] courtesy of BriskHeat Corporation)

Cure Cycle Recipes

The cure cycle recipe is used to drive the cure of the repair materials. It is programmed into the hot bonder. Many hot bonders can store multiple cure cycles in memory, allowing operators to select the desired cure recipe from a list and then check the graph to ensure that the right recipe has been selected. Cure recipes tell the hot bonder how fast to heat (ramp up), how long to hold the cure temperature (soak), and how fast to cool down (ramp down). Cure recipes will also specify the low vacuum (alarm) boundary. Most have a single soak temperature; however, special recipes may require multiple soaks.

A multiple-zone hot bonder (usually two zones) can run more than one cure at a time, or control a single repair cure over a larger than normal area using two or more zones. What constitutes a "large area repair" can range from an area greater than 8 square feet and is typically limited by the size of heat source available (e.g., heat blanket or heat lamps). Note: 120-volt heat blankets are limited to about 5 square feet on a 30-amp circuit; 240-volt blankets can cover about a 10-square-foot area. (Figures 13-52, 13-53)

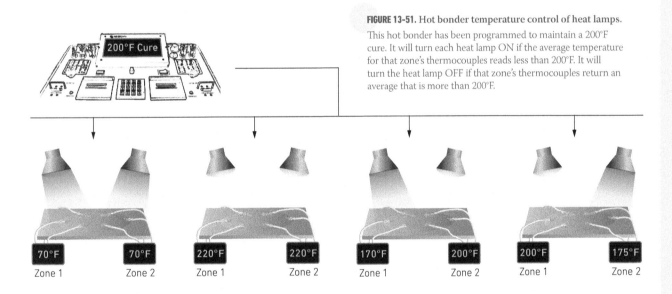

FIGURE 13-51. Hot bonder temperature control of heat lamps.

This hot bonder has been programmed to maintain a 200°F cure. It will turn each heat lamp ON if the average temperature for that zone's thermocouples reads less than 200°F. It will turn the heat lamp OFF if that zone's thermocouples return an average that is more than 200°F.

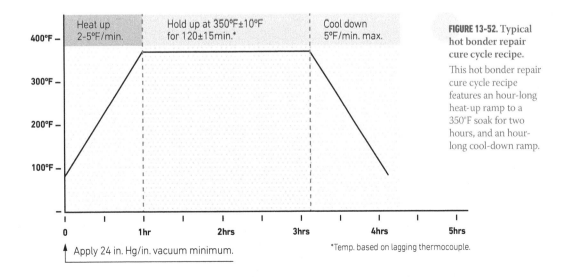

FIGURE 13-52. Typical hot bonder repair cure cycle recipe.

This hot bonder repair cure cycle recipe features an hour-long heat-up ramp to a 350°F soak for two hours, and an hour-long cool-down ramp.

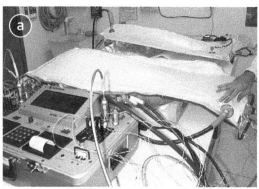

FIGURE 13-53. Dual-zone hot bonders.

[a] Repair of two composite parts using a two-zone hot bonder, [b] which prints a log record for each zone, i.e., a separate log record for each of the two parts being repaired.

In addition to a power source, many hot bonders also require a clean compressed air source to power a built-in venturi and generate vacuum for the repair during cure (not required when a separate vacuum pump is used for the repair).

Hot bonders are useful in controlling the curing process for a composite repair, but they are not cheap. Costs typically range from $4,000 to $40,000 or more. Also, by definition, using a hot bonder means an elevated cure temperature. This requires well-trained personnel who are familiar with the many issues involved in implementing and controlling elevated-temperature composite repairs. A hot bonder should always be monitored by the operator, at least within earshot of the alarms.

Hot Bonder Features

Keypad or touch screen display: The keypad or touch screen display enables the operator to input commands and program a variety of different features, including cure cycle parameters, alarms, and repair zones. The display guides the operator in monitoring the curing process. Many units will constantly display the cure status in both graphic and text form and will also provide text descriptions for alarms. (Figures 13-54, 13-55)

Chart recorder: The chart recorder prints a record of each repair cure cycle, providing evidence that the correct temperature and vacuum was used throughout the repair. This is required for FAA documentation. Print-out data can include time, date, cure cycle description, tag number,

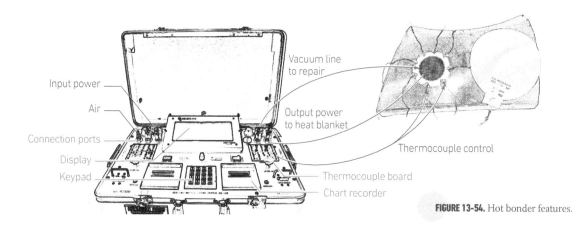

FIGURE 13-54. Hot bonder features.

FIGURE 13-55. A hot bonder display showing a graphic of the cure cycle being used.
(Photos courtesy of BriskHeat Corporation)

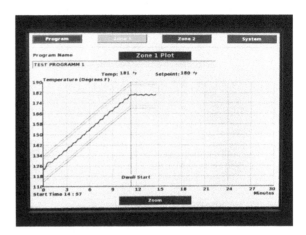

part number, operator ID, actual temperature and vacuum at selected intervals, graph of cure showing duration and temperature, description of alarms and alarm clear, and in-process program changes.

Digital record: Many new-generation hot bonders also provide for transfer of cure data via USB ports to thumb drives or directly to a database. The data can be read as Excel spreadsheet format or converted to other file types for future preservation.

Alarms: Alarms alert the operator when cure cycle parameters are not being met or when conditions arise which may threaten the repair. These include low temperature, low vacuum boundary, over-temperature, thermocouple failure, heat blanket failure, and power interrupt.

Vacuum Pump: Most bonders contain an internal venturi that transforms compressed air into a vacuum of 22"–28" Hg, depending upon altitude. Some hot bonder models use an internal electric pump instead of a venturi.

Connection Ports: Connection ports on a hot bonder typically include input power (power supply), output power (to heat source), air intake (compressed air or plant air), vacuum line (to repair), and thermocouple control board.

❱ HEAT BLANKETS AND OTHER HEATING SOURCES

Accurate temperature control and uniformity are difficult with heat lamps and heat guns. Temporary ovens, hot air machines, and heat blankets offer better control of curing temperature and even distribution of heat.

Heat blankets must be considerably larger (a minimum of 4 inches greater) than the area being cured. For example, an 8-inch diameter circular repair will require at least a 12-inch diameter heat blanket. Using this size blanket allows for thermocouple placement well inside the blanket edges.

Controlling thermocouples placed near the edge will make the blanket overheat, causing a very real fire risk and a damage risk to the component. Also, temperatures drop within 2 inches of the blanket's edge. The very edge of the blanket can have temperatures that are easily 100°F colder than at its center. This is often referred to as the heat blanket's "cold zone." (Figures 13-56, 13-57)

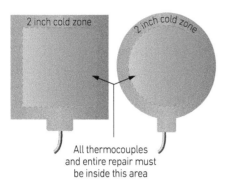

FIGURE 13-56. Thermocouple placement.

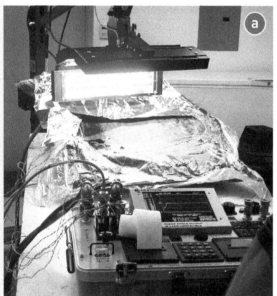

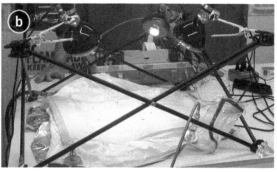

FIGURE 13-57. Repair heating methods.
Composite repairs may be cured using [b] heat lamps and [d] hot air machines or [a] by using a radiant heater; [c] however, heat blankets may provide the best control and distribution of temperature.

Smart Susceptor Technology

The Smart Susceptor system uses an inductive heating circuit within the heat blanket to uniformly distribute heat to cooler areas and self-limit heat to hot areas under the blanket. This can be particularly useful when heating solid laminate composite parts that incorporate elements such as stiffeners, stringers, or other underlying metal hardware in the structure.

The way it works is that the inductor circuit is wrapped with a susceptor alloy that absorbs electromagnetic energy and converts it to heat. To further increase thermal uniformity at the desired cure temperature, a susceptor alloy is selected with a Curie temperature that limits heating within the heat blanket circuit. When applied to thermally complex composite structures, this has a balancing effect that allows the Smart Susceptor heater to overcome the limitations of traditional resistive heat blankets, resulting in a uniform temperature gradient across the repair. (Figure 13-58)

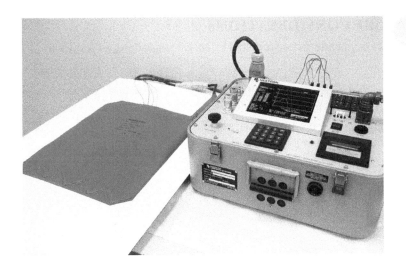

FIGURE 13-58. HCS9400-02 Smart Susceptor™ technology provides thermal uniformity across the heat blanket area to best deal with localized heat sinks during composite repair. *(Photo courtesy of Heatcon Composite Systems)*

❱ AUTOCLAVE PROCESSING

Autoclave processing of repairs is conducted in the same manner as that for manufacturing. The bagging schedule is likely to be different for the autoclave then that of vacuum bag only processes. Equipment manufacturers such as Heatcon Composite Systems (Tukwila, WA) make a semi-portable autoclave that relies on traditional heat blankets for the heat source. The autoclave controller works the same way as a hot bonder and requires little cross-training to use. These autoclaves are custom built with standard vessel lengths starting at 3 feet up to 30 feet, progressing in 3-foot increments. (Figure 13-59)

FIGURE 13-59. Heatcon HCS3100 RepairClave™ for repair processing. *(Photo courtesy of Heatcon Composite Systems)*

❭ CURE TEMPERATURE CONSIDERATIONS

There are five standard cure temperature requirements listed, with variations, as shown in Table 13.3.

For room-temperature-cured laminating resins, post-cures are often required in order to develop full strength within a reasonable time. Often, but not always, prepregs cured at higher temperatures are stronger than room temperature cured materials. However, high curing temperatures can cause problems:

- Creates steam in laminates and over cores,
- Blown skins,
- Excessive porosity in bondline,
- Uneven heating problems,
- Overheating/fire risks,
- Increased documentation,
- Additional training required.

TABLE 13.3 Common Cure Temperatures

Cure Temperature	Repair Materials
Room temp – 77°F (25°C)	Epoxy Wet Layup
Cure or post-cure at 150°F (66°C) – 200°F (93°C)	Epoxy Wet Layup
200°F (93°C) – 250°F (121°C)	Epoxy Prepregs
350°F (177°C)	Epoxy Prepregs
375°F (191°C) – 400°F (204°C)	BMI or Benzoxazine Prepregs

Often, repair instructions may offer a choice, especially with smaller repairs. In this case, go with the lowest-temperature repair allowed by the manual or guidelines. It may offer an easier repair with fewer risks associated with elevated temperatures. If there is no choice, then it is imperative to do exactly what is prescribed in the manual or guidelines, or as directed by engineering instructions.

❭ THERMOCOUPLE-RELATED ISSUES

Proper thermocouple placement is crucial for proper high-temperature repairs using hot bonders. With any heat source, multiple thermocouples are required. The general guideline is the more thermocouples, the better, because thermocouple failures are common and replacement during a running cure is virtually impossible.

Another reason for using numerous thermocouples is to control temperature spread across the repair. The goal is to supply uniform heat to the repair area and avoid hot or cold spots. Temperatures can vary widely across a repair during cure due to a variety of conditions. Spreads of 60°F–80°F are not uncommon and can be caused by the factors below.

Variations in Part Thickness and Conductivity

Most composite parts contain variances in thickness and/or thermal conductivity of the materials. Thicker laminates typically absorb heat while thinner surface skins will shed it rapidly. This can result in a wide temperature variation during cure.

Heat Sinks

There may also be heat sinks in the part or underlying tooling. For example, thick stringers or spars in a structure will take longer to absorb heat on the ramp up and will also retain heat longer on the cool down. Heat sinks may require an additional heat source to obtain the proper cure temperature.

Non-uniformity in Heat Blankets

Heat blankets are made by embedding wires between two layers of silicone rubber. As heat blankets are conformed to complex shaped parts, wires may be stretched around a corner. This area of stretched, thinner wire will create a hot spot in the heat blanket.

Thermocouple Placement—4 Rules

Here are four basic rules for installing thermocouples around a repair:

1. A thin aluminum (or CFRP composite) caul plate will help maintain uniform temperature over the repair. Place the caul plate over the repair/bleeder stack and below the heat blanket to equally distribute the heat from the heat blanket to the repair and the thermocouples. (Figure 13-60) Thermocouples should be taped to the backside surface of the caul plate. *Never place thermocouples under a caul plate.*

2. Always provide a heat blanket that is a minimum of 4 inches larger than the maximum repair size. Locate the thermocouples well inside the heat blanket. This will prevent errors in temperature readings and problems that can result from being in the 2-inch cool zone at the blanket's edge. However, if **not** using a solid caul plate, do not place thermocouples directly over the repair area as they will leave an imprint or "mark-off" in the repair that results in an ineffectual repair.

3. Always place a layer of flash tape under the head of each thermocouple around the repair to insulate the thermocouple from the conductive (carbon or metal) surface. This will prevent thermocouple cross-talk and small spikes in the thermocouple readings. Also, tape directly over the head with one layer of flash tape to secure each thermocouple in place.

4. Always use a uniform amount of bleeder/breather materials under the heat blanket, and insulation above the heat blanket (if applicable). Never double-up the insulation just to fit it inside the bag. This could provide more insulation over one thermocouple and less over another. Following these basic tips will reduce spreads in thermocouple temperature readings, help to prevent erroneous readings, alarms and runaways, and result in a better repair and documentation of the cure process.

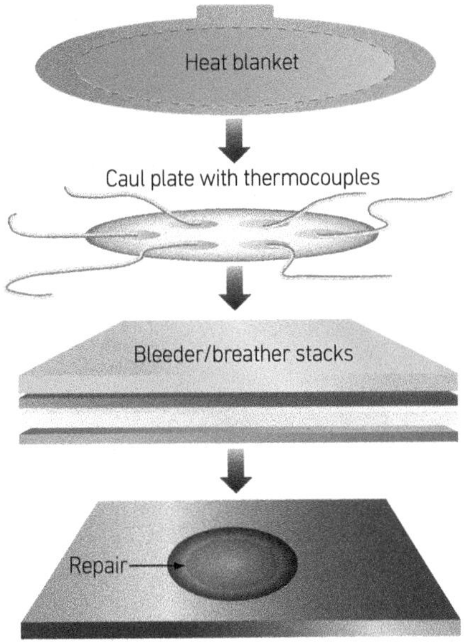

The next several sections will further explain why the above four rules are so important.

FIGURE 13-60. Illustration of Rule #1.

❱ CAUL PLATES

Caul plates are smooth metal or CFRP plates, slightly larger in size and the same basic shape as the composite repair. This is used to distribute uniform pressure and temperature across the repair during cure. (Figure 13-61, on the next page)

As was illustrated in Figure 13-60, caul plates are placed on top of the bleeder stack and provide a smooth surface for the finished laminate. They also enable placement of thermocouples closer to the center of the repair. (Figures 13-62, 13-63)

Aluminum caul plates can be difficult to form to complex contours. Also, metal caul plates can cause errors in temperature readings and false alarms as well, unless insulated with flash breaker tape (a.k.a. flash tape). Silicone rubber caul plates do not conduct electrical current like metal plates, but caution still must be taken to prevent thermocouple mark-off. CFRP caul plates can be molded to any contour and have a much lower CTE than aluminum or rubber, an important consideration for repairs to a leading edge or other protruding shape.

See "Cross-talk" for information about isolating thermocouples from conductive caul plates. (Figure 13-65)

FIGURE 13-61. Caul plates—two views of the same repair show caul plate size and thickness.

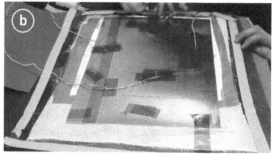
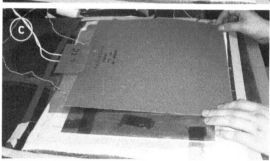

FIGURE 13-62. Caul plates.

[a] This repair begins with a repair patch, followed by [b] a sequence of vacuum bagging materials, including bleeder and peel ply, and then a rectangular caul plate. [c] A heat blanket sized larger than the repair area is then applied.

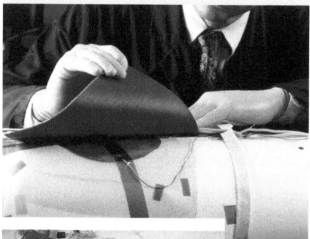

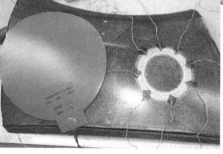

FIGURE 13-63. Illustration of Rule #2.

FIGURE 13-64. Illustration of Rule #4.

❱ CROSS-TALK

Thermocouple "**cross-talk**" occurs when very low voltage currents develop between thermo-couples that are taped directly to a conductive surface. The *cross-talk* current causes minor, very rapid fluctuations in the temperature readings, which then could spur the hot bonder to incorrectly add or subtract heat to the repair. This typically results in an alarm being triggered. In the worst case, this can cause a sub-standard repair. In almost all cases, it will cause alarms to print out on the log records, which then must be explained to quality assurance and regula-tory authorities. The best practice is to prevent cross-talk by insulating thermocouples on any conductive surfaces. (Figure 13-65)

Flash tape is used directly under the thermocouples to prevent them from contacting a conductive surface and promoting cross-talk between thermocouples. Place flash tape on top to secure the thermocouple to the part or caul plate. CFRP and aluminum surfaces are conduc-tive, glass and aramid surfaces are not. Isolating the thermocouple via flash tape also prevents resin from migrating up the thermocouple wire, thus minimizing clean-up afterward.

IMPORTANT: Put flash tape under the thermocouple to avoid spikes in temperature readings. If the thermocouple makes contact with an aluminum caul plate, it introduces a third alloy into the thermocouple's constantan-iron circuit, which may affect the calibration of the instrument and could cause erroneous readings.

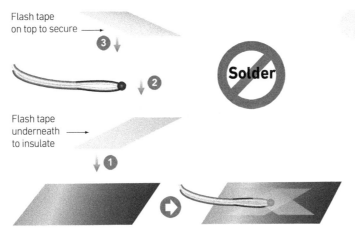

Flash tape on top to secure →
❸
❷
Solder
Flash tape underneath → to insulate
❶

FIGURE 13-65. Insulating thermocouples.

❱ REVERSE-WIRED THERMOCOUPLES

It could be possible to burn a part by wiring thermocouples backwards. Connecting the constantan wire to the positive plug (red wire to smaller plug) and the iron wire to the negative or J-plug (white wire to larger plug) will cause reverse polarity in the circuit. In other words, when the temperature goes up in the repair, the hot bonder will read it as going down.

This becomes dangerous very quickly, as some hot bonders might read a decreasing temperature—even though it is actually rising—and so it applies more heat, which makes the incoming readings look like the part is cooling down even further, so the hot bonder adds more heat, etc., until the part ignites. Most new-generation hot bonders have software and/or hardware protection against this event.

It is possible to purchase thermocouples that are already wired; however, if they are reused for multiple repairs they will eventually have frayed and broken wires and will need to be repaired or refabricated. Often, new thermocouple wire is purchased in long spools, and it will be up to the technician to cut them to length, weld the working end, and wire on the plugs. Thus, it is important to understand how to wire thermocouples correctly. (Figure 13-66)

❱ MICROWIRE SENSORS

As of this writing, development is underway for the use of microwire sensors that are embedded in the actual bondline between the adhesive and the repair. (Figure 13-67)

Microwire assemblies are small enough (0.010-inch diameter x 1.1-inch length) that they fall under the acceptable flaw level for most operations. Unlike thermocouples, these microwires have no wires running out of the repair area or through the vacuum bag seal. Instead, they use an antenna located outside of the bag that reads the magnetic field, discerning the **Curie temperature** of three separate wires that are in the sensor, and converting this information into common measurements of temperature (i.e., degrees C or F).

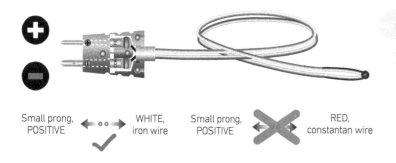

Small prong, POSITIVE ←○○→ WHITE, iron wire ✓

Small prong, POSITIVE ✕ RED, constantan wire

FIGURE 13-66. Reverse-wired thermocouples—a thermocouple can cause a repair to ignite if it is wired incorrectly.

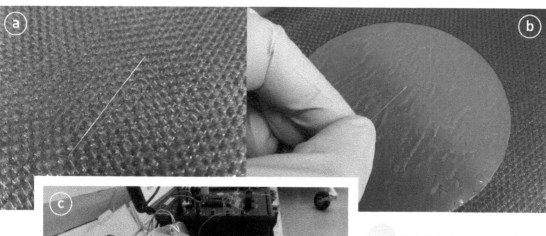

FIGURE 13-67. Embedded microwire sensor [a] on prepreg carbon fabric and [b] located on adhesive; [c] antenna located outside the vacuum bag reads the sensor and controls the process. *(Photos courtesy of AvPro, Inc., Norman, OK)*

Approach to a Repair

Below is a review of the general steps in performing a composite scarfed repair with a flush, bonded repair patch.

1. Get the best possible access to both sides if feasible.
2. Inspect for extent of damage:
   - visual.
   - tap, ultrasonic, thermographic, etc. (*see* Figure 13-68).

3. Remove all damaged and contaminated material , including core if applicable (*see* Figure 13-69):
   - remove damage in a circular or oval pattern, or radius corners as needed.
   - remove or treat contaminated material.

4. Determine the part's ply count, orientations, laminate thickness and materials in preparation for scarfing and repair. There are several choices depending on what information and materials are available:
   - Thorough repair instructions will detail exactly the number of plies, ply orientations and material specifications for the repair, as well as the scarf diameter for the repair area.
   - Ply count, ply orientations and material specifications may also be found in documentation from the original structure's manufacture (engineering drawings).
   - To determine plies, orientations and material practically, there are 2 options:
     – if a large enough piece of the damaged material is available, scarfing into that piece will reveal the number of plies, orientation, etc. The use of a magnifying glass is recommended to best view the materials.
     – otherwise, it will be necessary to taper sand into the area around the damage, roughly ½ inch for each ply of laminate thickness. Use calipers to determine the laminate thickness.

5. Taper-scarf the repair area, according to repair design instructions, to create a smooth, flat surface with high surface energy. *See* Figure 13-70.
6. Thoroughly dry the structure if moisture is present.
7. Develop a repair design, based on the damage and original structure information. Engineering support is usually required to ensure a successful repair.

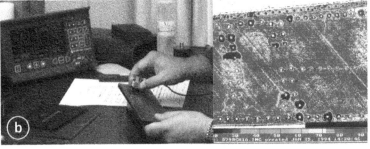

FIGURE 13-68. Nondestructive inspection.

[a] Tap testing and digital tap testing. [b] A-scan and C-scan ultrasonic scans can help to define the extent and degree of damage.

8. Replace materials. *See* Figure 13-71.

   a. Solid laminate:

     – adhesive layer first.

     – repair plies—match orientations with original structure.

     – extra plies—usually orientation matches original outer ply.

     – outer sanding layer if required (usually a film adhesive or one fiberglass fabric layer).

   b. Through-damaged sandwich structure:

     – adhesive layer.

     – one or more filler plies.

     – core material.

     – core splice adhesive around core plug.

     – inner skin plies.

     – outer skin plies (plus extra ply if required).

     – outer sanding layer if required.

Note: A two- or three-step process is recommended for sandwich panel structures where the outer skin is cured, the core is bonded then machined flush, followed by the inner skin repair and cure. The top and bottom skin repairs follow the solid laminate guidelines above.

9. Vacuum-bag and cure repair plies as required. *See* Figure 13-72.

10. Inspect repair. *See* Figure 13-73.

11. Sand and finish as required. Do not sand into fibers of repair plies.

FIGURE 13-69. Remove damaged material.
Damaged core is cut and scraped away. It may also be removed with an air-powered grinder.

FIGURE 13-70. Taper sanding/scarfing.

FIGURE 13-71. Replace materials.
[a] First, outer skin plies are replaced and cured, [b] and then core material.

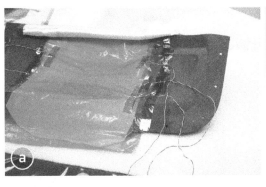

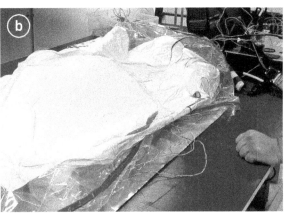

FIGURE 13-72. Vacuum bag and cure repair.

[a] Vacuum bagging materials are sequenced on top of the repair and [b] then sealed within the vacuum bag. The repair is then cured using heat and vacuum (atmospheric) pressure.

FIGURE 13-73. Inspect repair.

[a] Inner skin and [b] outer skin are inspected. Note use of repair tooling in restoring curvature of bottom skin surface.

14

Health and Safety Considerations

CONTENTS

Safety and the Industry

The subject of safety and advanced composite materials has always sparked tremendous debate. Since accurate data on these materials is often difficult to find, conjecture, rumor and hyperbole are often deeply rooted in an organization's safety policy where composites are concerned. To compound the problem, an organization may fall under the jurisdiction of more than one occupational health and safety entity, and these agencies may have differing opinions on regulations and priorities.

In the past decade, the industry has seen sweeping changes, particularly in the area of hazard communication. In 2012, the U.S. Occupational Health & Safety Administration (OSHA) began widespread implementation of the Globally Harmonized System of Classification and Labeling of Chemicals, or GHS. The GHS system of hazard communication is comprehensive and in use by over 70 countries worldwide.

It is important to understand the different aspects of safety to minimize the level of hazard employees are being exposed to. Key to this awareness are the roles of the industrial hygienist and the safety officer in setting policy that not only complies with the law, but also does not encumber the employee to the point where the quality of the work suffers.

The information contained here is not all-encompassing and should not be regarded as a replacement for safety information provided by the manufacturers of these products (such as a Safety Data Sheet, or SDS). Instead, this section is meant to give a broad overview of composite safety issues such as routes of exposure, and hazards associated with matrix systems, fibers, solvents, and nanomaterials. It will also discuss exposure limits and the role they play in establishing engineering controls, personal protective equipment, and setting organizational policy.

Routes of Exposure

Hazardous materials may enter the body in four different ways: absorption, inhalation, injection, and ingestion. Certain materials under the right circumstances may be able to enter the body via two or more of these routes. Conversely, one route of exposure may be used by two or more constituent materials to enter the body.

❱ ABSORPTION

The skin is a protective barrier that helps keep foreign chemicals out of the body. However, some chemicals can easily pass through the skin and enter the bloodstream. If the skin is cut or cracked, chemicals can penetrate through the skin more readily. Many chemicals, particularly organic solvents, dissolve the oils in the skin, leaving it dry, cracked, and susceptible to infection and absorption of other chemicals.

   The eye is particularly vulnerable to the absorption of chemicals as well. Many chemicals may burn or irritate the eye, but they may also be absorbed through the eye or the capillaries in the eyelid and then enter the bloodstream.

❱ INHALATION

An inhalation exposure occurs when a worker breathes a substance into the lungs. The lungs are made up of branching airways called bronchi. Clusters of tiny air sacs called alveoli are at the ends of these airways, and through the walls of these alveoli, gas exchange takes place. Gas exchange is the process whereby blood in capillaries outside the alveoli gives up carbon dioxide in exchange for oxygen. If other chemicals are present, they too may be absorbed into the bloodstream. If respirable fibers reach the alveoli, they can block the exchange of gas from the lungs to the bloodstream. *See* "Hazards Associated With Fibers" on page 387.

❱ INJECTION

This method introduces foreign material to the bloodstream or tissue by piercing the skin. It is relatively common in the field of composite repair since damage to structures often results in very sharp, stiff splinters. Wounds caused by splinters of this sort often get infected, but it is important to note that more often than not, contaminants on the splinter cause the infection and not the composite material itself.

   While limited, the use of hypodermic devices in both composite manufacturing and repair is also a consideration, because risk of accidental injection of a solvent, resin, or adhesive may exist. This area of risk and hazard is well documented and referenced in the medical community, which may be the best resource for information in this area.

❱ INGESTION

The least common source of exposure in the workplace is swallowing chemicals. Chemicals can be ingested if they are left on hands, clothing, or facial hair, or if they accidentally contaminate food, drinks, or cigarettes.

Hazards Associated with Matrix Systems

While there are ceramic as well as metal matrix materials, the vast majority of matrix systems used in the composites industry are polymeric. These can be further divided into two categories, thermoset and thermoplastic resins (*see* Chapter 2). Thermoplastics rely on heat to melt, or flow the resin in order to process it, while thermoset resins generally consist of two components that, when mixed and cured properly, form a solid.

〉 EPOXY

The most commonly used thermosets are the epoxy resins. They are made by reacting epichlorohydrin with a suitable backbone chemical, usually a bisphenol A or F resin, to produce diglycidyl ether of bisphenol A or F (DGEBA or DGEBF).

Chemical Irritants

These base resins are found in most epoxy formulations and are generally considered to have a low order of acute toxicity and are only slightly to moderately irritating to the skin. However, some resin formulations containing reactive diluents such as glycidyl ethers can range from moderate to severe as skin and eye irritants. In any case, direct contact with these materials in their liquid form, or their vapors, should be avoided.

Bisphenol-A (BPA) is classified as an endocrine-disrupting chemical due to its demonstrated ability to mimic estrogen and bind to estrogen receptors throughout the body. Estrogen mimicry has been shown to cause a variety of health issues. This has spurred an effort to develop chemical replacements for BPA in matrix materials that use it, including epoxies, polyesters, and polyinides.

Most of the hardeners for epoxies used today contain aromatic or aliphatic amine compounds. Aromatic amines such as methylene-dianiline (MDA), diaminodiphenylsulfone (DDS), and others are irritants to the eyes, skin, and respiratory tract. Symptoms of overexposure range dramatically from dizziness and headache to liver and possible retinal damage. Some amines, such as MDA, are also classified as Group 2 carcinogens or reasonably anticipated as being carcinogenic.

Like the base epoxy resins, these compounds have a very low vapor pressure and rarely present an airborne hazard unless the mixture is sprayed or cured at high temperatures. However, potential for dermal exposure remains high. Because of the toxicity of this class of hardeners, it is necessary to minimize or avoid exposure by all potential routes.

Several other types of curing agents are also used in the advanced composite industry. These include aliphatic and cycloaliphatic amines, polyaminoamides, amides, and anhydrides. The aliphatic and cycloaliphatic amines, such as diethylene triamine and triethylene tetramine, are strong bases and are usually severe skin and eye irritants. Many of these products can cause skin or respiratory sensitization. While the amide and polyaminoamide hardeners are generally less irritating to the skin, they are still considered potential skin sensitizers. There are two basic types of anhydride curing agents used in today's epoxies. Both types are severe eye irritants as well as strong skin and respiratory tract irritants. As with all materials mentioned in this section, the manufacturer's SDS should always be consulted prior to use.

Dust

Dust released from the sanding and machining of completely cured epoxy products is generally considered to be "nuisance" dust. The presence of filler materials such as fiberglass, calcium carbonate, powdered silica, lead, etc., may result in concentrations of airborne dust that are more toxic or irritating than nuisance dust. In addition, resins may not be completely cured for days after hardening. Therefore, inhalation of dusts from grinding, sawing, drilling, polishing, etc. of hardened but incompletely cured resins may cause allergic responses in individuals previously sensitized to uncured resins or curing agents.

Sensitization

Repeated exposure to certain constituents in resin systems may result in a condition known as *sensitization*. Sensitization, or allergic contact dermatitis, manifests as an allergic skin reaction characterized by red, itchy, irritated skin. In rare, severe cases, people who have been sensitized to the vapors of some constituents may exhibit asthma-like symptoms or worst case, full-blown anaphylaxis.

❱ POLYESTER AND VINYL ESTER

Polyester and vinyl ester resins are typically solutions of a solid resin in styrene (or methacrylate or vinyl toluene) monomer and may contain small amounts of other ingredients such as dimethylaniline (DMA), an accelerator, and cobalt naphthenate (CoNap), a promoter. Styrene is an acute eye, skin, and respiratory irritant, and can cause sleepiness and unconsciousness (narcosis) in high concentrations. The resin mixture is a viscous liquid with a strong, distinct odor.

These resins are cured by adding a small amount of an initiator (1 to 3 percent), typically an organic peroxide such as methyl ethyl ketone peroxide (MEKP), or benzoyl peroxide (BPO). These peroxides are strong irritants and can cause severe damage to the skin and eyes. Proper safety precautions must always be observed when using these materials.

Additionally, all promoters such as cobalt napthanate should be mixed thoroughly with the base resin before introducing an initiator like MEKP. Under no circumstances should a promoter be mixed directly with an initiator, as violent decomposition and fire may result.

❱ POLYURETHANE

These systems are formed by reacting a polyether or polyester polyol component with one of three isocyanate compounds, usually toluene diisocyanate (TDI), methylene diisocyanate (MDI), or hexamethylene diisocyanate (HDI). While the polyols generally have a low order of toxicity, the isocyanates are usually severe skin and respiratory irritants. Both skin and respiratory sensitization can readily occur from exposure to isocyanates. Additionally, respiratory sensitization can be caused by skin contact alone.

❱ POLYIMIDE AND BISMALEIMIDE

Since these resins have not been studied as extensively as the others, little information exists concerning their potential safety issues. Manufacturer's material safety data sheets indicate that prolonged or repeated contact with these resins may result in skin irritation or sensitization not unlike many epoxy formulations. Additionally, some polyimide chemistries use bisphenol-A (BPA)—as stated previously, BPA is classified as an endocrine-disrupting chemical due to its demonstrated estrogen-mimicking ability.

❱ BENZOXAZINE

Although discovered in the 1940s, serious research into benzoxazine (or, BZ) resin systems did not occur until the '80s and '90s. Recently, after decades of development, the composites industry has seen a dramatic uptick in the use and application of benzoxazine resin systems. This increased use stems primarily from the resins exceptional versatility and performance, but also because these materials offer some advantages in terms of industrial and environmental safety. While research is ongoing, early testing suggests that benzoxazines generally exhibit a reduced potential for acute toxicity, irritation and sensitization when compared to high performance legacy systems such as epoxy.

❱ PHENOLIC

This is a family of resins made by reacting one of at least seven different phenols with an aldehyde such as formaldehyde or furfural. Only exposure to phenol and formaldehyde vapors is likely to occur; however, both are potential eye, skin, and respiratory tract irritants. Additionally, phenol can be readily absorbed through the skin in toxic amounts, so both gloves and respiratory protection may be necessary.

Hazards Associated with Fibers

Handling dry carbon fiber, aramid, or fiberglass fabric is not an inherently hazardous activity. The potential to infiltrate and damage the human body is only present once the materials are processed with a matrix material and then machined. Machining a cured composite laminate generates a significant amount of dust. Although the dust created from machining appears to be homogeneous, it actually contains distinct resin and fiber components. The main safety concern that must be addressed is whether or not the particles, especially the fibers, can be considered *respirable*, according to OSHA. Respirable fibers can penetrate deep into the lungs and damage the alveoli. (Figure 14-1)

When composites undergo a machining process such as sanding, the relatively fragile matrix material shatters and exposes the fibers. The liberated fibers are then broken off at various lengths by the sanding process. It is important to note that as the fibers are abraded and broken, the diameter of the fiber remains largely unchanged. Since the vast majority of glass, carbon, aramid, and basalt fibers are greater than 6 microns in diameter, the percentage of fibers that fall into the respirable category according to OSHA is usually very small. However, that does not relieve an organization from accurately quantifying all exposures to hazardous materials in the workplace.

Respirable fibers are those whose shape and size allows them to potentially penetrate deep into the lungs and damage the alveoli, the tiny structures in the lung that are responsible for gas exchange. Generally, a fiber is considered respirable if it is 3 microns or less in diameter, 5 microns or more in length, with an aspect ratio greater than 3:1. Fibers larger than this are considered inhalable, but not respirable. Since inhalable fibers are too large to become permanently lodged in the alveoli, they are typically removed by normal lung clearance mechanisms such as coughing. (Figure 14-2)

FIGURE 14-1. Fiber and resin constituents from machining operation.

Because dust from machining contains resin and fiber components, the main safety concern is whether the particles, especially the fibers, are respirable. Respirable fibers can penetrate deep into the lungs and damage the alveoli.

FIGURE 14-2. Fibers and alveoli in the lung.

A fiber is considered respirable if it is 3 microns or less in diameter, 5 microns or more in length, and has an aspect ratio greater than 3:1. This size and shape allows the fiber particles to become lodged in the alveoli and damage the lung's gas-exchange ability.

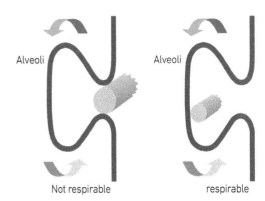

Exposure Limits

In the United States, exposure limits are established by OSHA to control exposure to hazardous substances. Collectively known as **Permissible Exposure Limits (PELs)**, they are set forth in OSHA regulations and employers who use regulated substances must control exposures to below the PELs for these substances. OSHA has a specific hierarchy in terms of the strategies employers may use to control employees' occupational exposure to hazardous materials. These control strategies for airborne particulates are required to be executed in the order shown in Table 14-1. PELs are usually expressed in parts per million (PPM) for liquids, gasses or vapors, and in milligrams per cubic meter (mg/m^3) of air for dusts and particulates. Particulates, including airborne fibers, are quantified in terms of the number of fibers per cubic centimeter (f/cm^3).

TABLE 14.1 OSHA Hierarchy of Strategies to Control Occupational Exposure to Hazardous Materials

Control strategy	Description
Elimination	Eliminate the hazardous constituent from the material if possible.
Substitution	Exchange the hazardous constituent for a non-hazardous or less hazardous version.
Engineering Controls	Implement local exhaust ventilation with HEPA filtration and/or closed process design (isolation).
Administrative Controls	Implement policy and/or procedural controls such as initial and annual medical screening, employee education and participation programs, and periodic evaluation of exposure data and engineering control efficacy.
Personal Protective Equipment	If exposure to particulate hazards is still above the PEL after employing the strategies above, use a respirator in accordance with 29 CFR 1910.134 or other appropriate respiratory standard.

The PEL found in most safety data sheets is expressed as an 8-hour time-weighted average (TWA). It is meant to characterize the maximum average exposure to a substance an employee can experience over an 8-hour period, based on industrial hygiene monitoring. The measured level may sometimes go above the TWA value, as long as the 8-hour average stays below. All chemicals with PELs have a TWA value. Only a few chemicals have ceiling and short-term exposure limits.

1. The Short-Term Exposure Limit (STEL) is a value that can be exceeded only for a specified short period of time (between 5–15 minutes). When there is an STEL for a substance, exposure still must never exceed the Ceiling Limit, and the 8-hour average still must remain at or below the TWA.
2. The Ceiling Limit is the maximum allowable level. It must never be exceeded, even for an instant.

Currently, OSHA classifies dust generated from composite machining operations as **particulates not otherwise regulated**, or PNOR (*see* below). The PEL for PNOR is 15 mg/m^3 total dust, where 5 mg/m^3 is allowed to be respirable. However, composite research is ongoing and safety personnel should check regularly for appropriate state and federal regulations changes.

❭ INDUSTRIAL HYGIENE REPORTS

In order to determine if a specific procedure or task does or does not generate exposures above the limits set forth by OSHA, a safety officer will usually enlist the help of an industrial hygienist. These specialists possess the knowledge and equipment necessary to quantify exposures in the workplace and will usually generate a report based on their findings. This industrial hygiene report is then used to determine what engineering controls or personal protective equipment should be used in the performance of that particular task.

❭ THRESHOLD LIMIT VALUE

Another figure that safety specialists should consider is the Threshold Limit Value, or TLV. This limit is established by the American Conference of Government Industrial Hygienists (ACGIH). It is used to express the degree of exposure to contamination in which most people can work consistently for 8 hours a day, day after day, with no harmful effects.

OSHA PELs are the law concerning the use of hazardous materials. TLVs are recommendations, scientific opinions based solely on health factors; there is no consideration given to economic or technical feasibility of implementing controls to keep worker exposure levels below this level. Notice that if they are included on an SDS, TLVs are often lower than PELs for a given substance. While the PEL is the legal standard to which an organization is held responsible, a prudent safety officer might err to the side of caution when determining the necessary engineering controls or personal protective equipment (PPE) to implement in order to control exposure. Consider the following example.

Carbon fiber dust is considered by OSHA to be PNOR. As such, its PEL is 15 mg/m^3 total dust, of which 5 mg/m^3 is allowed to be respirable. The ACGIH has assigned carbon fiber dust a TLV of 10 mg/m^3. If an industrial hygiene survey of a particular procedure showed exposure to be approximately 12 mg/m^3, a responsible and prudent safety officer might re-evaluate the engineering controls in place for that procedure to see if the exposure can be reduced below the TLV.

Hazards Associated with Nanomaterials

As discussed in Chapter 4, nanomaterials are tiny manufactured particles having dimensions between 1 and 100 nanometers (nm). A nanometer is one billionth of a meter. To provide some sense of scale, a human hair is between 80,000 and 100,000 nm in diameter and a strand of human DNA is about 2.5 nm across. Engineered nanomaterials are fabricated from nanoscale structures such as carbon nanotubes (CNTs), nanofibers and nanoparticles of clay, silica, polymers, metals and ceramics (*see* Chapter 4).

❭ RISK OF TOXICITY AND EXPLOSION

Assessing the true hazards associated with the use of any nanomaterial is difficult for two main reasons. First, nanomaterials are relatively new. As such, a robust body of health and safety research data for these materials is still being developed. However, in 2013 the National Institute for Occupational Safety and Health (NIOSH) reviewed 54 studies in which lab animals were exposed to CNT or carbon nanofibers (CNF). The studies indicated that respiratory exposure to CNT/CNF caused granulomas, fibrosis and inflammation of lung tissue. While respiratory studies found no evidence of mesothelioma, at least three studies found the presence of mesothelioma after CNT and CNF were injected into the body cavities of lab animals. Additionally, at least five of the studies found evidence of cellular damage at the DNA level. So while none of this is definitive, there is evidence suggesting that the use of CNT/CNF may have significant health risks associated with it.

The second problem is the number of different nanomaterials commercially available is already substantial and expected to continue growing rapidly for the next several years at least. Unlike composite constituents discussed earlier in this chapter, subtle changes in the geometry or chemistry of these nanomaterials can result in dramatic increases in their toxic effects on the body and/or the work environment. For example, due to their small size and large surface area, airborne particulate nanomaterials pose the risk of explosion. So, even if the particulate has been determined to be benign in terms of toxicity, the explosion threat certainly must be taken into account for occupational safety.

❭ MACHINING AND RESPIRABLE PARTICLES

When machining a traditional fiber reinforced composite structure, some of the airborne fibers generated may be of a certain size and shape to be classified as respirable. Research has shown that respirable fibers can enter the alveoli in the lung causing damage to the alveoli tissue, possibly rendering it unable to exchange oxygen for carbon dioxide. Carbon nanotubes behave similarly and for the same reason. However, some research indicates that certain nano-materials, particularly long, straight, multi-walled carbon nanofibers do not simply stop at the lung's alveoli and have been found to reach other organs and tissue including the brain, liver, heart, kidneys, and skeleton as well as other soft tissues.

At the nanoscale, a particle's behavior is governed by quantum effects and they cease to behave as they do in macro scale. Properties such as melting point, electrical conductivity, and chemical reactivity and others vary with the size of the nanoparticle, thereby making comprehensive assessments of a nanomaterial's relative safety in the workplace very difficult.

Though regulatory standards for nanomaterials are limited, OSHA has published exposure limits to nanomaterials based on NIOSH recommended exposure limits (RELs), but to date, these are limited to carbon nanotubes/nanofibers. They state: "worker exposure to respirable carbon nanotubes and carbon nanofibers should not exceed 1.0 micrograms per cubic meter ($1.0 \ \mu g/m^3$) as an 8-hour time-weighted average." While this is more restrictive than some carbon nanotube manufacturers have suggested (between 2.5 and 50 $\mu g/m^3$), it is orders of magnitude more restrictive than OSHA's PEL for airborne carbon fibers, which is 15 mg/m^3 total dust, 5 mg/m^3 of which is allowed to be respirable—demonstrating the profound difference size makes in terms of particulates.

TABLE 14.2 Published Exposure Limits for Nanomaterials vs. Carbon Fibers

Substance	Occupational exposure limit (8-hr time-weighted avg)	Source
Respirable CNT/CNF	1.0 $\mu g/m^3$	OSHA/NIOSH
	2.5-50 $\mu g/m^3$	CNT manufacturers
Airborne Carbon Fibers	15 mg/m^3 total dust 5 mg/m^3 respirable	OSHA/NIOSH

As stated earlier, OSHA's hierarchy for protecting workers exposed to these hazards is as follows:

1. Elimination
2. Substitution
3. Engineering controls
4. Administrative controls
5. Personal protective equipment (PPE), such as a respirator Personal protective equipment (PPE), such as a respirator

NIOSH states that "...current respirator performance research suggests that its traditional respirator selection tools apply to nanoparticles. NIOSH-certified respirators should provide the expected levels of protection, consistent with their assigned protection factor (N95 and N100)." However, a NIOSH-rated P100 (HEPA) filter or respirator is only for particles 0.3 microns in diameter (300 nm) and above. A National Institute of Health (NIH) study found the most penetrating particle size (MPPS) generally ranged from 0.05 to 0.2 microns (50 to 200 nm) for P100 filters. Thus, there is a risk that respirators may do little to eliminate all particles below 50 nm in diameter from being inhaled by workers. However, since high-aspect ratio fibers are the more prevalent concern, in general, it would seem the agglomeration of such fibers combined with their length helps NIOSH-rated filters arrest these particulates despite their small diameter.

Additionally, a 2011 report by researchers at Lund University on behalf of the Swedish Work Environment Authority (Arbetsmiljöverket)[1] asserts that "airborne nanoparticles behave like gas molecules, therefore established safety and protection devices for gases should also work in order to protect workers from exposure to nanoparticles."

❯ MACROSTRUCTURES (ARTICLES) VS. LOOSE POWDERS (PARTICLES)

Nanocomp Technologies argues that its Miralon® CNT fibers and sheet products are classified by the EPA as articles—macro-scale bulk constructions rather than a collection of free tubes or particles. This is supported by two factors: first, Miralon® products emerge from their production reactors fully-formed as sheets and yarn; also, its CNT fibers are formed as a network of interlocked, intertwined bundles that are millimeters long, not as single or small aggregates of nanotubes with micron lengths, as in common CNT, CNF, graphene and nanoparticle powder products.

In collaboration with NIOSH, Nanocomp also tested Miralon® products by cutting, tearing, puncturing, and submitting them to other common industrial processes—and then checked for the release of nanomaterials. Though the materials did shred, the pieces produced were large and easily visible. A variety of particle detectors were used to check for nanomaterial release and confirmed there was none.

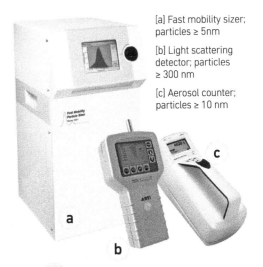

[a] Fast mobility sizer; particles ≥ 5nm

[b] Light scattering detector; particles ≥ 300 nm

[c] Aerosol counter; particles ≥ 10 nm

FIGURE 14-3. Particle detectors used in Nanocomp Technologies safety testing. *(Photos courtesy of Huntsman)*

1   "Carbon nanotubes—Exposure, toxicology and protective measures in the work environment," Per Gustavsson, Maria Hedmer, Jenny Rissler, Lund University (and Arbetsmiljöverket), Jan. 2011. www.av.se

❱ RECOMMENDED PROTECTION MEASURES

The recommendations below are from the same 2011 report for the Swedish government cited above.

- To minimize risk when handling CNTs, it is advisable to minimize:
  - number of workers potentially exposed to CNTs
  - level and duration of exposure (through training and handling instructions)
  - volumes used
  - handling of the CNTs
- If possible, keep the CNT materials wet or damp to reduce the risk of them becoming airborne.
- Keep all processes using nanoparticles contained within enclosures designed for gaseous substances. If this is not possible, then local ventilation should be used such as exhaust ventilation with a particulate filter (e.g. HEPA—note that cyclones do not capture nanoparticles effectively) along with extensive personal protective equipment (respiratory equipment, protective gloves, protective clothing, safety shoes).
- Power-assisted respiratory protective equipment is reported as offering the best protection, including a half-face mask with FFP3 class particle filter. Respiratory protective equipment should be rated with a protection factor ≥ 40.
- Disposable gloves made of nitrile are reported to be suitable for work with nanoparticles. However, how the gloves are put on and taken off, as well as their overlap with protective clothing, is more relevant for possible skin exposure than the glove material's permeability.
- Protective clothing should prevent dermal exposure and should preferably be made of membrane materials (e.g., polypropylene or polyethylene). Studies show that woven or knitted fabrics made of cotton or wool provide less protection.
- CNTs are stable compounds that have no natural biodegradability. Thus, CNTs that have been emitted into the work environment will remain until cleaned or removed. The work environment should be regularly cleaned, and thus it is important to have easily cleaned surfaces when handling nanoparticles. Vacuuming is reported to be an effective cleaning method, as long as guidelines with regard to preventing dust explosion are followed.
- Waste containing CNTs must be classified and labeled as hazardous waste and sealed carefully using double layers of polyethylene bags. This should be completed in a fume cupboard that has a HEPA filter, or by using process ventilation such as local exhaust vents with HEPA filters. If waste containing CNTs is to be eliminated by combustion, pyrolysis >500°C is preferred because it oxidizes the CNTs completely.

Much more research is needed, including further studies on toxicity, the factors that affect toxicity, how long CNTs and other nanomaterials stay in the body after exposure (bio-persistency) and the long-term effects of exposure. This research will require many more personal exposure measurements as well as a standardized measurement methodology for the quantification of exposure. Because of this, combined with the fact that the type of nanomaterials used in composites continues to change rapidly, any organization considering their use should consult with an industrial hygienist knowledgeable in nanotechnology to establish recommended exposure levels and aid in developing the necessary engineering and procedural controls.

Solvents

Solvents may be used sparingly or extensively depending on the type of components being manufactured and the type of manufacturing process being employed. They might be used at many stages from the impregnation of dry fabrics (prepregging), to the clean-up of tools, parts, and the application of release agents on the molds. Generally speaking, most of the solvents used in the composites industry fall into one of two chemical families: ketones, and chlorinated solvents.

❱ KETONES

The ketone family includes acetone, methyl ethyl ketone (MEK) and methyl isobutyl ketone MIBK), and methyl propyl ketone (MPK). In addition to being extremely flammable, these chemicals pose a significant inhalation hazard. Over-exposure to vapors can cause irritation to the nose and throat, headache, dizziness, nausea, shortness of breath, and vomiting. Higher concentrations may cause central nervous system depression and unconsciousness.

The ketones absorb readily through the skin producing mild to moderate irritation, redness, itching, and pain. Chronic or prolonged skin contact can cause dermatitis and may also cause depression of the central nervous system. Vapors may irritate the eyes and contact with the liquid will produce painful irritation and possibly eye damage. Persons with pre-existing skin disorders, eye problems, or impaired respiratory function may be more susceptible to the effects of chemicals in this family.

❱ CHLORINATED SOLVENTS

The family of chlorinated solvents is quite large containing hundreds of chemicals. Two that are used in the composites industry are trichloroethane (and its variants—trichlorotrifloroethane and trichloroethylene), and methylene chloride.

Trichloroethane and its variants can be absorbed by intact skin. They cause central nervous system (CNS) depression, and are known to damage the kidneys and liver. Additionally, trichloroethylene is a suspected carcinogen. It can adversely affect the lungs, and can cause rapid and irregular heartbeat that can cause death. It also causes skin, eye, and mucous membrane irritation. Toxic effects are increased when combined with alcohol, caffeine, and other drugs.

This substance has been used extensively in vapor degreasers. Many members of this family of chemicals are similar to Halon, and thus displace oxygen, which makes knowing the exact chemical involved critical in choosing a respirator for the task.

Methylene chloride is listed by the ACGIH as a potential carcinogen. As a result, OSHA has recently lowered the PEL for this chemical. Overexposure causes CNS depression, liver and kidney damage, and can cause elevated blood carboxyhemoglobin (also caused by exposure to carbon monoxide). Contact of the liquid with skin or eyes causes an extremely painful irritation similar to a burn. In the aerospace industry, it has long been used as the active ingredient in paint stripper.

❱ SOLVENTS IN RELEASE AGENTS

Exposure to solvent-borne release agents may be one of the most overlooked hazards in the workplace. Engineering controls for proper ventilation and containment of the Volatile Organic Compounds (VOCs) found in these chemistries are paramount to maintaining a safe working environment.

Hydrocarbon solvents used in release agents are usually aliphatic or aromatic. At one time chlorinated solvents were very popular because most are not flammable. However, they have been mostly regulated out of use, because they are Ozone Depleting Compounds (ODCs) and suspected carcinogens.

Aromatic hydrocarbons are the most common solvent for release agent systems. They range in flash point from combustible down to flammable. Aromatic solvents are VOCs; they contain Hazardous Air Pollutants (HAPs) and most are suspected carcinogens. To reduce VOCs, manufacturers will use low-density solvents. Lower weight per gallon equals fewer VOCs per gallon. However, these lighter weight solvents are more flammable.

Aliphatic hydrocarbons range from combustible to flammable and are relatively low in VOCs. Many of them are HAP-free and are not known carcinogens. These are usually a safer choice for the end-user; however, non-VOC chemistries may ultimately be the safest and the most environmentally friendly option for today's composite manufacturing factory.

Personal Protective Equipment

When dealing with the chemicals and other hazards in the advanced composite industry, it is important to use Personal Protective Equipment (PPE) that is appropriate for the task. The guidance offered in this section is general in nature and not meant to be specific to, or suggestive of the PPE that the reader may require for a given task.

It is important to note that while PPE is important, OSHA considers it to be the last in a series of defenses against hazards in the workplace. Exposure to hazardous materials in the workplace should first be mitigated by engineering controls, then administrative controls. If these are impractical, or simply do not reduce the exposure below acceptable limits, then the use of personal protective equipment should be explored.

Engineering controls are apparatus designed to remove hazardous materials as close to their point of origin as possible. This includes equipment such as downdraft tables, exhaust hoods, paint booths, etc.

Administrative controls are rules implemented by an organization to regulate how long any employee may be engaged in activities where they are exposed to hazardous materials.

Personal protective equipment may still be necessary to supplement engineering and administrative controls in order to provide the best protection for the operator from the four primary routes of exposure, as outlined below.

❱ PROTECTION FROM ABSORPTION

Gloves are the most common protection against absorption hazards, worn to protect the hands while touching resins, adhesives, or solvents. The proper glove should be selected based on the materials being handled. For instance, when using a small amount of isopropyl alcohol,

FIGURE 14-4. Clayton Tools has a full line of tools that vacuum away particulates at the point of generation, such as the random orbital sander seen here. *(Photos courtesy of Clayton Associates, Inc.)*

nitrile gloves may be appropriate. However, if using acetone, nitrile gloves will break down quickly and expose the operator to the solvent. The actual use of the glove must be considered as well as the degradation and permeation ratings of the glove material. Always consult a reputable supplier of safety equipment before selecting gloves for a specific task.

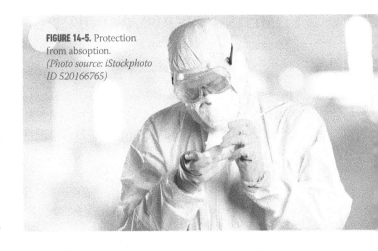

FIGURE 14-5. Protection from absoption. *(Photo source: iStockphoto ID 520166765)*

Since the category of absorption includes exposure of the eyes, safety glasses, goggles, or face shields may also be required if the materials are prone to splashing. In that case, other items of protective clothing such as plastic aprons or even Tyvek™ suits may be warranted. Much like glove selection, the material selected for this protective gear is determined by the hazards involved.

❱ PROTECTION FROM INHALATION

For respiratory hazards, the primary protection should be derived from engineering controls in the work area. Ventilation hoods, exhaust fans, etc. may be available that can bring exposure levels down below permissible limits. This is not only the preferred solution according to OSHA, but in the long run, it is likely to be the most economical because it is a one-time expense.

If engineering controls are not sufficient, are impractical, or are unavailable, then appropriate respirators are necessary. Respirators are a continuing expense that not only includes the equipment, but also the recurrent training for end-users. This may involve medical examinations, screening, and regular health monitoring of the employees. Establishing a respiratory protection program can be relatively simple or quite complex, depending on your workplace and the number of people involved. Your local safety equipment vendor will likely have a list of companies in your area who are qualified to conduct this sort of training. Whether your workplace is large or small, it is important that respirators be used in compliance with OSHA regulations (in particular, §1910.134) and in accordance with the equipment's NIOSH approvals.

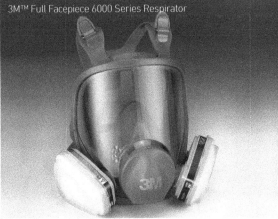

FIGURE 14-6. Respirators protect against inhalation. *(Photos courtesy of 3M)*

❱ LIMITING THE RISK OF INJECTION

Protection against injection is difficult because the type of equipment that would be needed to guard against a sharp object penetration tends to be stiff and difficult to work in. New materials and technologies in protective gear are making this more attainable. Obviously, the best protection is not rushing the work, and paying close attention to the task at hand. Proper first aid supplies and training are another aspect of protection in this category. As mentioned earlier, infection is as big a concern as actual chemical hazards in this mode of injury.

❱ PREVENTING INGESTION

We all know better than to eat or drink the chemicals we work with in this industry. What we don't always think of is the trace amounts that might be on our hands when we grab a quick snack in the workplace, or the little bit that is on our fingers if we have a cigarette at break time.

The main two ways to protect against this method of entry is good personal hygiene (washing hands before eating/drinking/smoking) and the use of gloves. If gloves are worn and then removed before taking a break, it is that much less likely that chemical hazards will be ingested.

Glossary

FOR READERS' CONVENIENCE, some of the term entries in this glossary are highlighted in bold red type where they occur in the main text, in order to facilitate lookup whenever an unfamiliar term may be encountered. This list represents just a few of the terms and definitions used daily by the advanced composites industries.

3D fabrics. *See* three-dimensional fabrics.

accelerator. A material that, when mixed with a catalyst or a resin, will speed up the chemical reaction between the catalyst or hardener and the resin.

additive manufacturing (AM). The process of joining materials to make objects from 3D model data, usually layer upon layer, as opposed to subtractive (machining) manufacturing methodologies.

adherend. A surface or object to which an adhesive is applied or which is bonded by an adhesive.

adhesive. A substance capable of holding materials together by attachment to each surface. Structural adhesives are those which are capable of transferring considerable structural loads between materials.

adhesive (or liquid) shim. Typically, paste adhesive used as a gap-filling material between two mechanically fastened substrates.

adhesively bonded. Two or more materials that are attached to each other with an adhesive, typically resulting in a permanent bond.

advanced composites. Generally considered to be made of fibers with lower density, higher strength or modulus than E-glass, combined with a matrix resin that has significantly better thermal and mechanical properties than unsaturated polyester or vinyl ester resins.

agglomerate (tendancy of nanomaterials). The tendency of particles to form into a mass or grouping.

allotropes. Two or more different physical forms in which an element can exist. Graphite, charcoal, graphene, carbon nanotubes, fullerenes, carbon nanobuds and diamond are all allotropes of carbon.

amide curing agents. Basically, ammonia with a hydrogen atom replaced by a carbon/oxygen and organic group; used as a hardener for epoxy resins.

amine curing agents. Basically, ammonia with one or more hydrogen atoms replaced by organic groups. Aliphatic amines and polyamines are common curing agents for epoxy resins.

anhydride curing agents. Agents such as Dodecenyl Succinic Anhydride (DDSA), Methyl Hexahydro Phthalic Anhydride (MHHPA), Nadic Methyl Anhydride (NMA), and Hexahydrophthalic Anhydride (HHPA). They provide for good electrical and structural properties and exhibit a long working time. Anhydrides require an elevated temperature cure.

amorphous carbon. A non-crystalline form of carbon. often related to coal or coke carbon foam products.

aramid fiber. A structural fiber made of aromatic polyamide, often recognized by the brand name Kevlar®.

A-scan. A non-destructive ultrasonic pulse-echo inspection method where ultrasonic sound waves are transmitted and received by a single transducer through a composite laminate.

A-stage prepreg. Early stage in the reaction of a thermosetting resin in which the resin is still soluble and fusible. (*See also* B-stage and C-stage prepregs)

ASTM. American Society for Testing and Materials; an internationally recognized standards organization.

asymptote/asymptotic. In this context, a line or curve with very little change in shape, usually related to a vertical or horizontal axis.

atmospheric pressure. The pressure exerted on all things by the weight of the earth's atmosphere. The accepted standard-day pressure at sea level is 29.92 inches (760 mm) of mercury, or 14.7 psi. This pressure varies inversely with altitude and decreases by approximately 1/2 pound for every 1,000 feet in altitude gained.

autoclave. A pressure vessel used for processing composite materials at pressures exceeding one atmosphere. Typical autoclaves will have heating, cooling, pressurization, and vacuum/vent systems. They are usually pressurized with an inert gas such as nitrogen.

Automated Ply Verification (APV) device. A high-resolution camera designed to work with a laser projector to verify and document fiber alignment compared to preset fiber axial tolerances. Patented by Aligned Vision.

Automated Fiber Placement (AFP). An automated fiber placement system where the mold or mandrel moves on multiple axes as well as the robotic fiber placement head. The machine lays fiber bundles in engineered axial patterns against the mold or mandrel, to produce a composite layup.

Automated Tape Layup (ATL). An automated tape layup system. An automated tape layup head, usually on a moving gantry, is used to lay up unidirectional tape widths on a fixed mold or mandrel in an automated tape laying process.

balanced laminate (in-plane). A laminate that has an equal number of "minus" and "plus" angled plies for torsional equilibrium.

basket weave fabric. Fabrics with interwoven multiple yarns, bundled side by side (for bulk) and woven in a plain weave format. (*See also* woven roving)

bending stiffness. The relative stiffness of a beam or plate of a material when a bending moment is applied, usually measured with a three-point bending apparatus. Resistance to out-of-plane flexure.

benzoyl peroxide (BPO). A chemical in the organic peroxide family, consisting of two benzoyl groups joined by a peroxide group. Typically used as a catalyst for reacting polyester or vinyl ester resin systems.

biaxial-stitched reinforcement. A fabric consisting of one layer of unidirectional tows stitched to another layer of unidirectional tows, in which each layer has a different fiber-axis angle; typically 0/90 or +45/-45.

bismaleimide (BMI). A type of thermoset polyimide resin that cures by an addition reaction rather than condensation reaction, thus minimizing volatiles, and therefore void formations during cure. BMIs are tougher, and serve a higher temperature range than epoxies.

bisphenol A. Condensation product formed by the reaction of phenol with acetone. This polyhydric phenol is the standard intermediary that is then reacted with epichlorohydrin in the formulation of epoxy resins.

bisphenol F. Condensation product formed by the reaction of phenol with formaldehyde, which is then reacted with epichlorohydrin in the formulation of epoxy resins. (Also known as a novolac epoxy formulation.)

blank. "Blank" is a broad description used to identify any material form that is to be used for a subsequent process. It can be anything from a dry preform to a pre-consolidated stack of thermoplastic sheets awaiting the next step in a cutting, kitting, or molding process. (*See also* charge)

blasting. The process of using alumina, sand, plastic, wheat starch, or any other media through a high pressure nozzle to abrade the surface of a substrate for the purpose of paint or coating removal, or surface preparation for painting or bonding.

bleeder. An absorbent layer used in the vacuum bag schedule to allow resin to move or "bleed" into the layer during processing. Typically used in non-autoclave processes.

bond. To adhere to a material and provide substantial chemical and/or mechanical adhesion to the material at the bonded interface.

bond assembly fixture (BAF). A tool used to locate and bond materials and/or components together in a primary, secondary, or cobond process.

bondline. The line of adhesive between two substrates.

bond strength. The stability of a bond, usually described in terms of bond energy, which is the energy required to "fail" the bond, or the amount of energy that the bonded joint can take without failure.

boron fiber. A high-modulus fiber made by vapor deposition of boron gas onto tungsten filament, producing a strong, stiff fiber.

bow-tie coupon. A tensile or compressive test coupon resembling a bow tie, that is narrow at the mid section and wider at the grip ends to facilitate failure in the mid-section of the specimen. (*See also* dog bone coupon)

braid, braiding. Braiding intertwines three or more yarns so that no two yarns are twisted around one another. In practical terms, "braid" refers to fabrics continuously woven on the bias. The braiding process incorporates axial yarns between woven bias yarns, but without crimping the axial yarns in the weaving process. Thus, braided materials combine the properties of both filament winding and weaving, and are used in a variety of applications, especially in tube construction.

breather. A breather layer is used in the vacuum bag schedule to maintain a pathway for air and gas to escape from the processing laminate to the port or vents within the bag.

bridging. Refers to a condition where the fibers in a laminate span or "bridge" across a radius or curve in a part, as a result of a lack of contact or compaction to the previous layer during layup or processing of the panel.

B-scan. An ultrasonic inspection technique rarely used on composite materials because it sends an ultrasonic sound wave along the laminate, in-plane, along the width or length of the laminate (rather than through the cross-section).

B-stage prepreg. An intermediate cure stage of a thermosetting resin that is between completely uncured and completely cured. All thermosetting prepregs are supplied in a B-stage condition. The degree of B-staging will advance as a prepreg material ages. (*See also* A-stage and C-stage prepregs)

buckling. Compressive deformation of fibers typically combined with matrix failure in a laminated panel, beam, or channel.

bulking materials. Typically, a non-woven core such as Soric® used within the laminate to create bulk or thickness.

bulk molding compound (BMC). Thermoset resin mixed with short fiber reinforcements and fillers into a viscous compound for use in compression molding processes.

carbon fiber. Carbonized polyacrylonitrile (PAN) or pitch-based fiber with higher stiffness (flexural or tensile modulus) than that of glass and many other fibers. Also referred to as graphite fiber.

carbon matrix/carbon-carbon composites (CCC). Carbon-carbon composites consist of carbon fibers embedded in a carbon matrix. Typically fabricated by building up a carbonized matrix on a fiber preform through a series of resin impregnation and pyrolysis steps.

carbonize/carbonization. The process of pyrolyzing a precursor material to a carbon form, usually at temperatures around 1,315°C (2,400°F), in an inert atmosphere. Actual carbonization temperature is dependent on the precursor materials, manufacturer's specifications, and the desired material properties of the end-product.

catalyst. A substance that changes the rate of a chemical reaction without itself undergoing permanent change in composition or becoming a part of the molecular structure of the product. It markedly speeds up the cure of a compound when added in minor quantity as compared to the amounts of primary reactants.

caul plate. A smooth metallic, elastomeric, or composite sheet or plate used on the bag-side of a composite layup or repair to provide a smooth, wrinkle-free surface to a localized area of the finished laminate.

centipoise. A measurement of viscosity; 1 centipoise is 1/100 Poise, equal to the viscosity of water. (*See also* viscosity)

ceramic fiber. Continuous fibers made from metal oxides or refractory oxides which are resistant to high temperatures (2,000–3,000°F / 1,093–1,649°C). This class of fibers includes alumina, beryllia, boron, magnesia, zirconia, silicon carbide, quartz and high silica (glass) reinforcements. Ceramic fibers are created by chemical vapor deposition, melt drawing, spinning and extrusion. Their main advantage is high strength and modulus. (Although glass and carbon fibers are also ceramic materials, they are not generally included in this list.)

ceramic matrix composites (CMC). Ceramic matrix composites combine ceramic reinforcing fibers with a ceramic matrix to create superior high-temperature properties.

CFRP. Carbon fiber reinforced polymer.

charge. A "charge" is typically a tailored or cut piece of material made from a larger blank, which is subsequently loaded onto a mandrel or into a mold for processing. (*See also* blank.)

chemical vapor deposition. Precursor gases such as elemental boron are introduced into a reaction chamber at ambient temperature. As they come into contact with the heated substrate, (such as a tungsten filament in the manufacture of boron fiber), they decompose, forming a solid material which is deposited onto the substrate.

chopped strand mat. A material form made with randomly placed chopped strand, typically in a tissue or mat form, designated by the fiber type and areal weight.

clockwise (CW) symbol. Refers to the fiber orientation symbol, a mirror image of the counter-clockwise symbol used on composite manufacturing drawings to designate the fiber orientation for production of the composite part. Useful in repair instructions where applicable.

closed molding. Any molding process in which Volatile Organic Compounds (VOCs) are either contained or controlled from release into the atmosphere.

closed cell foam. Any solid foam product that has a closed cell structure where the material is not porous between cells in the structure.

cobond, cobonding. The process of bonding a precured detail to an uncured laminate, requiring surface preparation and the use of an adhesive at the bondline. Cobonded composite repairs fall under this definition.

cocure, cocuring, cocured. The practice of curing all active thermoset materials in one process.

coefficient of thermal expansion (CTE). The change in length or volume per unit length or volume produced by a 1° rise in temperature.

cold mixed resin. In this context, comparatively, a "cold" mixed batch of polyester or vinyl ester resin could be mixed with less peroxide so that it reacts slowly and provides more working time. With epoxy resins, a cold mix using less hardener is not an option, as epoxies require precise measurements of each polymer component to achieve proper structural properties.

combination fixture, or jig. A tool designed to provide more than one tool or operational function.

compaction of prepreg, or laminate. Refers to interim or continuous compaction of combined resin and fiber forms in the mold, upon the mandrel, or in-situ at the machine head in an automated material placement process.

compaction shoe. A mechanical device used at the head of an automated material placement machine that provides in-situ compaction pressure to the material-form being applied in the automated process.

composite. A composite is a material comprised of two or more individual materials in which each retains its own unique properties, but when combined, the resulting material has superior properties to that of its constituents.

compression molding. A molding process using matched cavity dies mounted in a press, where bulk molding compound (BMC) is introduced into the open die set and is compressed at high pressure and temperature while being redistributed throughout the cavity as the press/die is closed. The die remains closed at elevated temperature for the duration of the cure cycle.

compressive strength properties. The ability of a material to resist a force that tends to crush or buckle; the maximum compressive load sustained by a specimen divided by the original cross-sectional area of the specimen.

compressive stress. The normal stress caused by forces directed toward the plane upon which they act; the compressive load per unit area of original cross-section carried by the specimen during the compression tests.

computer controlled processing. Controlling process parameters using a computer software program and a computer control interface to a PID Loop Controller (PLC.)

conformal shield. A surface or coating that provides electromagnetic interference (EMI) shielding.

constituent. A single element in a larger grouping. For example, in advanced composites, the principal constituents of a composite laminate are the fibers and the matrix resin.

contact molding. A spray-up or layup process where resin and fiber are placed in a mold or on a mandrel, rolled or squeegeed in place; but no additional pressure is applied to the layup in the process as it cures, typically at room temperature.

contamination. An impurity or extraneous substance in a material, on a surface, or in the environment that affects one or more properties of the material, for example, adhesion.

continuous fiber reinforcement. Refers to a fiber form or fabric where the fibers run continuously throughout the form of the material (as opposed to chopped or discontinuous fiber reinforcements).

continuous strand mat. Non-woven reinforcement made with randomly oriented continuous fibers. They are typically used as surface ply materials in structural laminates and as reinforcements in semi-structural applications.

controller. Equipment used to control temperature, pressure, vacuum, or other aspects related to composite processing.

copolymer. A long-chain molecule formed by the reaction of two or more dissimilar monomers.

core forming jig, or fixture. A tool specifically used for forming a core material to a desired shape using a vacuum bag, press, or other mechanical clamping mechanism at an elevated process temperature.

core potting compound. A syntactic paste used to pot core close-outs, core ramps, hard-points, etc.

core splice, core splice adhesive. Edgewise splice of core materials, typically using a foaming core-splice adhesive at butt-joints or by overlapping 2–4 honeycomb core cells, using a "crush core" process where the overlapped core cells are driven or "staked" together.

counterclockwise (CCW) symbol. This fiber orientation symbol is shown from the manufacturing point of view, where the fiber orientation is viewed from the inside of the panel looking toward the tool surface (as would be the case during layup). The symbol controls the primary fiber orientation where it is shown on the structure on the drawing.

couplant. A liquid or gel-like material used to "couple" to a solid surface to allow sound waves to travel from the transducer, through the couplant to facilitate an ultrasonic inspection.

coupon test program. Refers to a program in which coupons made from composite materials are tested to destruction in order to validate material properties.

cosmetic repair. A repair to a scratch, dent, or a ding of a size and/or location that is accomplished using filler putty or other non-structural material simply to restore the appearance of the part.

course (in stitched fabrics). Number of stitches of yarn per inch in a stitched fabric in the longitudinal/warp direction.

covalent bond. A chemical bond formed by sharing one or more electrons, particularly pairs of electrons, between atoms.

creel. A tension-controlled spool, designed to control the release of fiber from an attached spool of tow, usually arranged in a multi-creel array, supporting an automated fiber placement or filament winding operation.

crosslinking. A process in which polymer molecules or chains are chemically attached or "linked" to one another, resulting in a single complex polymeric molecule. Refers to a reacting thermoset resin system.

cross-talk. Small changes in amperage measured by a thermocouple due to contact with metal surfaces that also contact other thermocouples in close proximity, causing erroneous reading of temperature.

C-scan. A through-transmission ultrasound inspection technique.

C-stage prepreg. The final stage of the evolution of a thermosetting resin in which the material has become infusible and insoluble in common solvents. Fully vitrified thermosets are in this stage. (*See also* A-stage and B-stage prepregs)

cure, curing (or, fully reacted). To permanently change the properties of a thermosetting resin system as a result of a controlled chemical reaction, usually involving heat and pressure. This is a nonreversible process as the resin goes from a viscous or liquid state to a solid or elastic state.

cure cycle. Typically refers to a time/temperature recipe prescribed to cure a thermoset resin system. Technically, it is the time and temperature required to achieve the desired properties of a thermoset resin.

cure temperature. The temperature required to vitrify or fully cure a thermoset resin system.

cure time. The time required to vitrify or fully cure a thermoset resin system.

Curie temperature. The temperature at which certain magnetic materials undergo a sharp change in their magnetic properties, changing the inherent magnetic field of the material.

curing agent. The hardener or catalyst that activates the chemical reaction in a thermoset resin system.

cyanate ester. A high performance thermoset resin characterized as having a low coefficient of thermal expansion, good hot-wet properties, and a low dielectric constant.

cycle time (or, production time). The total time that it takes to process (heat, soak, and cool) a part or load of parts in a process vessel such as an oven or autoclave.

damage tolerance. The resistance of a component or structure to failure when damaged.

debulking, vacuum debulking. Intermittent compaction steps during a layup of a composite panel to achieve good consolidation of materials, especially in a complex shape. Typically this is done under a vacuum bag for a short period of time. A longer debulk may be prescribed to remove gas and air and further compact a laminate prior to final processing.

decitex. Unit of weight measure of strand or yarn in grams per 10,000 meters in length.

degas. The removal of static or evolving gas from a substance (resin).

delamination. The term used to describe a disbond between layers or plies in a laminate.

denier. Unit of weight measure of strand or yarn in grams per 9,000 meters in length.

desiccant. A substance that promotes drying (for example, calcium oxide).

die, die set (or, heated die, matched die). A tool used for forming sheet material in a press, or a die used for forming a shape in a pultrusion process. Also used to define tools used for injection molding, RTM, and compression molding processes (the latter are closed-cavity dies).

dielectric constant. Dielectric means a nonconductor of electricity; therefore dielectric constant is the measurement of the ability of a material to resist the flow of an electrical current.

differential scanning calorimetry (DSC). A thermo-analytical method measuring the amount of energy absorbed (endotherm) or produced (exotherm) during the cure of a thermoset resin system.

disbond. Refers to a lack of a bond between two substrates. Typically, a composite skin to a core disbonds in a sandwich panel.

dog bone coupon. A tensile or compressive test coupon resembling a bow tie, which is narrow at the mid-section and wider at the grip ends in order to facilitate failure in the mid-section of the specimen. (*See* bow-tie coupon)

double bias reinforcement. A fabric form, typically stitched, in which two layers of fibers are oriented at equal angles (i.e., +45° and -45°).

drape, drapeability, drapable. The ability of a fabric to drape or form over a complex shape without cutting and tailoring.

drape formed, drape forming. A process where many layers of prepreg or bulk fiber reinforced thermoplastic materials are heated and formed over a mandrel.

drill fixture, or jig. A tool used for drilling holes in a metallic or composite part, typically designed with slip-renewable bushings to minimize wear.

dry pack. Refers to a dry pack or preform of materials placed in a mold, vacuum bagged, and infused in a resin infusion process.

dry T_g. The initial thermal glass transition temperature (T_g) of a cured thermoset resin or adhesive, prior to exposure to moisture or hot-wet conditions.

dynamic mechanical analysis (DMA). Measures the stiffness and damping properties of a material, typically through a change in temperature. DMA can be used to measure the viscoelastic properties of a curing thermoset resin during the cure cycle. Also useful in determining the T_g.

Dyneema® fiber. An ultra-lightweight, high-strength thermoplastic fiber made from ultra-high molecular weight polyethylene (UHMWPE), also known as high modulus polyethylene (HMPE) or high performance polyethylene (HPPE). Dyneema® was developed by the Dutch company DSM in 1979.

edge closure. The edge closure design for a sandwich panel. Design methods range from simple potted edges to tapered close-outs. Also plastic or metallic channels may be used for edge closure design.

edge distance. The distance from the center of a fastener to the net-edge of a panel. Fastener pitch is the distance between fasteners in the same row, measured from center-to-center. Transverse pitch is the distance between rows of fasteners in the same joint.

electromagnetic (EMF) shielding. The process of limiting the penetration of electromagnetic fields into a space, by blocking them with a barrier made of conductive material. Typically it is applied to enclosures, separating electrical devices from outside access. Electromagnetic shielding is also used to block radio frequencies (also known as RF shielding) and electromagnetic radiation.

electrospinning. A fiber production method using electric force to draw charged threads of polymer solutions/melts up to larger fiber diameters, up to 100 nanometers in diameter.

elongation (to break). Measured amount of elongation of a material loaded in tension to failure. Typically measured with an extensiometer or strain gauge.

end grain balsa. Balsa wood core material cut in sheets where the grain runs perpendicular to the plane of the sheet, exhibiting excellent compressive properties.

emissions. In this context, emissions are volatile organic compounds (VOC) that are emitted into the atmosphere.

encapsulate. To surround, encase, or protect, in or as if in a capsule (such as fibers encapsulated in a polymer matrix).

environmental conditions. Refers to the environment that the structure will be exposed to in service.

exfoliation. To shed or cast-off scales, lamina, or splinters when abraded.

epoxy. A high performance thermoset resin that has good thermal, structural, and adhesive properties.

exothermic reaction. A common chemical reaction producing a slight increase in temperature when a thermoset resin evolves during crosslinking. Large masses of resin can produce an out-of-control exotherm, which can cause excessive heat and toxic smoke.

expanded aluminum, or copper. Refers to a metallic "mesh" used for lightning strike protection or EMF/EMI protection.

extraction (of volatiles, air, and gas). The inherent removal of trapped air and gas when the composite laminate is vacuum bagged and "debulked" for a period of time. Routes for extraction are provided in the vacuum bag schedule for debulking.

face skin. The face sheet or skin on a cored sandwich panel or structure. Also known as "facing."

fastener footprint. The area covered by the head of a fastener in a composite or metallic panel. Typically composite panels require a larger footprint than metallic structures.

fatigue. The failure or decay of mechanical properties after repeated applications of stress; the ability of a material to resist the development of cracks, which eventually bring about failure as a result of a large number of cycles. Failure is also the event leading to fracture under repeated or variable stresses having a maximum value less than the tensile strength of the material. Fatigue fractures are progressive, beginning as miniature cracks that grow under the action of the fluctuating stress. In composites, the effect of cyclic damage (fatigue) is different than metals, i.e., splitting and delamination, matrix cracking, and fiber breakage.

fatigue life. The number of cycles of deformation required to produce failure of the test specimen under oscillating (stress or strain) conditions.

faying plies/surfaces. Faying surfaces are those that make contact in a joint. Faying plies are those in direct contact with each other. Generally, the contacting surfaces, plies or faces of two similar or dissimilar materials placed in tight contact.

fiber angle callout. Primary fiber angles for each ply are controlled by the ply orientation symbol and the ply table callouts on the composite drawing.

fiber buckling. Refers to buckling of reinforcement fibers in a laminate loaded edgewise, on-axis in compression beyond the strength of the matrix that is binding the fibers together.

fiber-dominated properties. The subset of physical properties that are dependent on the type of fiber used in a laminate versus the type of matrix. Tensile strength and flexural stiffness (modulus) are examples of fiber-dominated properties.

fiber placement. *See* Automated Fiber Placement (AFP).

fiber reinforcement. Any fiber used for the purpose of reinforcing a matrix material to gain composite properties.

fiberglass. A common term that refers to fibers made of glass. (Fibers made of silica compositions, *see* glass fiber.)

fiber volume, fiber volume fraction. The amount of fiber in a composite laminate; usually expressed as a percentage volume fraction.

filament. Describes the basic structural fibrous element. It is either extruded or spun from the precursor materials that are used to make the specific type of fiber desired.

filament winding. A process of continuous fiber placement on a mandrel where both helical and hoop wraps produce the desired fiber axial strength and stiffness.

fill face. Refers to the fill-yarn dominant surface of a harness satin weave fabric. (Opposite of warp face.)

filled resin. Any resin that has been filled with any of a number of fillers such as silica or milled fibers. (*See* syntactic foam)

fillers. In this context, fillers are physical additives such as silica microballoons or milled fibers, used to thicken or enhance other properties within the resin or adhesive.

fill yarn (or, weft yarn). Yarns that run across or cross-wise to the length of a roll of woven cloth.

film adhesive. Flat, "B-staged" structural adhesive with a backing film applied. Typically furnished in rolls not unlike prepreg fabrics. Often used with other prepreg materials or along the bondline in an adhesively bonded structural joint.

finish. A coating applied to a yarn or fabric that provides chemical compatibility with the fiber so that the matrix resin will easily wet and distribute along the fiber during processing.

flame/flammability resistance. The ability of a material to resist burning and releasing heat or flame and toxic smoke.

flame treatment. Refers to using a flame to treat a surface to change the surface free energy prior to bonding.

flash breaker tape. A heat resistant tape used to secure ancillary materials within a vacuum bag for processing. The term "flash breaker" refers to the ability to seal off a layer of film at a constant distance from the edge of a laminate to contain the amount of resin flash around the periphery of a laminate during processing.

flat braid. Flat braids use only one set of yarns, and each yarn in the set is interwoven with every other yarn in the set in a zigzag pattern from edge to edge.

Flex Core®. A Hexcel trademarked core design that allows for the core to flex or bend over complex contours.

floating mold. A mold set that floats in water or other liquid used for hydraulic pressure and/or heat transfer.

flow, in resin. The ability of a resin to flow when processed. This is particularly important to prepreg materials that are approaching or have exceeded their initial out-time.

flexural properties. Flexural stiffness, in this context, is related to on and off- axis flexural loading of a composite laminate.

fracture toughness. The ability of a material to resist fracture when loaded or impacted. Also refers to the measure of damage tolerance of a material containing initial flaws or cracks.

functionalization, of nanomaterials. Introduction of organic molecules or polymers as catalysts on the surface of nanoparticles. This coating determines many of their physical and chemical properties such as stability, solubility, and functionality.

fusion bonding. The ability to fuse two thermoplastic materials together either through a solvent process or by welding. (*See* thermoplastic welding)

gage section (gage length). The part of the specimen over which strain or deformation is measured. (*See* bow-tie coupon)

galvanic corrosion. Corrosion due to galvanic activity between two dissimilar materials, where one is anodic and one is cathodic. The material on the anodic side of the scale experiences the corrosion.

gauge. Number of stitches of yarn per inch in a stitched fabric in the transverse/fill direction.

gauge pressure. Gauge measurement of positive pressure, typically expressed numerically as psig or barg. Gauge pressure is not absolute pressure, which starts at 0 absolute (below atmospheric pressure).

gel point, gelation, gelled. Measured change in viscosity of a thermoset resin by means of rheometric analysis. Typically the gel point is designated at half way up the transition slope before the viscosity peaks at vitrification.

gel coat, gel coated. A surface coat of filled resin used at the face of a laminate for cosmetic reasons (color, etc.), or for void filling purposes in a composite tooling application.

generative design. When a part's geometry is not predefined but instead computer-generated, by applying advanced algorithms to data derived from the application's requirements and constraints— size envelope, loads, strength, stiffness, impact, deflection, cost, etc.

GFRP. Glass fiber reinforced polymer.

glass fiber. Fibers made from one or more formulations containing silica, of which the resulting spun filament has excellent strength to weight, as well as electrical and thermal insulation properties.

glass mat thermoplastic (GMT). Glass fiber reinforced thermoplastic sheet used for compression and press molding processes.

glass transition temperature (T_g). The temperature at which a material goes through a transition from a glassy-solid state to a rubbery, flexible state (or vice-versa).

graphite fiber. "Graphite" is used in pencils, as a lubricant, and in electrodes. Carbon atoms are strongly bonded together into graphite sheets (like mica flake). The term is mistakenly used to describe high modulus carbon fiber, as they are atomically the same but extremely different in form; carbon fibers are manufactured in tension, aligning the molecular structure, providing for a functional fiber that is not as brittle or soft as graphite. Although there is a distinct difference in the form, the industry still refers to carbon fiber as graphite.

hardener. A material added to a polymeric mixture to promote or control the curing action by becoming an integral part of the cured resin.

harness satin (HS) weave fabrics. In this weave patter, a fill yarn floats over a number of warp yarns and "hooks" under one yarn in the fabric, and then repeats the pattern. With all harness-satin weaves, there is a dominant warp and fill face to consider in layup, which effects symmetry.

high modulus carbon fiber. Carbon fibers that have a tensile modulus greater than 50 msi/345 GPa.

high temperature. Generally considered to be temperatures greater than 212°F/100°C, or the boiling point of water at sea level.

heat distortion temperature (HDT). The temperature at which a standardized test specimen deflects a specific amount under a stated load. Also called deflection temperature under load (DTUL).

HOBE (HOneycomb Before Expansion). The HOBE (HOneycomb Before Expansion) is the block of bonded sheets before they are expanded into honeycomb patterns. When made from paper they are expanded and dipped in resin (typically phenolic). Metallic HOBE can be expanded and made into sheets or sold in HOBE form for ease of machining.

honeycomb nodes (or, node lines). These are the bonded interfaces between the honeycomb cells, denoting the "ribbon" direction.

honeycomb ribbon direction. The line, which is parallel to the bonded nodes in the honeycomb core, called out and oriented on the composite panel drawing.

honeycomb web. The wall of the honeycomb cell; much like mini I-beams placed throughout the core sandwich panel.

hot melt prepreg. Prepreg manufactured with heated resin typically coated on a drum and then rolled onto a backing paper prior to the dry fabric being impregnated into the resin in the process. Hot-melt prepregs have little to no volatile content as a result of the process.

hot mix. Comparatively, a "hot" mixed batch of polyester or vinyl ester resin could be mixed with more peroxide so that it reacts quickly and shortens the working time. With epoxy resins, a hot mix using more hardener is not an option, as epoxies require precise measurements of each polymer component to achieve the proper structural properties.

hot-wet service conditions, hot-wet properties. Referring to hot, humid (including salt spray) environments to which composite structures are exposed, and the ability to resist these conditions.

hybrid fabrics. Fabrics woven with two or more different fiber types, such as carbon and aramid fiber hybrid fabric.

hybrid laminates. Laminates with layers of different fiber types such as a carbon fiber laminate with an aramid fiber outer layer in a laminate used for abrasion or impact resistance.

hydrostatic. Refers to the equilibrium of liquids and the pressure exerted by liquid at rest.

impact resistance. The ability of a material to resist damage from impact. Toughened thermoset resins or thermoplastic resins provide good impact resistance in a composite structure.

impregnators. A resin bath or other resin impregnation machine where fibers are wet-out with liquid resin.

infill. A repetitive geometric structure used to take up space within a greater structure, such as a 3D mold or part.

infrared radiation. Relating to the invisible part of the electromagnetic spectrum with wavelengths longer than those of visible red light but shorter than those of microwaves. Infrared radiation from the sun is measured as heat.

Infusion, resin infusion, vacuum infusion. A fabrication process that involves the introduction of liquid resin into a dry-pack or fiber preform. (*See* resin infusion)

inhibitor. A chemical added to a liquefied monomer to inhibit polymerization or solidification.

initiator. Technically not a catalyst, but often referred to as such, the initiator starts the reaction that allows the chemistry to evolve to a cured solid. The most commonly used materials for this purpose are Benzoyl Peroxide (BPO) and Methyl Ethyl Ketone Peroxide (MEKP).

initiated resin systems. Technically all thermosets are initiated, however in this context, polyester and vinyl ester resin systems are initiated with a peroxide initiator.

in-plane shear strength. Resistance to in-plane surface distortion of a material.

in-situ consolidation (ISC). In-place consolidation of fiber reinforced thermoplastic materials on a mandrel or mold where no additional processing is required to produce the laminate.

intercalation. The reversible inclusion or insertion of a molecule (or ion) into materials with layered structures.

interlaminar shear. Through-the-thickness distortion of a material.

interleaf channel. A channel system or path for resin distribution between layers or objects in a layup.

intermediate modulus carbon fiber. Carbon fibers that have a tensile modulus greater than 42 msi/290 GPa, and less than 50 msi/345 GPa.

Invar (tooling). Invar is an iron-nickel alloy with very low thermal expansion properties. Typically, Invar 36 or 42 (36 or 42% nickel) is used for tooling.

isothermal. Relating to a process at a constant temperature.

isotropic. Having the same mechanical properties in all directions within the material.

Kevlar®. DuPont's brand name for aramid fiber.

kitting. Refers to pre-cutting and packaging "kits" of materials to be used to manufacture a composite component.

lagging thermocouple. Referring to a single thermocouple out of a group of thermocouples, which lags in response to temperature changes when compared to the rest of the group.

lamina. Refers to a single layer or ply of material.

laminae. Plural of lamina, refers to more than one layer or ply over another.

laminate. Refers to multiple layers or plies bonded together; i.e., composite laminate or laminate panel.

laminating resins. Typically, liquid resins that have relatively low viscosity and surface tension values that allow for ease of wetting (of fibers) in a laminating process.

lap shear strength. The calculated strength of a bonded lap joint when the substrates are pulled in tension. Typically lap shear test coupons utilize a 1/2 inch long x 1 inch wide lap joint, and the breaking strength is multiplied by 2 to provide pounds per square inch values. These values can be used for material qualification but are not applicable to actual lap shear values for use in design.

laser positioning. Refers to the use of a laser projector to outline ply locations or other details in layup or assembly.

laser shearography. Shearography is a nondestructive test using an image-shearing interferometer to detect and measure local out-of-plane deformation on the test part surface, as small as 3 nanometers, in response to a change in the applied engineered load.

layup. The act of laying-up a single ply or layer at a time on a mold, plate, or mandrel. The resulting laminate may also be called a layup.

layup assembly approach to manufacturing. A cocure manufacturing approach where all of the panel layers, stiffeners, etc. are laid up and compacted on separate molds or mandrels and then assembled in the primary mold for final cure.

layup mold or mandrel. A tool used to layup and process the composite laminate, panel, cocured or cobonded structure.

legacy cure cycle. A cure profile or recipe based on legacy time and temperature specifications rather than actual material state feedback processing.

lightning strike protection (LSP). A metallic or conductive surface on a composite structure to dissipate lightning strike energy along the outer surfaces of the structure in order to minimize damage to the structure. (*See* expanded aluminum or copper)

liquid compression molding (LCM). Compression molding a charge of fiber wet-out with a liquid resin prior to or during the molding operation.

long-fiber reinforcement. A fiber form or fabric where the longitudinal fibers in the roll, bolt, or spool runs full length (as compared to chopped or discontinuous fiber reinforcements). Suited for structural applications where long-fiber load transfer is required.

long fiber thermoplastic (LFT). Thermoplastic pellets with fiber lengths ranging from 0.5–1.0 inch (13–25 mm) in length, used for injection molding processes.

Long-Fiber Reinforced Thermoplastic (LFRT). Thermoplastic prepregged material containing long fiber reinforcements, typically unidirectional tape used for tape placement or other fabrication processes.

longitudinal properties. Properties along the primary axis of load in a composite panel or structure.

low observable (L.O.). Engineering and material technologies designed to reduce a craft's signature in the visual, radar, sonar, infrared, and audio ranges. Often referred to as "stealth," these technologies are used primarily to enhance the survivability of military aircraft and watercraft.

low pressure molding compounds (LPMC). Low pressure sheet molding compounds (SMC) that process in the range of 150 psi.

low-energy impact. Impact to a structure where the impacted area and subsequent damage may be difficult to detect.

material state conditions, material state management of cure. Material state conditions refer to the actual viscoelastic state of the material being processed. Material state management refers to controlling the cure cycle based on the actual viscoelastic properties of the material, as opposed to using antiquated legacy cure methods.

mandrel. A tool used to layup and process the composite laminate, panel, cocured or cobonded structure.

mat. Any non-woven tissue, veil, or mat made with random short or continuous fibers in a thin sheet form.

matched die forming. A press-forming process using a set of dies (a cavity die-set) to control the inner and outer surfaces of a reinforced thermoset or thermoplastic composite part.

matrix, matrix resin, polymer matrix. A material used to encapsulate or bond fibers together in a fiber reinforced composite structure.

matrix degradation. Degradation of the matrix material, typically from exposure to excessive heat or from chemical attack.

matrix digestion test method (or, acid digestion). Used to remove the matrix material from a laminate specimen in order to interpret the fiber-resin-void volume of a test panel or laminate coupon. (ASTM D3171 standard test method.)

matrix dominated properties. The subset of physical properties that are dependent on the type of matrix used in a laminate versus the type of fiber. Examples are service temperature, interlaminar shear strength, and compressive strength.

matrix-fiber interface. The interfacial bond of the matrix to fiber at the interface between fibers or layers of fibers in a laminate.

matrix ignition loss test method (or, burn-out). Used to remove the matrix material from a laminate specimen in order to interpret the fiber-resin-void volume of a test panel or laminate coupon. (ASTM D2584 standard test method.)

mechanical fasteners. Screws, bolts, blind bolts, rivets, or a number of different mechanical fasteners designed to be used to fasten together a structure.

mechanical properties. Compressive, tensile strength, and modulus of materials associated with elastic and inelastic reaction when force or load is applied; the individual relationship between stress and strain. (*See* structural properties)

mechanically fastened. Refers to a fastened joint where all loads through the joint are transferred through mechanical fasteners.

meta-aramid. A type of aramid fiber that inherently has a non-aligned molecular chain structure. It is very resistant to high temperatures and is used as a substitute for asbestos in industrial applications. It is good for use in making fire-resistant racing suits, and aramid paper for honeycomb core materials. However, it is not a good chemistry for making structural fibers. First developed and patented in the mid-1960s by DuPont and sold under the trade name NomexTM.

metal matrix composites (MMC). Composites containing non-metallic (ceramic) fibers bound with a metal-matrix; usually aluminum, titanium, magnesium or copper.

methacrylate. Refers to any of several polymer substances formed by polymerizing esters of methacrylic acid to make an adhesive or sealant.

methylene chloride. A solvent that will attack and degrade most plastic materials. It is a nonflammable liquid typically used as a paint remover for metallic structures.

Methyl Ethyl Ketone Peroxide (MEKP). Highly reactive organic peroxide used to produce a chemical reaction with a promoter such as cobalt napthenate to start the polymerization of a polyester, vinyl ester, or acrylic resin system, causing it to evolve to a solid and crosslink with a functional monomer to produce a cured thermoset polymer.

microballoons. Typically micro-fine precipitated silica filler (Cab-O-Sil or Visco-Fill) used to thicken a resin system so that it will fill a gap or hang on a vertical surface. Microballoons are non-uniform in size.

microbeads. Usually silica beads manufactured to a very precise diameter, used as spacers along a bondline to maintain a constant bondline thickness.

microcracking. Microscopic cracking of the matrix material between fibers primarily caused by thermal stress fracture, but may also be related to flex or impact stress.

micro-focused CT scanning. The use of a micro-focus x-ray source that can accurately identify internal and external geometry of a scanned part in 3D.

mid-plane. The mid-section at the center (cross-section) of a laminate. Laminates are normally designed to be symmetric about the mid-plane to ensure dimensional stability when exposed to a change in temperature.

mix ratio. The ratio of resin to hardener. (*See* parts by weight and parts by volume)

mold (or, tool). A layup mold, mandrel, or any of many different tools used in the production of composite parts. (*See* tooling)

modulus, tensile modulus. Modulus of elasticity is the ratio of stress or applied load to the strain or deformation produced in a material. Modulus of rigidity is the ratio of shearing stress to angular deformation within the elastic region for shear or torsional loading.

moisture contamination of prepreg. Usually moisture contamination occurs when the moisture barrier is removed from the frozen prepreg roll prior to the roll thawing to room temperature. When this happens, condensation brings moisture to the exposed prepreg.

moisture ingression/intrusion. The movements of water or other fluid contaminants into a composite structure, particularly into honeycomb sandwich structure.

monocoque. A type of construction in which the skin or outer shell bears all or most of the load or stress on the structure.

multiaxial fabrics. Typically, a stitched fabric made up of long fibers running in three or more axis of orientation.

natural fibers. Fibers such as sisal, hemp, flax, cotton, etc. used in the production of composite parts.

neat-resin. Resin as to which nothing (fibers, fillers, etc.) has been added. Usually the base resin for further formulation.

nesting. Orientation of harness-satin weave plies so the faying or mating surfaces "nest" or have the dominant (warp or fill) fibers running on the same axis of orientation at the interface.

Non-Destructive Testing (NDT), Non-Destructive Inspection (NDI). One of many different test or inspection techniques that do not require destruction of a specimen or part.

non-woven core material. A lightweight, non-woven material typically used for the dual purpose of both a core and an interlaminate resin distribution media for a resin infusion process.

non-woven materials. A compressed tissue or mat made up of non-woven short or continuous fibers that are bound to each other mechanically or with a binder. Also applies to stitched materials.

oil-canning. In this context, refers to the ability of an asymmetric laminate, that when loaded from one side, it curls or warps about the dominant axis on the loaded side of the panel. When loaded from the opposite side of the panel, the laminate "flips" and curls along the opposite axis, making a noise similar to that of an old-fashioned oil can.

open cell foam. Any solid foam product that inherently has an open cell structure where the material is porous between cells in the structure.

open time. The time that a resin or adhesive is left open and exposed to the atmosphere before is covered or joined.

open molding, open mold. A mold that controls a single surface. Today, an open mold is typically used for vacuum bagged, oven or autoclave cured composite parts, however some non-vacuum bag open molding is still done in the industry.

operational service temperature. A service temperature of a composite that is set conservatively lower than the wet T_g of a thermoset or thermoplastic resin matrix. (*See* service temperature)

out of autoclave (OoA). Processes that are performed in an oven, press, or by other means, outside of an autoclave environment.

out-of-plane shear. Shear stress effected through the cross-section of a composite laminate. Typically tested via a short beam shear test. (ASTM D2344 standard test.)

out-time. The accumulative time that a prepreg is out of the freezer. Limited to a specified amount of time.

oven, oven cure. A non-pressurized process vessel capable of heating. Process ovens normally have ramp programmable controllers for processing (curing) a variety of materials.

overbraid. This is a technique where fibers are braided directly onto cores or tools that will be placed within molding processes. It is often used when a preform with very high bias angles or a contoured triaxial design is required, or when it is desirable to include circumferential windings in a preforms architecture.

overmolding. In overmolding, injection molding is used to mold a thermoplastic polymer (unreinforced or chopped-fiber filled) over a continuous-fiber-reinforced composite blank or insert.

PAN. Polyacrylonitrile. Used as a precursor for making carbon fiber. (*See* polyacrylonitrile)

para-aramid. A type of aramid fiber that has a highly aligned molecular chain structure, making it a good material for making structural fibers. First developed in 1973, patented by DuPont and sold under the trade name Kevlar®.

particulates not otherwise regulated (PNORs). Dusts from solid substances that have particles without specific occupational exposure standards.

paste adhesive. A thermoset adhesive that usually contains fillers, making it paste-like, allowing for gap filling and the ability to hang on a vertical surface.

P.B.V. (parts by volume). Refers to a resin mix ratio expressed in parts by volume.

P.B.W. (parts by weight). Refers to a resin mix ratio expressed in parts by weight.

peel ply. A layer of fabric, either porous or non-porous, applied to the backside of a laminate to provide for a protective layer that peels away from the laminate upon removal.

permeability. The rate of passage of a gas, liquid, or a solid through a barrier without chemically or physically affecting it.

Permissible Exposure Limits (PELs). The legal limit that an employer can expose any employee to a specified chemical substance over an 8-hour period. It also applies to high-level noise.

phased array inspection. An ultrasonic inspection technique using an array transducer that has many elements, which allows for a variety of wide area and localized point inspection capabilities within a composite panel.

phenolic. A thermosetting resin that is produced by condensation of an aromatic alcohol with an aldehyde (phenol with formaldehyde). Phenolic functions at higher temperatures than epoxies; however, it is generally considered to be a low strength resin.

pitch carbon fiber. Carbon fiber made from petroleum pitch through a high temperature production process, resulting in intermediate, high, and ultra-high modulus carbon fibers for use in structural applications.

plain weave. Consists of yarns interlaced in an alternating fashion, one over and one under every other yarn. The plain weave fabric provides stability, but is generally the least pliable, and least strong, due to a lower yarn count and a higher number of crimps than other weave styles. Plain weave fabrics tend to have significant open areas at the numerous yarn intersections. These intersections will be either resin rich pockets or voids in a laminate made with this fabric style. There is no warp- or fill-dominant face on a plain weave fabric.

plasma treatment. Plasma surface treatment is used to raise the surface free energy of a polymer surface in preparation for bonding.

pleats (in a vacuum bag). Pleats or "dog ears" are placed along the periphery of a vacuum bag and sealed with sealant tape. This allows for plenty of bag materials for pleating at the base of core transitions, or changes in geometry within the bagged part area.

ply/plies. A ply refers to a single layer of material and plies refer to multiple layers of materials.

ply layup table. The ply layup table is the master tabulation of ply numbing, sequencing, orientation, material callout, splice requirements, symbol relevance, and revision information located on the face of the composite panel drawing.

ply orientation, ply angle callout. Refers to the alignment of the fibers of each layer in the laminate, on the desired axis, inspected at the point designated by the location of the universal ply orientation symbol and the ply table on the drawing.

ply orientation shorthand. Refers to ply orientation shorthand code, typically used in engineering and design for communication of complex laminate layups in a simplified manner.

ply orientation symbol. The primary orientation symbol on the plan view of the composite panel that controls the fiber axial direction relative to the symbol, applicable only where the symbol is located on the panel. If there are multiple symbols used on the drawing, then that information is coordinated with the ply table.

Poise. A measurement of viscosity. From Poiseuille (named for the French physician, Jean Louis Poiseuille, who researched flow resistance of liquids extensively in 1846). The common unit for expressing absolute viscosity is centipoise (1/100 Poise, 1/1000 of a Poiseuille).

Poisson's ratio. The ratio of transverse contraction strain to longitudinal extension strain in the direction of stretching force.

polyacrylonitrile (PAN). A synthetic resin prepared by the polymerization of acrylonitrile. An acrylic resin, it is a hard, rigid thermoplastic material that is resistant to most solvents and chemicals and of low permeability to gases. Most polyacrylonitrile is produced as acrylic and modacrylic fiber used as a precursor for making carbon fiber.

polyamide (PA). A thermoplastic polymer containing repeated amide groups. A type of condensation polymer produced by the interaction of an amino group of one molecule and a carboxylic acid group of another molecule to give a protein-like structure. The polyamide chains are linked together by hydrogen bonding.

polyamideimide (PAI). A thermoplastic amorphous polymer that has exceptional mechanical, thermal and chemical resistant properties. Polyamide-imides are produced by Solvay Advanced Polymers under the trademark Torlon®.

polycarbonate (PC). Polycarbonates have high impact strength and exceptional clarity. These unique properties have resulted in applications such as ballistically resistant windows and shatter resistant lenses, etc. More recently however, low flammability of polycarbonate is of interest to industry. This material is formed by a condensation polymerization reaction of bisphenol A and phosgene.

polyester resin (unsaturated). A thermoset resin made of unsaturated polyester mixed with a vinyl-active monomer such as styrene. The cure is affected through polymerization using a peroxide catalyst along with a promoter and an accelerator. The end result is a crosslinked polymer (to the styrene monomer).

polyester resin, (saturated) thermoplastic (PET). Polyethylene terephthalate (PET) is a polymer resin from the polyester family and is used in synthetic fibers, thermoforming applications, and engineering applications often in combination with glass fiber. (*See* polyethylene terephthalate).

polyetheretherketone (PEEK). PEEK is a crystalline thermoplastic, which makes it a very chemically resistant product. PEEK exhibits excellent chemical resistance including organic solvents, aqueous reagents. The heat distortion temperature for PEEK is 600° F/316°C, with continuous service above 450°F/232°C. PEEK has high mechanical properties, very low smoke emission and self extinguishing characteristics. Therefore it is used for many demanding applications, for example: automotive engine parts, aerospace components, compressor valve parts, etc.

polyetherketoneketone (PEKK). PEKK is a high-temperature thermoplastic that has high stiffness and strength and excellent chemical resistance. The term PEKK actually represents a variety of copolymers with different ratios of terephthalate (T) and isophthalate (I) moieties. By varying the T/I ratio, it is possible to control the crystallization rate and ultimate crystallinity and mechanical properties of the polymer, without substantially reducing its end-use temperature. Despite its many attractive properties, the commercial application of PEKK has been limited by its high cost and its relatively poor melt processability.

polyetherimide (PEI). PEI is an amorphous, amber-to-transparent thermoplastic with similar characteristics to PEEK. Relative to PEEK, it is cheaper, but less temperature-resistant, and lower in impact strength. It is prone to stress cracking in chlorinated solvents. Available under the trade name Ultem®, a registered trademark of SABIC Innovative Plastics.

polyethylene (PE). Any of the polymers of ethylene, i.e. high-density and ultrahigh-molecular-weight polyethylene (HDPE and UHMWPE, respectively), or branched to a greater or lesser degree (low-density and linear low-density polyethylene; LDPE and LLDPE, respectively). The branched polyethylenes have similar structural characteristics (e.g., low crystalline content), properties (high flexibility), and uses. HDPE has a dense, highly crystalline structure of high strength and moderate stiffness. UHMWPE is made with molecular weights 6–12 times that of HDPE; it can be spun and stretched or aligned into stiff, highly crystalline fiber with a tensile strength many times that of steel. Uses include bulletproof vests and protective armor.

polyethylene terephthalate (PET). A saturated, thermoplastic, polyester resin made by condensing ethylene glycol and terephthalic acid. It is extremely hard, wear-resistant, dimensionally stable, resistant to chemicals, and has good dielectric properties.

polyimide (PI) thermoset. Polyimides include condensation and addition polymers. Addition-type polyimides based on reacting maleic anhydride and 4,4'-methylenedianiline are processable by conventional thermoset transfer and compression molding, film casting and solution fiber techniques.

polyimide (PI) thermoplastic. A synthetic polymeric resin originally developed by DuPont that is very durable, easy to machine and can handle very high temperatures. Polyimide is also highly insulative and does not outgas. Vespel® and Kapton® are examples of thermoplastic polyimide products from DuPont.

polymer. A high molecular weight organic compound, natural or synthetic, typically a thermoset or thermoplastic polymer, or elastomeric material.

polymer matrix composites. Composites made with polymeric matrix materials, either thermoset or thermoplastic.

polymerization. A chemical reaction in which the molecules of a monomer are linked together to form large molecules whose molecular weight is a multiple of that of the original substance. When two or more monomers are involved, the process is called copolymerization.

polyphenylene sulfide (PPS). PPS is a high-strength, highly crystalline engineering thermoplastic that exhibits good thermal stability and chemical resistance. It is polymerized by reacting dichlorobenzene monomers with sodium sulfide at about 480°F/250°C in a high-boiling, polar solvent. It is used for many fiber-reinforced parts such as battery boxes and covers, etc.

polypropylene (PP). Polypropylene (PP) shares some of the properties of polyethylene, but it is stiffer, has a higher melting temperature, and is slightly more oxidation-sensitive. A large proportion goes into the production of fibers, where it is a major constituent in fabrics used in industry.

polysulfone (PSU). Polysulfone (PSU) is a rigid, high-strength, semi-tough thermoplastic that has a heat deflection temperature of 345°F (174°C), and maintains its properties over a wide temperature range.

polyurethane. A thermoset resin made from the reaction of diisocyanates with polyols, polyamides, alkyd polymers, or polyether polymers.

porosity. In this context, porosity refers to surface voids in a laminate.

post-cure. A post cure is necessary when a thermoset resin has been vitrified at a temperature below the target service temperature. Therefore it is "post-cured" at temperature intervals or at a very slow rate so as to "push" the T_g to higher temperatures without allowing distortion of the matrix, typically during a free-standing post-cure.

pot life. The time at which a freshly mixed thermoset resin has at 25°C (77°F), in a known amount, at a designated thickness, before it starts to gel or rapidly change viscosity (not necessarily the working time).

potting compound. A lightweight syntactic or filled resin used to "pot" or fill the edges or other hard-point locations in honeycomb core to prevent core crushing during processing.

precursor, carbon fiber. A precursor refers to the material used as the primary ingredient in the manufacture of a product. In the case of carbon fibers, pitch and PAN are precursor materials. (*See* pitch and PAN)

preform. Preforms are fiber reinforcements that have been oriented and pre-shaped to provide the desired contour and load path requirements of the finished composite structure. They are most commonly used in resin transfer molding (RTM), where their complex fiber orientations provide multiple flow paths for the injected resin, enabling good fiber encapsulation and higher properties.

prepreg. A reinforcement material that is pre-impregnated with resin. The reinforcement can be made from any type of fiber, in a unidirectional, stitched, or woven form. With thermosets, the resin is already mixed prior to saturating the reinforcement. The amount of resin in the prepreg is strictly controlled to achieve the desired resin content. It is stored in a moisture barrier and kept frozen until needed to preserve out-time.

press. A hydraulically-driven, heated set of platens, or special set of heated dies. It has high-pressure capabilities used for processing composites and other materials.

pressure (or, equilibrium pressure). Refers to when the hydraulic pressure of the resin in a vacuum bag comes to equilibrium with the atmospheric pressure, causing a loss of down-force or compaction on the underlying laminate.

promoter, promoted resin systems. Typically refers to polyester or vinyl ester resins that have had the promoter (Cobalt Napthenate) added to the resin formula at the manufacturer. This eliminates the hazard of storing and mixing three-part systems.

primary structure. Also known as a "fracture critical" structure that carries or transfers primary loads throughout a major, heavily loaded structure. Failure of a primary structure typically is catastrophic. (*See* secondary structure)

primer. Typically, a corrosion-inhibiting layer sprayed or otherwise applied to a metallic surface prior to painting or structural adhesive bonding. The term also applies to an interface coat of resin used for wetting a surface of a composite prior to applying a paste adhesive.

processing. The act of following a process to achieve consistent results; for example, processing thermoset resins.

production rate. The number of parts or assemblies being manufactured within a certain amount of time.

pulse-echo ultrasonic inspection. Also known as an A-scan, a non-destructive, ultrasonic pulse-echo inspection method where ultrasonic frequency sound waves are transmitted through, and received by, a single transducer through a composite laminate. (*See* A-scan)

pultrusion. An automated production process in which all of the raw materials are pulled in tension through a die or die-set in order to form a shape. The shape is cured as it exits the pultrusion machine and is cut to length.

pyrolization/pyrolysis. The thermal decomposition of materials at elevated temperatures in an inert (no air or oxygen) environment.

quadraxial reinforcement. A four-layer stitched fabric, usually with one layer in each of the four primary directions: 0°/+45°/90°/-45°.

quartz fiber. Quartz fiber is produced from quartz crystal, which is the purest form of silica. The crystals, which are mined mainly in Brazil, are ground and purified to enhance chemical purity. The quartz powder is then fused into silica rods, which are drawn into fibers and coated with size for protection during further processing. Quartz fiber has the best dielectric constant and loss tangent factor among all mineral fibers. For this reason, along with its low density (2.2 g/cm³), zero moisture absorption, and high mechanical properties, quartz fiber has been used frequently in high-performance aerospace and defense radomes. Quartz fiber also has almost zero coefficient of thermal expansion in both radial and axial directions, giving it excellent dimensional stability under thermal cycling, and resistance to thermal shock.

quasi-isotropic. Refers to the in-plane isotropic properties achieved in laminates using multi-axial orientations of the fibers to attain similar properties when the laminate is loaded in-plane, in flexure, compression or tension, on any axis.

R-value. Refers to a material's resistance to heat transfer. The higher the "R" value, the better the insulation properties.

ramp rate. The rate at which the temperature is rising or falling in a ramp cycle. (*See* ramp cycle)

ramp/soak cycle. The portion of a cure recipe in which the temperature is ramping up, holding at a specified temperature, and cooling down. (*See* cure cycle)

rate of cure. Refers to the time compared to temperature required to cure a thermoset resin at elevated temperatures; the "rate of cure" (or rate of chemical reaction) is accomplished faster at temperatures above the glass transition temperature than below it. Therefore, final cure temperatures are typically engineered to be as near as practical to the final desired T_g in order to minimize the cure time or rate of cure.

REACH. A regulation of the European Union, adopted to improve the protection of human health and the environment from the risks that can be posed by chemicals.

recertification/requalification. Refers to materials that have exceeded their shelf life but may be deemed usable through testing. Usually prepreg materials will be recertified or requalified for use after a flow/gel test has been conducted to verify that the resin still meets original certification parameters. The material then is recertified for a short duration and again can be used in production of components.

reconfigurable/adaptive tooling. Tooling that can be reshaped or adapted to a variety of different shapes.

release film. A low-energy thermoplastic film such as Teflon® or some other low-energy polymer film used in the vacuum bag schedule for the purpose of releasing from the resin, blocking, or restricting resin flow.

release agents/release systems. Semi-permanent polymer coatings that are used on the layup tool or mandrel to provide release from the tool after processing. Also refers to polyvinyl alcohol (PVA) and paste wax used in tooling and other room-temperature composite manufacturing operations.

resin. A solid or pseudo-solid organic material, generally of a high molecular weight. In fiber reinforced plastics, it is the resin that binds the fibers.

resin chemistry. Refers to the specific chemical makeup of a selected resin or resin system.

resin content (or, matrix content). The amount of resin in a composite laminate, usually expressed as a percentage by weight or volume.

resin/fiber ratio. The amount of resin to fiber expressed as a percentage of each; i.e., 40/60 resin to fiber by volume or by weight.

resin infusion. A fabrication process that involves the introduction of liquid resin into a dry-pack or fiber preform.

resin pyrolization. The carbonation of a resin matrix, usually phenolic, by ultra-high temperature exposure, in the manufacture of carbon-carbon composites.

resin transfer molding (RTM). RTM is a process in which mixed resin is injected into a closed mold containing fiber reinforcement. It produces a part with good surface finish on both sides and is well-suited to complex shapes and large surface areas.

reticulated film adhesive. According to Webster's Dictionary, reticulation is defined as "resembling or forming a net or network." Therefore, reticulated adhesive forms a network of fillets around the honeycomb core nodes when it is heated during processing.

rheometer. A torsion plate rheometer is used to measures the stiffness and damping properties of a composite material through a change in temperature. A rheometer can be used to measure the viscoelastic properties of a curing thermoset resin during the cure cycle. Also useful in determining the T_g. (*See* DMA)

ribbon direction. Refers to honeycomb core; the direction of the bonded nodes between layers in the manufacture of core.

room temperature cure. A thermoset resin system that cures at "room temperature" (77°F/25°C); or, the time it takes a thermoset resin to cure to desired properties at room temperature.

roving. A roving is a number of strands, yarns, or tows collected into a parallel bundle with little or no twist. Single-end rovings are smaller in diameter and are more expensive to produce due to the higher precision process required for the smaller diameter. Multi-end rovings are larger in diameter and generally less expensive, but can be more difficult to process in filament winding and pultrusion applications. Rovings are most commonly used for constructing heavy fabrics. (*See* woven roving)

Safety Data Sheet (SDS). A document that contains information on the potential hazards (health, fire, reactivity and environmental) and how to work safely with a chemical product.

sanding ply or layer. A sacrificial ply, typically applied to the outer layer of a repair, designed to allow for post-process sanding of the repair to blend or "feather" the repair to a suitable finish prior to primer and paint application.

sandwich construction. Sandwich panel laminates are designed with an integral lightweight core material to provide web space between laminate skins, increasing the stiffness and strength-to-weight ratio of the panel. Sandwich panels are processed at lower process pressure than solid laminate panels to prevent core crushing. Typically, sandwich laminate panels are less damage-tolerant than solid laminate structures.

scarf. A tapered or otherwise-formed end on each of the pieces to be assembled using a scarf joint.

scarf joint, scarf joint repair, scarfed repair, scarfed structural repair. A joint design that consists of a uniformly tapered angle machined through a composite laminate. This tapered angle is matched in a corresponding laminate that is then secondary-bonded in place, or matched ply-for-ply with equivalent layers in a cobond process to create a very robust bond joint in the composite structure. (*See* tapered or scarf joint)

sealant, sealer, sealing system. Using an adhesive to seal around fasteners, external doublers, and various repairs in order to prevent liquid or gas ingress into a structure. Usually a polysulfide or polythioether based material, but many different materials can be used for this purpose.

secondarily bonded, secondary bonding. In secondary bonding, two or more already-cured composite parts are joined together by the process of adhesive bonding, during which the only chemical or thermal reaction is the curing of the adhesive itself.

secondary structure. Generally considered to be less "fracture critical" than a primary structure, yet still carries significant loads throughout a structure. Failure of a secondary structure typically does not lead to catastrophic failure. (*See* primary structure)

semipreg. A material that has been partially impregnated with a resin layer, typically on one side of a dry fabric.

semi-rigid tooling. Usually refers to localized tooling (cauls) made with a combination of flexible and rigid materials. These tools are frequently used to control steep core-ramp angles, detail interface locations, and edgeband smoothness along the bag-side surface of a cocured composite panel.

separator layer/film, in vacuum bagging. The separator layer is used between the bleeder layer and the subsequent breather layer to restrict or prevent resin flow. This is usually a solid or perforated release film that extends to the edge of the layup, but stops slightly inside the edge of the bleeder layer, to allow a gas path to the vacuum ports.

service environment. The environmental conditions to which a composite structure is exposed during its service life.

service temperature. The service temperature of a composite is set at a temperature that is conservatively lower than the wet T_g of a thermoset or thermoplastic resin matrix.

service-life. The designed or actual service life of a structure.

S-2 glass. "S" glass refers to structural grade glass with higher tensile strength and modulus than that of "E" glass. Pure S glass is no longer manufactured and S-2 glass has replaced it industry-wide as it has similar properties and is less expensive to manufacture.

shear properties. Properties related to shear stress, strain, modulus, etc., resulting from applied forces that cause two adjacent parts of a bonded laminate or joint to slide relative to each other in the direction parallel to their plane of contact. In interlaminar shear for example, the plane of contact is mainly resin.

sheet molding compound (SMC). A fiber reinforced thermosetting compound, manufactured into sheet form, rolled up into rolls separated with a plastic backing film and used to manufacture a variety of different composite parts and panels.

shelf life. The amount of time that materials such as prepreg, liquid resins, or adhesives have at their specified storage temperature and still meet specifications or maintain suitable function.

shrinkage. The relative change in dimension from the length measured on the mold at room temperature, compared to the length of the molded object at room temperature 24 hours after it has been removed from the mold.

signal attenuation. The diminution of a sound signal over time or distance, such as when performing ultrasonic inspection on composite materials.

silane finish. Silane is a known couplant agent for polyester, vinyl ester, and epoxy resins. It is used as a finish on glass fibers/fabrics to better wet the fiber and facilitate bonding. (*See* finish)

size (applied to fibers). A chemical coating applied to fibers during initial manufacturing. Size is typically added at 0.5 to 2.0 percent by weight and may include lubricants, binders and/or coupling agents. The lubricants help to protect the filaments from abrading and breaking. Coupling agents tend to cause the fiber to have an affinity for an individual resin chemistry. Binder, size and sizing are often used interchangeably in the industry, size is the correct term for the coating that is applied, and sizing is the process used to apply it.

smoke toxicity. Composite materials produce smoke when they burn that consists of a potentially toxic mixture of gases, airborne fibers, soot particles, etc. The degree to which it is toxic is determined through testing and analysis of the specific material.

soak. The isothermal hold for a specified period of time in a cure recipe designed to allow for a "full cure" of a thermoset resin system.

solid laminate. A "solid" laminate made of all fiber-reinforced resin layers and no added core material. Distinguished from sandwich panel construction, solid-laminate panels can be processed at much higher pressure than sandwich panel laminates as there is no risk of crushing a core material. Typically, solid laminate panels are more damage-tolerant than sandwich structures.

solvent. A substance, usually a liquid, capable of dissolving another substance; i.e. acetone, alcohol, methyl ethyl, ketone (MEK), methylene chloride, etc.

solvolysis. A chemical reaction occurring between a dissolved substance and its solvent.

Spectra® fiber. An ultra lightweight, high-strength thermoplastic fiber made from ultra high molecular weight polyethylene (UHMWPE), also known as high modulus polyethylene (HMPE) or high performance polyethylene (HPPE).

splice. A butt-splice, overlap, or other specific splice joint may be required for the fabrication or repair of composite structures. Typically splice requirements are controlled using a process specification. These specifications will be noted on the associated part or assembly drawing.

springback. A geometric change to the angular dimensions of a part after the initial molding or forming process when the part has been released from the forces of the tool.

stacking. In this context, this term refers to "stacking" harness satin (HS) weave fabrics as opposed to "nesting" these materials. A stacked laminate has different performance characteristics than a nested laminate. (*See* nesting)

stacking sequence. The engineered sequence in which the layers or plies of a laminate are arranged to meet design performance requirements.

state of cure. The actual material state condition of a thermoset resin during a cure cycle. Typically measured with analytical equipment reading real-time sensor data, parallel to the actual cure of a composite panel.

static-dissipative. Refers to the ability of a material to conduct or dissipate static electricity to ground. Measured in ohms, this classification refers to a material with an inherent resistivity range between 1x104 ohms and 1x1011 ohms.

stealth. *See* low observable.

stiffness-to-weight ratio. Comparative stiffness to weight of different materials; also known as the specific stiffness.

strand. A number of filaments bundled together to form a strand; for example, a single glass strand is comprised of anywhere from 51 to 1,624 filaments.

strength-to-weight ratio. Comparative strength to weight of different materials; also known as the specific strength.

structural properties. Compressive, tensile strength, and modulus of materials associated with elastic and inelastic reaction when force or load is applied; the individual relationship between stress and strain. (*See* mechanical properties)

styrene. A colorless aromatic liquid that can be easily polymerized, which is used in organic mixtures, specifically in the manufacture of synthetic rubber and plastics.

surface coat. A thin layer of filled or unfilled resin applied to the surface of a tool or part laminate for cosmetic reasons.

Surface Free Energy Value (SFEV). Denotes the measurement of the total energy content of a solid surface in equivalence to the surface tension of a liquid. Values are expressed in SI units of measure: mN/m and mJ/m^2.

Surface Tension Value (STV). Denotes the measurement of the total energy content of a liquid in equivalence to the surface free energy of a solid surface. Values are expressed in SI units of measure: mN/m and mJ/m^2.

symmetry. Refers to symmetry of the fiber axial orientations about the mid-plane of a laminate. Laminate symmetry is required for dimensional accuracy and elimination of in-plane and out-of-plane behavioral coupling.

syntactic foam. Compounds made by mixing hollow microspheres of glass, epoxy, phenolic, etc. into fluid resin systems to form a lightweight, formable paste. (*See* filled resin)

tack. Referring to the stickiness, or the lack thereof, of a resin or prepreg material.

tailored blank. A preform blank made up of layers of materials, typically with different axial orientations of the layers, or layers of different materials designed to have specific mechanical properties as a result of tailoring the construction.

tapered or scarf angle. The angle of taper of a scarf joint; i.e., the taper ratio of length to thickness.

tapered or scarf diameter. The outer diameter of a circular shaped tapered or scarf repair removal area. This corresponds to the diameter of the largest original repair ply.

tapered or scarf distance. The calculated distance from the edge of the removed damage to the outermost edge of the tapered angle.

tapered or scarf joint. A joint design that consists of a uniformly tapered angle machined through a composite laminate. This tapered angle is matched in a corresponding laminate, which is then secondary-bonded in place, or matched ply-for-ply with equivalent layers in a cobond process in order to create a very robust bond joint in the composite structure.

tap testing. A simple non-destructive test involving a coin or small tap-hammer used to lightly tap the surface of a composite laminate and listen for subtle changes in the auditory response, indicating the possibility of an inclusion, delamination, or disbond in the area.

tensile properties. The measured tensile stress and strain (and related calculated properties, such as modulus) of a given material.

tex. A unit for expressing linear density equal to the mass or weight, in grams, of 1,000 meter increments of filament, strand, yarn, or other linear fiber forms.

TFNP. Teflon® Fabric Non-Porous.

TFP. Teflon® Fabric Porous.

T_g. *See* glass transition temperature.

thermal analysis (TA). Any one of many analytical techniques developed to examine physical or chemical changes of a sample of material that occurs as the temperature of the sample is increased or decreased.

thermal conductivity. The ability of a material to conduct heat between two surfaces of a given material with a temperature difference between them. The thermal conductivity is the heat energy transferred per unit time and per unit surface area, divided by the temperature difference. It is measured in watts per degree Kelvin.

thermal cycle. Refers to a cycle where the material experiences a change in temperature from ambient to a specified temperature (or set of temperatures), eventually returning to ambient.

thermal expansion. The change in dimension related to a change in temperature for a given material.

thermal imaging (thermography). A non-destructive inspection method involving the use of an infrared camera in order to look at the thermal signature of a panel or structure, in an effort to locate defects or anomalies within the panel or structure.

thermal conductivity coefficient (TCC). The ability of a material to transfer heat; measured as heat energy transferred per unit-time and per unit-surface area, divided by the temperature difference from one surface to the other.

thermocouple. A temperature measurement sensor made up of two dissimilar wires where a change in temperature at the welded or joined end creates a change in voltage, which is measured and converted into a temperature reading.

thermocouple mark-off. The molded imprint of thermocouple wires when they are run across the back side of a laminate during fabrication or repair.

thermoform, thermoforming. The processes of heating and forming a sheet (or pre-consolidated stack) of thermoplastic materials across a mold or mandrel.

thermo mechanical analysis (TMA). Thermo mechanical analysis (TMA) measures material sample displacement as a function of temperature, time, and applied force. TMA is typically used to characterize thermal expansion, glass transitions, and softening points of materials.

thermoplastic resins. Polymers or copolymers that essentially have no crosslinks and can be melted or reshaped when heated to melt point or above the T_g.

thermoplastic welding. The melding together of thermoplastic materials by taking the materials to a temperature above the melt point in order to weld them together.

thermoset resins. A polymer that cures by the application of heat or by chemical reaction resulting in a crosslinked insoluble molecule; considered an irreversible process.

three-dimensional (3D) fabrics. Fabrics that have structural fibers running in the Z-axis, generally between two plies of biaxial fabric.

through-transmission ultrasonic inspection. A non-destructive inspection technique where ultrasonic waves are sent from a transmitter to a receiver through the part (usually coupled with water). The signal is then processed and shown on a multi-color display so that the operator can determine if there are any reductions in the signal, indicating an anomaly.

tooling, tools. Molds, fixtures, jigs, templates, and other equipment used in the manufacture of other tools, parts, and assemblies.

tool sealer. A polymeric surface sealer used as part of a semi-permanent release system that is designed to fill micro-pores at the tool surface and minimize the possibility for mechanical attachment to the surface.

tool or mold surface. The primary surface of a composite part or assembly controlled by tooling for the purpose of aerodynamic smoothness, cosmetic or aesthetic requirements, or other necessity.

torsion load. The twisting stress applied to a structure.

toughness. The ability of a material to absorb shock or impact without damage.

tow. A unidirectional bundle of filaments for use in filament winding or fiber placement applications. Also used for manufacturing unidirectional tape.

towpreg. A tow or bundle of tow that is prepregged with either a thermoset or thermoplastic resin.

tracers. Contrasting colored threads woven into fabric to identify the warp and fill directions; also used to show the fiber-dominant surfaces of harness satin weave fabrics.

transesterification. Conversion of an organic acid ester into another ester of that same acid. This conversion process can have a profound effect on certain cyanate ester resins.

transverse properties. Properties measured in the transverse direction, in-plane and perpendicular to the longitudinal direction of a laminate or structure.

triaxial reinforcement. A fiber form with three primary axes of fiber reinforcement.

trim fixture, or jig. Tooling used to hold and provide net-trim off-set dimensions, designed to be used for the purpose of trimming the component with a hand-held router or grit-edge rotary saw.

Twaron®. The trademarked name of an aramid fiber reinforcement used in manufacturing and repair operations.

twill weave. A weave style that allows for additional drapeability and conformance to complex geometric shapes.

twin sheet forming. A process in which two sheets of FRTP are processed at the same time in a dual-die mold-set, resulting in a completed, hollow-cored structure.

twisting in laminates. Twisting of an FRP laminate, in a post-processed laminate or panel, is typically due to an asymmetric layup. (*See* symmetry)

ultimate service temperature. The maximum temperature that a material can sustain for a short period of time with little to no degradation.

ultimate stiffness. The highest flexural or bending load that a material can sustain before failure.

ultimate strength. The highest tensile or compressive stress that a material will sustain before failure.

ultra high modulus (UHM) carbon fiber. Made from a pitch precursor, this type of fiber is processed at very high temperatures within an inert gas atmosphere to achieve the desired properties.

ultrasonic inspection/testing. A non-destructive inspection method using ultrasonic frequency sound waves to detect flaws in a structure.

ultrasonic leak detector. An ultrasonic receiver that "listens" to a high frequency band for noise emitted from a leak in a vacuum bag that may not be detectable to the human ear.

ultraviolet (UV) light, or radiation. Radiation from the ultraviolet light spectrum. UV light can degrade certain materials over time. Composite parts are usually painted with opaque colored paint to resist UV degradation.

unidirectional reinforcement. A fiber form or fabric where 80% or more of the fibers are running in one direction.

unidirectional tape. Dry or prepreg tows placed side by side, so that all of the structural fibers are running the length of the roll. Dry "tape" is usually held together with a zigzagging thread lightly bound to the backside of the tape with an inert binder. Prepreg tape is supported on a coated paper backing.

unreinforced plastics. Refers to a molded or cast thermoset resin or a thermoplastic sheet having no reinforcing fibers or structural fillers within the material.

unsaturated polyester. Condensation polymer formed by the reaction of polyols and polycarboxylic that contain double bonds, the typical polyester thermoset resin used in the composite manufacturing industry.

vacuum bag, vacuum bagging. The application of a flexible plastic or elastomeric membrane, sealed to the mold about the perimeter, from which vacuum is drawn to allow atmospheric pressure to be applied for the purpose of compacting and/or bonding materials together. Can be used in an autoclave to allow many atmospheres of pressure to be applied to a part or bonded assembly.

vacuum bag only (VBO). Processing a composite laminate using vacuum bag pressure only, without additional forces applied.

vacuum bag schedule, or sequence (of vacuum bag materials). Refers to the layers between the part and the vacuum bag, i.e. peel ply, release film, breather materials, etc. These materials can vary from part to part and are not always the same, particularly in non-autoclave, resin-bleed applications where the bleed is controlled to maintain resin content, yet allow for gas and air movement (with the resin) during processing.

vacuum bag sealant tape. Rubberized putty used to seal the vacuum bag film to the mold or fixture surface.

vacuum forming. A thermoforming process where a heated sheet of thermoplastic is draped over a vacuum form mold, pulled to the surface and into detail areas through a series of holes in the mold drawing vacuum.

vacuum gauge. A gauge that measures vacuum in millibars or inches of mercury.

vacuum leaks. Refers to leaks in a mold, fixture or vacuum bag during processing. Typically, there is specified leak rate that is deemed acceptable for processing.

vacuum port. A piece of hardware designed to penetrate either the vacuum bag or the mold/fixture, through which vacuum is drawn on the vacuum bag for processing.

vacuum pump. An air pump designed to pull vacuum at the inlet instead of pump air at the exhaust. A vacuum pump may be a portable unit, or a large pump attached to a tank and plumbing system for use by an entire facility.

Van der Wals force. A weak force of attraction between non-polar molecules caused by a temporary change in dipole moment due to a shift of orbital electrons from one side of one atom or molecule to the other, stimulating a similar shift in adjacent atoms or molecules and maintaining a weak attraction or bond. Named after Dutch physicist Johannes Diderik van der Waals (1837-1923).

vinyl ester. Vinyl esters are chemically similar to unsaturated polyester resins with the backbone of an epoxy resin. Like polyesters, they are relatively low cost and have similar thermal and mechanical properties but not necessarily the adhesion properties of an epoxy.

viscoelastic. The change in a thermoset resin from a liquid (visco) to a solid (elastic) state.

viscosity. The resistance of a material to flow; measured in Poise (P) or more commonly, centipoise (cP). The viscosity of water is a reference at 1 cP at 20°C/68°F.

vitrification. A solid or glassy phase of a material. When a thermoset vitrifies, it has become solid. (However it is not necessarily cured to the temperature capability desired.)

vitrimer. Derived from thermoset polymers comprised of molecular, covalent polymer chain (crosslinked) networks. Sometimes referred to as "reversible" or "self-healing" resins, these polymers allow covalent bonds to be exchanged and rearranged at elevated temperatures.

voids, void content. A void is a location in a laminate where there is no resin or fiber in that space. The void content is usually calculated based on mass to volume measurements. Refer to ASTM D 2734 for void volume calculation.

Volan finish. A Hexcel Schwebel proprietary surface finish used on glass fabrics to enhance wettability and bonding to polyester, vinyl ester, and epoxy matrices.

Volatile Organic Compound (VOC). Compound that has a high vapor pressure and low water solubility, typically controlled from direct release into the atmosphere for environmental and health concerns.

volatiles, volatile content. Materials such as water and alcohol, inherent to a resin formula that can be turned into vapor under vacuum, at ambient or low temperature, allowing for evacuation from the resin. The volatile content is the percentage of volatile that is contained in the resin.

volume ratio (or, fiber/resin ratio). Refers to the volume percentage ratio of fiber to resin contained in a composite laminate.

warp face. The face of a harness satin weave fabric that has predominately warp-yarns visible on the surface.

warp yarns, warp fiber direction. The longitudinal yarns in a roll of woven fabric; i.e., the yarns that run the length of the roll parallel to the selvedge. The primary yarns in a fabric used for axial orientation purposes.

weave, weaving, woven fabric. Any bidirectional fabric has been woven in a loom, with warp fibers alternating position at certain intervals in the weaving process so the fill yarns can be placed across the loom, thereby producing either a plain, twill, harness-satin, or other weave pattern.

weight ratio (or, fiber/resin ratio). Refers to the weight percentage ratio of fiber to resin contained in a composite laminate.

wet layup. A fabrication process where both freshly mixed resin and dry fiber forms are laid up in a mold or fixture.

wet-out. The action that takes place when a liquid (resin) is applied to a material (fiber) and the material either "wets out" or the liquid "beads up." This wetting action is crucial for good distribution of resin throughout the fiber form (tow, mat or fabric) and obtaining a good bond between the resin and fiber. A finish is usually applied to a fiber form or fabric to enhance wetting or wet-out.

wet T_g. The thermal glass transition temperature of a wet or water-saturated composite matrix.

wet winding. Using a liquid resin bath to wet-out dry fiber prior to winding on a mandrel.

working time. The actual working time of a mixed thermoset resin system, dependent upon the volume mixed and the amount of time left in mass in the bucket or cup while working, before spreading into a thinner cross-section. The actual time to gel in the final form.

woven roving. Fabrics with interwoven multiple yarns, bundled side by side (for bulk) and woven in a plain weave format. (*See* basket weave fabric)

yarn. A bundle of filaments or strands twisted together for use in a weaving process.

yarn count. The number of yarns per square inch in both the warp (W) direction and the fill (F) direction.

Zylon® fiber. A Hexcel Schwebel trademark name for their polyphenylenebenzobisoxazole (PBO) fiber. It is comparable to aramid in properties and does not hydrate at the same rate as aramid. It also can service a higher temperature range than aramid fibers.

Bibliography and Acknowledgments

WORKS QUOTED OR RESEARCHED and/or companies acknowledged for their permission as to use of their materials, photos, or diagrams are listed here in the order presented in the main chapter text. The authors and publisher acknowledge and thank the following for their contributions to this edition. (All of the following photographs are copyright of the individual owners listed below and are used with permission.) Drawings throughout are credited to ASA Staff unless otherwise noted; sources of the information for the drawings and diagrams created by the ASA artists are also listed and acknowledged here.

In addition, the authors acknowledge *CompositesWorld* magazine *(www.compositesworld.com)* as a valuable resource for a wide range of material presented in this textbook. For their support of this third edition the authors wish to thank *CompositesWorld*, as well as the many participants from the composites industry overall.

CHAPTER 1

Photo set: **Aligned Vision** Chelmsford, MA; **Marine Plastics, Ltd.** Langley, BC.

ASM Handbook Volume 21, Composites, ASM International, 2001; Material Park, Ohio.

Stork Fokker Papendrecht, the Netherlands; **FACC Operations GmbH** Ried im Innkreis, Austria *(www.facc.com)*

Photo set: **Select Charter Services** Villefranche sur Mer, France; **Kockums AB** Malmö, Sweden; **H2X Yachts & Ships** La Ciotat, France; **Brødrene Aa** Hyen, Norway; **SouthernSpars** Auckland, NZ; **Placid Boatworks** Lake Placid, NY.

H160 photos used by permission **Airbus Helicopters**, and are copyright *(left)* Productions Autrement Dit, and *(right)* J.Deulin/Airbus Helicopters.

Photo set: **Siemens Renewable Energy** (Siemens Corporation, Washington DC); **SSP Technologies** Stenstrup, Denmark *(ssptech.com)*. *Diagrams:* **Gurit** Zurich, Switzerland; Siemens Renewable Energy.

James Webb Space Telescope (JWST) and other photos, National Aeronautics and Space Administration (NASA).

Photo set: **Mubea Carbo Tech GmbH** Salzburg, Austria *(www.carbotech.at)*.

Photo sets: **Audi of America** Herndon, VA; **Voith Group** Heidenheim, Germany *(www.voith.com)*; **BMW Group** Munich, Germany *(www.bmwgroup.com)*; **Saint Jean Industries SAS** Saint Jean d'Ardiéres, France *(www.st-ji.com)*; **SGL Carbon** Wiesbaden, Germany.

Photo set: **Bianchi** Treviglio, Italy, *and* **Countervail®** Greenville, SC; **Innegra Technologies** Greenville, SC; **BAUER Hockey, LLC*** Exeter, NH; **Lingrove** San Francisco, CA; TeXtreme; North Kiteboarding – **Boards & More GmbH** Munich, Germany.

*© 2016 BAUER Hockey, Inc. and its affiliates. All rights reserved. NHL and the NHL Shield are registered trademarks of the National Hockey League. NHL and NHL team marks are the property of NHL and its teams. © NHL 2016. All Rights Reserved.

Photo set: **Groupe Médical Gaumond** *(www.groupemedicalgaumond.com)*; **Ferno-Washington Inc.** Wilmington, OH *(www.ferno.com)*; **Markforged** *(markforged.com)*; Invibio.

Photo sets: **Composite Panel Systems LLC** Eagle River, WI; **Sika Corporation** Lyndhurst, NJ *(usa.sika.com)*; Felix Weber, **Arup** *(www.arup.com)*, and San Francisco Museum of Modern Art; **9T Labs** Zürich, Switzerland; **Moi Composites s.r.l.** Como, Italy; McNAIR Center for Aerospace Innovation and Research (funded by TIGHITCO), University of South Carolina; **Stratasys** Eden Prairie, MN; **AREVO, Inc.** Milpitas, CA; Markforged.

"Timeline" photo set: Pages 14–16, and top of Page 17, © **Depositphotos.com**, licensed and used with permission *(Detail credits for Depositphotos.com— p.14 bottom, Gino Santa Maria; p.16 top, Celso Diniz, and bottom, Paul Fleet; p.17 middle, Sascha Burkard).* Page 16 *(middle)* courtesy of **TeXtreme** (Borås, Sweden). Page 17 *(bottom)*, courtesy of GE Aviation; Page 18 *(top)*, courtesy of **Kreysler & Associates** (American Canyon, CA) and **SFMOMA** (San Francisco Museum of Modern Art), *(middle)* courtesy of Stratasys, and *(bottom)*, courtesy of **Covestro** (Leverkusen, Germany); Page 19 *(top)*, courtesy of **Joby Aviation**, © Joby Aero, Inc.

CHAPTER 2

Wageningen University, and Wageningen Research Foundation, The Netherlands *(www.wur.nl)*.

Sara Black, "Automotive composites: Thermosets for the fast zone," *CompositesWorld* Sept. 2015; reproduced with permission of Gardner Business Media, Cincinnati, OH.

Connora Technologies Hayward, CA *(connoracomposites.com)*.

"Ceramic Matrix Composites—an Alternative for Challenging Construction Tasks," Friedrich Raether (Fraunhofer Center for High Temperature Materials and Design HTL), 2013, in *Ceramic Applications* 2013 (excerpts reprinted with permission).

Photo set: **Brembo SGL Carbon Ceramic Brakes (BSCCB) S.p.A.** Stezzano, Italy *(www.brembo-sgl.com)* and **SGL Group; GE Aviation** Evendale, OH, and **Composite Horizons** Covina, CA *(chi-covina.com)*.

C. A. Litzler Co., Inc. Cleveland, OH *(calitzler.com)*.

Photograph and drawing of prepreg machines, **Century Design, Inc.** San Diego.

CHAPTER 3

Clauss, B., "Fibers for Ceramix Matrix Composites" *Ceramic Matrix Composites*, 2008 Wiley-VCH Verlag GmbH & Co. KGaA, Weinhelm, Germany; *and*

Jawaid, M., HPS Abdul Khalil, "Cellulosic/synthetic fibre reinforced polymer hybrid composites: a review" *Carbohydrate. Polymers* 2011 (Vol. 86).

Two textile photographs, Wikimedia Creative Commons "CC by SA 3.0" *(Damieng, 2004; Edal Anton Lefterov, 2010).*

Ginger Gardiner, "The Making of Glass Fiber" *Composites Technology Magazine,* April 2009, Gardner Publications, Inc., Cincinnati, OH.

HexForce® Technical Fabrics Handbook, **Hexcel** website *(www.hexcel.com).*

Sara Black, "Can basalt fiber bridge the gap between glass and carbon?" *CompositesWorld* blog; *www.compositesworld.com.*

Photo set: **Coyote Design** Boise, ID; **Mafic** Shelby, NC; *Basalt Today (www.basalt.today).*

Ceramic Fibers and Coatings: Advanced Materials for the Twenty-First Century, 1998, The National Academies Press, National Academy of Sciences, 500 Fifth St., NW, Washington, DC.

Omar Faruk, Andrzej K. Bledzki, Hans-Peter Fink, and Mohini Sain, "Biocomposites reinforced with natural fibers: 2000-2010," *Progress in Polymer Science 37;* 2012 Elsevier.

Yan, Libo, Chouw, Nawawi and Jayaraman, Krishnan, "Flax fibre and its composites—A review," *Composites Part B: Engineering, An International Journal* (Vol. 56); Elsevier 2013 (10.1016/j.compositesb.2013.08.014); *also*
W.D. Brouwer, "Natural Fibre Composites in Structural Components: Alternative Applications for Sisal?" Delft University, The Netherlands.

Photo set: **Bcomp Ltd.** Fribourg, Switzerland; **Cretes** Gullegem – Wevelgem, Belgium; CELC (European Confederation of Linen and Hemp), Paris.

JELUPLAST® products from **JELU-WERK** – Josef Ehrler GmbH & Co., Rosenberg, Germany.

Michael R. Legault, "Natural fiber composites: Market share, one part at a time," *CompositesWorld,* posted March 2016.

"Sound and Vibration Damping Properties of Flax Fiber Reinforced Composites," Procedia Engineering, ScienceDirect website *(www.sciencedirect.com).*

Product photos: Lingrove, Bcomp, Lineo; **FlexForm Technologies** Elkhart, IN; **Composites Innovation Centre** Winnepeg, MB Canada; **Swift** Gravenhurst, ON Canada.

F.C. Campbell, "Structural Composite Materials," ASM International, 2010 *(www.asminternational.org).*

Ronald F. Gibson, *Principles of Composite Material Mechanics,* 4th Edition; CRC Press, 2016.

Photo set: **Tri-Mack Plastics Manufacturing** Bristol, RI, and **TxV Aero Composites; DSM Engineering Plastics** Heerlen, Netherlands; *CompositesWorld;* SGL Carbon.

Photo sets: **3D-LightTrans** *("A project co-founded by the European Commission under the 7th Framework Program within the NMP thematic area")* Vienna, Austria; **Poly Lining Systems, Inc.** Plymouth, NH; **BMW Group;** TeXtreme; North Thin Ply Technology, and Hexcel; *diagrams:* Hexcel PrimeText.

David J. Spencer, *Knitting Technology,* 1983 Pergamon Press *(www.elsevier.com/books).*

Vectorply Corporation Phenix City, AL

A&P Technology Cincinnati, OH

Stephen W. Tsai, Edward M. Wu, "A General Theory of Strength for Anisotropic Materials," *Journal of Composite Materials* 1971, (Vol. 5).

Herzog GmbH Oldenburg, Germany

Photo sets: Porsche AG; GE Aviation; Munich Composites; NASA Ames; **Bally Ribbon Mills** Bally, PA; **Albany Engineered Composites** Rochester, NH.

"3D Woven Fabrics," Pelin Gurkan Unal, Associate Professor, Namik Kemal University Department of Textile Engineering, Tekirdağ, Turkey. *InTechOpen (www.cdn.intechopen.com/pdfs/36903/InTech-3d_woven_fabrics.pdf).*

Photo sets: **Saint Gobain Performance Plastics** Puyallup, WA *(www.plastics.saint-gobain.com);* **Preform Technologies Ltd.** Derby, U.K.; textilestudycenter.com; **Carbon Conversions** Lake City, SC; Connora Technologies; **ELG Carbon Fibre** Birmingham, U.K.; **Vartega** Golden, CO.

Table photos: **CFK Valley Stade Recycling GmbH & Co. KG** Wischhafen, Germany; **Technical Fibre Products Ltd.** Cumbria, U.K.;

Photo set: ELG Carbon Fibre; CFK Valley Recycling GmbH; Technical Fibre Products (Optiveil®); Carbon Conversions.

CHAPTER 4

"Nano Graphene Platelets (NGPs)" article, *azonano.com*.

Bien Dong Che, et al., "The impact of different multi-walled carbon nanotubes on the X-band microwave absorption of their epoxy nanocomposites" (Table 3); *Chemistry Central Journal* 2015. *(https://www.ncbi.nlm.nih.gov/pmc/articles/PMC4353877/)*.

N12 Technologies, and Dr. Enrique Garcia, **necstlab** (MIT).

"CNT Composites: Processing of CNT/Polymer Composites"; website of Marcio R. Loos, PhD, Case Western Reserve University, Cleveland OH (site accessed September 2018). *(https://sites.google.com/site/cntcomposites/processing-of-cnt-polymercomposites)*

Nanocomp Technologies Merrimack, NH *(www.miralon.com)*.

Dr. Brian Wardle, Department of Aeronautics and Astronautics, MIT, Cambridge, MA. *(www.sciencedirect.com/science/article/pii/S026635381630687X)*

Shaohua Jiang, "Electrospun Nanofiber Reinforced Composites: Fabrication and Properties," *University of Bayreuth*, 2014 (pp. 54-55). *(https://epub.uni-bayreuth.de/1699/1/thesis---final%20 version---Shaohua%20Jiang.pdf)*

Dr. Ir. Lode Daelemans, researcher at Centre for Textile Science and Engineering, Dept. of Materials, Textiles and Chemical Engineering (MATCH), Ghent University, Belgium.

Arindam Chakrabarty, Yoshikuni Teramoto, "Recent Advances in Nanocellulose Composites with Polymers: A Guide for Choosing Partners and How to Incorporate Them," *Polymers,* 10(5) 2018, p. 517 *(doi.org/10.3390/polym10050517)*; *and*

Masaya Nogi, et al, "Transparent Conductive Nanofiber Paper for Foldable Solar Cells," *Scientific Reports* 2015, Vol.5, Article# 17254 *(nature.com/articles/srep17254)*, also "Printed Electronics on NanoPaper" 2005, Osaka Univ. *(www.nogimasaya.com/research/en/)*

3M St. Paul, MN

CHAPTER 5

Photo sets: Hexcel; I-Core Composites; **Goodrich** Engineered Polymer Products; **Livrea Yachts.**

Jeff Sloan, "Core for composites: Winds of Change," *CompositesWorld* June 2010.

DIAB Laholm, Sweden *(www.diabgroup.com)*.

Photo set: **Tricel Honeycomb Corporation** Gurnee, IL *(www.tricelcorp.com)*; **EconCore N.V.** Leuven, Belgium *(www.econcore.com)*.

Photo sets: Albany Engineered Composites; **Saertex GmbH & Co.KG** Saerbeck, Germany; **Seriforge** San Francisco, CA *(www.seriforge.com)*.

Lantor BV Veenendaal, The Netherlands *(www.lantor.com)*.

Photo set: Moi Composites; MarkForged; Stratasys; *CompositesWorld* magazine.

CHAPTER 6

Photo sets: Ginger Gardiner; **Compsys** Melbourne, FL; **Covestro** Leverkusen, Germany; SGL Carbon.

American Composites Manufacturers Assoication (ACMA) website, "Corrosion Resistance"; also, "Different FRP Resin Chemistries for Different Chemical Environments," Don Kelley, Jim Graham and Thom Johnson (Ashland Performance Materials), online paper at *metatron.cl.*

Engineering Toolbox *www.engineeringtoolbox.com*

Photo sets: Countervail®, Innegra Technologies, Lingrove; **CompoTech** Sušice, Czech Republic.

Photo set: **Dielettrika Ligure, s.r.l.** Genova, Italy; CompoTech.

R. Amacher, J. Botsis, J. Cugnoni, G. Frossard, Th. Gmür, and S. Kohler, "Strength and fracture of thin ply laminate composites: experiments and analysis," *École Polytechnique Fédérale de Lausanne* (EFPL), Comptest 2017, Leuven, Belgium.

C. Dransfeld, R. Amacher, J. Cugnoni and J. Botsis, "Toward aerospace grade thin-ply composites," June 2016, presented *17th European Conference on Composite Materials* (ECCM17, Munich, Germany), Publication #304659240 at *www.researchgate.net.*

University of Cambridge, Department of Mechanics, Materials, and Design *(www-mech.eng.cam.ac.uk/mmd).*

I.A. Tsekmes, et al, "Thermal conductivity of polymeric composites: A review," 2013 *IEEE International Conference on Solid Dielectrics*, Bologna, Italy.

Mahadev Bar, et al, "Flame Retardant Polymer Composites," *Fibers and Polymers* 2015, Vol.16, No.4.; *photo:* **CFP Composites** Solihull, West Midlands, U.K.

Kandola and Krishnan, "Fire Performance Evaluation of Different Resins for Potential Application in Fire Resistant Structural Marine Composites," Institute for Materials Research and Innovation, University of Bolton, U.K.; *Fire Safety Science – Proceedings Of The Eleventh International Symposium,* 2014 International Association for Fire Safety Science; Kandola, et al, "Development of vinyl ester resins with improved flame retardant properties for structural marine applications," *Reactive and Functional Polymers* 129, August 2017; Belén Redondo, Composites Department of AIMPLAS, "Composites and fire: developments and new trends in flame-retardant additives," Plastics Technology Centre, 2018.

Christophe Swistak, et al, "Flame Retardant Nanocomposites," *École Polytechnique Fédérale de Lausanne*, Switzerland, September 2009 (online); Giovanni Camino, "Flame retardants: Intumescent systems," *Plastics Additives* 1998 (part of *Polymer Science and Technology Series* Springer Science + Business Media).

Photo sets: **Dexmet Corporation** Wallingford, CT *(dexmet.com)*; **nVent ERICO** Solon, OH *(nvent.com).*

CHAPTER 7

Photo sets: **BriskHeat Corporation** Columbus, OH; **Electrotherm Industry** Migdal HaEmek, Israel; **Aligned Vision** Chelmsford, MA; **Eastman Machine Company** Buffalo, NY; **Fives Cincinnati** Hebron, KY; **Electroimpact** Mukilteo, WA; **MTorres** Everett, WA.

"ElectroImpact" photo, copyright and courtesy of *The Seattle Times*, by Mike Siegel, staff photographer; licensed and used with permission 2016.

Photo set: **Trelleborg** Trelleborg, Sweden (*formerly* Automated Dynamics); Electroimpact; **Coriolis Composites** Quéven, France; *also:* ISC thermoplastic fuselage demo photo, courtesy NAVAIR Public Release 11-1897 ("Approved for public release—distribution unlimited").

"Consolidating thermoplastic composite aerostructures in place, Part 1" and "Part 2", *CompositesWorld,* February and March, 2018.

FIDAMC Madrid, Spain; **American Autoclave Co.** Jasper, GA.

"The Thermoforming Process," April 2006 *Composites Technology,* Gardner Publications, Inc.

Photo sets: **Gurit; Propex Fabrics** Chattanooga, TN; **Broetje Automation** Rastede, Germany; **Cetim** Senlis, France; Ginger Gardiner; **Compositence** Leonberg, Germany; **Dieffenbacher-Fiberforge** Eppingen, Germany.

Photo sets: **Litespeed Racing** Los Angeles, CA; **LyondellBasell** Houston, TX; Mubea Carbo Tech; **Clear Carbon & Components** Bristol, RI.

Peggy Malnati, "Lower cost, less waste: Inline prepreg production," March 2016 *CompositesWorld.*

Photo sets: **Huntsman Advanced Materials** The Woodlands, TX; **Mitsubishi Rayon Co., Ltd.** Tokyo, Japan; **CTC Stade (an Airbus Company)** Stade, Germany; **AZL** – *The Aachen Center for Integrative Lightweight Construction* Aachen, Germany; **XELIS GmbH, a member of AVANCO Group** Markdorf, Germany; **Radius Engineering** Salt Lake City, UT; **Piercan USA, Inc.** San Marcos, CA; Heidi Dorworth.

CHAPTER 8

CompositesWorld supplied many of the photographs in this chapter, in cooperation with several companies listed below.

Photo sets: Voith Group; **DLR** Stade, Germany; **Teijin Carbon Europe GmbH** Wuppertal, Germany; Dieffenbacher.

IPF Liebnitz-Institut für Polymerforschung, Dresden, Germany (and photographers Emanuel Richter, J. Läsel).

Photo sets: **ShapeTex** Witney, Oxfordshire, U.K.; **TPI Composites**, Inc. Warren, RI; **DD-Compound GmbH** Ibbenbüren, Germany; **Superior Signal** Old Bridge, NJ; **Airtech International** *instruments (photos by Abaris).*

Source info. for diagrams: **Vacmobiles** Auckland, New Zealand; *also:* "Composite Propeller Construction," Andy Gunkler and Dr. C. Mark Archibald; © 2011 *American Society for Engineering Education,* Proceedings of the 2011 ASEE NC & IL/IN Section Conference, Grove City College, PA.

Photos, and info. for diagrams: **Bravolab** Fall River, MA *(www.bravolab.com)*; **Molded Fiber Glass Companies** Ashtabula, OH; **KrauseMaffei** Munich, Germany.

Photo sets: **Smooth-On** Macungie, PA; **Jasper Plastics** Syracuse, IN; **Matrix Composites** Rockledge, FL; SGL Carbon; Audi; Voith Group; SAERTEX.

Photo sets: **VEC Technology** Greenville, PA; Matrix Composites; **Israel Aerospace Industries (IAI)** Ben Gurion Arpt, Israel; Airbus Bremen; FACC; Albany Engineered Composites; **Elbit-Cyclone** Haifa, Israel.

Photo sets: **Protectolite Composites** Toronto, Ontario CAN; BMW Group; SGL Carbon; Mubea Carbo Tech.

Source info. for diagrams: Dow Automotive Systems; Huntsman Advanced Materials.

Photo sets: **Connova | *new ways in composites*** Villmergen, Germany; **Mikrosam A.D.** Prilep, Rep. of Macedonia; **Northrop Grumman** Promontory, UT; **MF Tech SAS** Argentan, France; **Cygnet Texkimp Ltd.** Northwich, Cheshire, U.K.; **CIKONI** Stuttgart, Germany; **Murata Machinery Ltd.** *and* Institut für Textiltechnik (ITA) der RWTH, Aachen University, Germany.

Photo sets: **Daimler AG** Stuttgart, Germany; **Automotive Management Consulting GmbH** Penzburg, Germany; **Sumerak Pultrusion Resource International** Bedford, OH; **Shape Corp.** Grand Rapics, MI; Romeo RIM *and* KrauseMaffei; Dieffenbacker.

Source info. for diagrams: **Thomas Technik & Innovation** Lower Saxony, Germany; **Exel Composites** Vantaa, Finland; (Pulpress™) **Evonik** Parsippany, NJ; (centrifugal casting) **ACMA**, American Composites Manufacturers Association *(compositeslab.com)*; **RIM Mfg.** *(reactioninjectionmolding.com).*

CHAPTER 9

This chapter's photography is used with the kind permission of the following individuals and companies (in order of appearance).

DUNA USA Baytown, TX

CFoam, LLC Triadelphia, WV

Compotool, LLC Monroe, WA

LaminaHeat, LLC Greer, SC

Ascent Airspace – Coast Composites Santa Ana, CA

Weber Manufacturing, Inc. Midland, Ontario CAN

Radius Engineering, Inc. Salt Lake City, UT

Alpex Technologies GmbH Mils, Austria

Hunter Brankamp, *photographer*

Stratasys Eden Prairie, MN

Oak Ridge National Laboratory, U.S. Dept. of Energy Oakridge, TN

Surface Generation Ltd. Oakham, U.K.

Regloplas St. Joseph, MI

CurveWorks BV Zoetermeer, The Netherlands

Torr Technologies, Inc. Auburn, WA

Advanced Ceramics Manufacturing Tucson, AZ

CHAPTER 10

The Composite Materials Handbook Organization, *Composite Materials Handbook-17* (CMH-17): CMH-17-1G, Vol. 1, Chapter 2: "Guidelines for Property Testing of Composites."

Tests from organizations: ASTM International, International Standards Organization (ISO), Suppliers of Advanced Composite Materials Association (SACMA), and Titles 14 and 49 of the Code of Federal Regulations, Part 25.

Several test fixtures and specimens photos throughout the chapter: **Wyoming Test Fixtures** Salt Lake City, UT

NPL Report Measurement Note CMMT(MN) 067, "Environmental and Fatigue Testing of Fibre-Reinforced Polymer Composites," December 2000, by W. R. Broughton and M. J. Lodeiro, Industry and Innovation Division, National Physical Laboratory, Middlesex, U.K.; used with permission.

Jared Nelson, "Passenger Safety: Flame, Smoke Toxicity Control," *Composites Technology Magazine,* December 2005; excerpts copyright of and reproduced with permission of *Composites Technology Magazine,* Gardner Publications, Inc., Cincinnati, Ohio.

Photo sets: **Wichitech Industries, Inc.** Randallstown, MD *(wichitech.com)*; **Marietta NDT, LLC** Marietta, GA.

"X-Rays for NDT of Composites," *CompositesWorld,* Jan. 2017.

Photo set: GE Inspection Technologies; **DolphiTech AS** Raufoss, Norway.

Autodesk, Inc. Moldflow® simulation software; San Rafael, CA (corporate HQ).

NTS (National Technical Systems) Fremont, CA.

TNDT photos: Dr. Steven Shephard, **Thermal Wave Imaging, Inc.** Ferndale, MI.

Laser shearography photos: John Newman, **Laser Technology, Inc.**; and "Application and interpretation of a portable holographic interferometry testing system in detection of defects in structural materials," R.B. Heslehurst, PhD, University of South Wales, 1998.

Agilent Technologies Santa Clara, CA.

CHAPTER 11

CompositesWorld, and 3M.

Brighton Science Cincinnati, OH.

CHAPTER 12

This chapter's photography is used with the kind permission of the following individuals and companies (in order of appearance).

CMS North America Caldonia, MI

Abrasive Technology Lewis Center, OH

Synova Advanced Laser Systems − Duillier (Nyon), Switzerland

Monogram Aerospace Fasteners Commerce, CA

Traditional Woodworker Richardson, TX (became part of *WoodWorld* in 2017)

Novator Swånga, Sweden

Arconic Fastening Systems (Arconic, Inc.) Torrance, CA *(www.arconic.com)*

CHAPTER 13

Photo sets: The University of Auckland, New Zealand; **LUSAS** Surrey, U.K.; **ElectraWatch, Inc.** Charlottesville, VA.

Photo set: **Archer Daniels Midland Company (ADM)** Chicago, IL; **Composition Materials, Inc.** Milford, CT *(www.compomat.com)*;

Photo sets: **AGM Container Controls** Tucson, AZ; **Moisture Register Products (JD Instruments, Inc.)** Houston, TX; **Heatcon Composite Systems** Tukwila, WA;

André Cocquyt, *VIP: Vacuum Infusion Process,* GRPguru.com

Photo set: **DMG MORI/Sauer GmbH** Bielefeld, Germany; **Lufthansa Technik Ag** Hamburg, Germany *(headquarters).*

Photo set: BriskHeat Corporation, Wichitech, Heatcon.

CHAPTER 14

Per Gustavsson, Maria Hedmer, Jenny Rissler, "Carbon nanotubes—Exposure, toxicology and protective measures in the work environment"; Lund University (and Arbetsmiljöverket), Jan. 2011 *(www.av.se).*

Photo sets: Nanocomp Technologies *and* Huntsman Advanced Materials; **Clayton Associates, Inc.** Lakewood, NJ; iStockphoto/carlosphotos; 3M.

Index